Natural Resources and the Environment Series

Volume 16.2

**Ecology in Practice**
**Part II: The Social Response**

**NATURAL RESOURCES AND THE ENVIRONMENT** SERIES

**Series Directors:** *Margaret R. Biswas & Asit K. Biswas.*



Vol 1: **Environmental Impacts of Production and Use of Energy.**
*Essam El-Hinnawi,* United Nations Environment Programme.

Vol 2: **Renewable Natural Resources and the Environment: Pressing Problems in the Developing World**
*Kenneth Ruddle & Walther Manshard,* United Nations University.

Vol 3: **Assessing Tropical Forest Lands: Their Suitability for Sustainable Uses**
*Richard Carpenter,* Editor. East-West Center.

Vol 4: **Fuelwood and Rural Energy Production and Supply in the Humid Tropics**
*W. B. Morgan & R. P. Moss,* United Nations University.

Vol 5: **Economic Approaches to Natural Resource and Environmental Quality Analysis**
*Maynard M. Hufschmidt & E. Hyman,* Editors. East-West Center.

Vol 6: **Renewable Sources of Energy and the Environment**
*Essam El-Hinnawi & Asit K. Biswas,* Editors.

Vol 7: **Global Environmental Issues**
*Essam El-Hinnawi & Manzur-Ul-Haque Hashmi,* Editors. United Nations Environment Programme.

Vol 8: **The World Environment 1972-1982**
*Martin Holdgate, Mohammed Kassas & Gilbert White,* Editors; *Essam El-Hinnawi,* Study Director. United Nations Environment Programme.

Vol 9: **Energy Alternatives in Latin America**
*Francisco Szekely,* Editor. United Nations Environment Programme.

Vol 10: **Integrated Physical Socio-Economic and Environmental Planning**
*Yusuf J. Ahmad & Frank G. Müller,* Editors. United Nations Environment Programme.

Vol 11: **Third World and the Environment**
*Essam El-Hinnawi & Asit K. Biswas,* Editors.

Vol 12: **Development without Destruction: Evolving Environmental Perceptions**
*Mostafa K. Tolba.*

Vol 13: **Climate and Development**
*Asit K. Biswas,* Editor.

Vol 14: **New and Renewable Sources of Energy: Future Prospects**
*Essam El-Hinnawi, Margaret R. Biswas & Asit K. Biswas,* Editors.

# Ecology in Practice

## Part II: The Social Response

*Edited by*

F. DI CASTRI
F. W. G. BAKER
M. HADLEY

*Based on papers presented at the conference "Ecology in Practice — Establishing a Scientific Basis for Land Management", September 22 to 29, 1981, organized jointly by Unesco and ICSU to mark ten years of the Man and Biosphere (MAB) Programme*

TYCOOLY INTERNATIONAL PUBLISHING LIMITED, DUBLIN
UNESCO, PARIS

First published 1984 by the
**United Nations Educational, Scientific and Cultural Organisation**
7 Place de Fontenoy, 75700 Paris, France
and
**Tycooly International Publishing Limited**
6 Crofton Terrace, Dun Laoghaire, Co. Dublin


Published by arrangement with Unesco
and sold throughout the world by
Tycooly International Publishing Limited
6 Crofton Terrace, Dun Laoghaire, Co. Dublin

ISBN 0 907567 80 0 Hardback
ISBN 0 907567 81 9 Flexicover
ISBN 0 86346 043 7 Boxed Library Edition

Typeset and printed by Cahills Printers Limited, Dublin

# Contents

# Preface

THE main aim of *Ecology in Practice, Parts I and II* is to provide a review and evaluation of some critical problems of land management, and to report the findings and experience of a selection of concrete field activities where an attempt has been made to conduct ecological research which responds to the needs of planning and management. In doing so, some more far-reaching questions are examined, such as: "To what extent has ecological research during the last ten years been sufficiently problem-oriented? How to present research results in a form which can be better understood and used by decision-makers? To what extent are the real social demands perceived by the scientist? How can ecologists better fulfil their responsibility to help the general public to better understand environmental problems? What are the most effective means to bring about interdisciplinary as well as international cooperation in the ecological sciences? To what extent can the results and experience obtained in a given situation be useful and applicable elsewhere? What are the emerging issues and trends for the 1980s?"

*Ecology in Practice* is based on papers presented at an international scientific conference: "Ecology in practice establishing a scientific basis for land management", held in Paris from September 22 to 29, 1981. Some 400 scientists and planners from 67 countries took part in the conference, which was organized jointly by the United Nations Educational, Scientific and Cultural Organization (Unesco) and the International Council of Scientific Unions (ICSU).

The conference was convened as one of a series of activities to mark ten years of the Man and Biosphere (MAB) Programme, an international programme of research and training launched by Unesco in 1971. In planning the programme of the conference, a deliberate effort was made to avoid generalized discussion of theory or strategies which had not been backed by practical experience. The conference was based on invited papers, which analyzed some of the technical outputs of specific projects or groups of projects. Authors were encouraged to highlight shortcomings as well as accomplishments.

The sixty-one contributions and the field experience on which they are based, do not offer any simple solutions to the major world problems relating to man-environment interactions. But they do illustrate how a certain approach to land use problems — involving the participatation of planners, local populations and scientists from a range of disciplines — can be made to work in different regions of the world, given the right mix of ingredients and the necessary good will on the part of all concerned. As such, it is hoped that the book will prove of general interest to those concerned with seeking to apply the approaches and techniques of ecological research to the needs of land use planning and resource management.

Divided into two parts and following an introductory contribution in *Part I* there are six main sections. The first part (three sections) deals with the use and management of ecosystems, special attention being given to the problems of (a) the humid and subhumid tropics, (b) grazing and marginal lands and (c) the issue of seeking new approaches for conserving the earth's genetic diversity in the face of increasing demands for land and natural resources.

This second volume, also divided into three sections, is focussed on society's response to some of the principal environmental and resource management problems of today and tomorrow. A first group of contributions treats the issue of ecological approaches for improving the planning of urban systems, those areas where a rising proportion of the world's population is concentrated. The final two sections are concerned respectively with the use of scientific information for environmental education purposes and with providing the types of information needed for decision-making on land management. They thus provide a bridge between ecological research on the one hand and education and decision-making on the other.

Each of the six sections of the book has a short introduction which tries to place the contributions that follow within the broader context of the topic being examined. The first five themes have a similar structure. Each is introduced by a review and evaluation paper, for which one or more specialists were invited to identify: the key scientific problems and prospectives involved in that topic (e.g. what are the key scientific problems in the long-term management of humid tropical ecosystems); the principal constraints and obstacles encountered within the MAB Programme in trying to overcome these problems; and future research and training directions. This review and evaluation paper is followed by one or more invited papers on selected issues in the theme being treated. There follow about six case studies which report on experience gained in particular geographical situations of the land use problem or topic under discussion. These case studies are largely, but not entirely, drawn from the MAB Programme.

The editors thank the contributors for responding so enthusiastically to invitations to present papers to the conference and for putting their papers in a form for publication. They also express their gratitude to the moderators of the various conference sessions on which the book is based: Messrs A. Gómez-Pompa (Mexico), F. Bourlière (France), O. V. Lindquist (Finland), P. Zaremba (Poland), A. El-Hili (Tunisia), H. Sene (Senegal), P. Nogueira-Neto (Brazil), D. Sastrapradja (Indonesia). Thanks are also due to Jacqueline Baker, Alison Clayson, Michèle de Maistre, Lydie Guillaud, Brian Payne and Jane Robertson Verhnes for their help in the tasks of editing, translating and typing of manuscripts.

All contributions to the book are presented by the authors as individuals and do not necessarily represent the views of Unesco, ICSU, or any other organization or jurisdiction.

*F. di Castri*
*F. W. G. Baker*
*M. Hadley*

*March, 1983*

# List of Contributors


Howard F. Andrews
The Child in the City Programme
University of Toronto
455 Spadina Avenue, Suite 424
Toronto, Ontario M5S 2G8
Canada

Serge Antoine
Ministère de l'Environnement et du Cadre de Vie
14, boulevard du Général Leclerc
92521 Neuilly sur Seine
France

Mohammed Ayyad
Alexandria University
P.B. 589
Alexandria
Egypt

Agadjan Babaev
Desert Research Institute
Academy of Sciences
Gogol St. 15
Ashkabad 74400
Turkmen SSR
USSR

F. W. G. Baker
International Council of Scientific Unions
51, boulevard de Montmorency
75016 Paris
France

Paul T. Baker
Department of Anthropology
Pennsylvania State University
409 Carpenter Building
University Park, Pennsylvania 16802
USA

Thelma S. Baker
Department of Anthropology
Pennsylvania State University
409 Carpenter Building.
University Park, Pennsylvania 16802
USA

Mirilia Bonnes
Istituto di Psicologia
Università di Roma
Via Monte di Casa, 65
00138 Rome
Italy

Tahar Bouhaouach
Institut des Régions Arides
4119 Medenine
Tunisia

Stephen Boyden
Human Ecology Group
Centre for Resource and Environmental Studies
Australian National University
Canberra
Australia

Harold Brookfield
Department of Geography
University of Melbourne, Parkville
Victoria 3052
Australia

Sandra Brown
Department of Forestry
University of Illinois
110 Mumford Hall, 1301 W. Gregory
Urbana, Illinois 61801
USA

Eduardo Bustos
Facultad de Medicina
Universidad de Chile
Casilla 13898
Santiago
Chile

F. El Cadi
Association pour la Sauvegarde de l'île de Djerba
Houmt Souk
Tunisia

Francesco di Castri
Division of Ecological Sciences
Unesco
7, Place de Fontenoy
75700 Paris
France

Milka Slavia Castro Lucic
Facultad de Medicina Norte
Universidad de Chile
Casilla 9183
Santiago
Chile

Albert Chapalay
Association MAB-Pays d'Enhaut
CH-1837 Château d'Oex
Switzerland

E. J. Costello
Tussock Grasslands and Mountain Lands Institute
Lincoln College
Canterbury
New Zealand

Charly Darbellay
Directeur du Projet MAB-Pays d'Enhaut
1906 Charrat
Switzerland

R. P. Dhir
Central Arid Zone Research Institute
Jodhpur 35203
India

J. C. I. Dooge
Dept. of Civil Engineering,
University College
Upper Merrion Street
Dublin 2, Ireland

H. Dosso
Institut d'Ecologie Tropicale
BP 109
Abidjan 08
Ivory Coast

Neville Dyson-Hudson
International Livestock Centre for Africa (ILCA)
Box 46847
Nairobi
Kenya

Sir Otto Frankel
Division of Plant Industry
CSIRO
PO Box 1600
Canberra 2601
Australia

Yona Friedman
42 Bd Pasteur
75015 Paris
France

Eduardo Fuentes
Laboratorio de Ecología
Instituto de Ciencias Biológicas
Universidad Católica de Chile
Casilla 114-D
Santiago
Chile

Madhav Gadgil
Indian Institute of Science
Bangladore 560012
India

Gisbert Glaser
Division of Ecological Sciences
Unesco
7 Place de Fontenoy
75700 Paris
France

Monica W. Goigel
University of Georgia
Athens, Georgia 30602
USA

Frank Golley
Institute of Ecology
University of Georgia
Athens, Georgia 30602
USA

Fernando Gonzales Bernaldez
Catedratico de Ecología
Departamento de Ecologia C-XU
Facultad de Ciencias
Universidad Autonoma
Madrid—34
Spain

William P. J. Gregg Jr.
National Park Service
Office of Science and Technology
Washington, DC 20240
USA

Mohamed Sahbi Hadjej
Institut des Régions Arides
Medenine
Tunisia

Abdelhamid Hadji
Institut des Régions Arides
Medenine
Tunisia

Malcolm Hadley
Division of Ecological Sciences
Unesco
7, Place de Fontenoy
75700 Paris
France

Gonzalo Halffter
Instituto de Ecología
Cerrada Monte Kamerum
Mexico 10, DF
Mexico

Rafael Herrera
Centro de Ecología
Instituto Venezolano de
Investigaciones
Científicas (IVIC)
Apartado Postal 1827
Caracas 101
Venezuela

Alexander von Hesler
Umlandverband Frankfurt
Regional Planning Association
Fliederweg II
638 Bad Homburg
Federal Republic of Germany

Martin Holdgate
Department of Environment and
Transport
2 Marsham Street
London SW1P 3EB
United Kingdom

Indonesian MAB National Committee
Indonesian Institute of Sciences (LIPI)
J1. Teuku Chik Ditiro No. 43
Jakarta
Indonesia

John N. R. Jeffers
Institute of Terrestrial Ecology
Merlewood Research Station
Grange-over-Sands
Cumbria
United Kingdom

Jan Jeník
Botany Institute
Czechoslovak Academy of Sciences
37989 Třeboň
Czechoslovakia

Carl F. Jordan
Institute of Ecology
University of Georgia
Athens, Georgia 30602
USA

Kuswata Kartawinata
Herbarium Bogoriense
Lembaga Biologi Nasional
(National Biological Institute)
Jalan Raya Juanda 22-24
Bogor
Indonesia

Jiro Kawakita
School of Environmental Sciences
University of Tsukuba
Meguro-ku, Tokyo
Japan

I. G. C. Kerr
Tussock Grasslands and Mountain Lands Institute
Lincoln College
Canterbury
New Zealand

Hans Klinge
Max-Planck-Institut für Limnologie
AG Tropenökologie
2320 Plön
Federal Republic of Germany

Jan Květ
Botany Institute
Czechoslovak Academy of Sciences
37989 Třeboň
Czechoslovakia

Bruno Lefèvre
Etudes et Planification des Communications
16, rue Chardon Lagache
75016 Paris
France

Gilbert Long
Ecothèque Méditerranéenne
Centre National de la Recherche Scientifique
BP 5051
Route de Mende
34033 Montpellier
France

Thomas E. Lovejoy
World Wildlife Fund-USA
1601 Connecticut Avenue, NW
Washington, DC 20009
USA

Ariel Lugo
Institute of Tropical Forestry
US Forest Service
PO Box AQ
Rio Piedras
Puerto Rico 00928
USA

Bengt Lundholm
Forskningådsnämnden
Box 6710
S-13 85 Stockholm
Sweden

Walter J. Lusigi
Integrated Project in Arid Lands
Unesco Regional Office for Science and Technology for Africa
P.O. Box 30592
Nairobi
Kenya

Michel Maldague
Programme Interdisciplinaire en Aménagement du Territoire et Développement Régional (ATDR)
Université Laval
Tour des Arts, Cité Universitaire
Sainte-Foy, Que
Canada

H. S. Mann
Central Arid Zone Research Institute
Jodhpur 3520003
India

Ernesto Medina
Centro de Ecología
Instituto Venezolano de Investigaciones Científicas (IVIC)
Apartado Postal 1827
Caracas 101
Venezuela

Bruno Messerli
Institut de Géographie
Université de Berne
Hallerstrasse 12
3012 Bern
Switzerland

Paul Messerli
Institut de Géographie
Université de Berne
Hallerstrasse 12
3012 Bern
Switzerland

Hector Luis Morales
Instituto de Investigaciones Sobre Recursos Bióticos (INIREB)
Apartado Postal 63
Xalapa
Veracruz, Mexico

A. Moroni
Institute of Ecology
University of Parma
Parma, Italy

Justin Murphy
CSIRO
Division of Land Resources Management
Private Bag, PO
Wembley, W. Australia 6014
Australia

N. Netchayeva
Desert Research Institute
Academy of Sciences
Gogol St. 15
Ashkabad 74400
Turkmen SSR
USSR

Ken Newcombe
Electricity Commission of PNG
PO Box 2352
Konedobu
Papua New Guinea

Kevin O'Connor
Joint Centre for Environmental Studies
Lincoln College
Canterbury
New Zealand

Tokio Okino
Shinshu University
Suwa Hydrobiological Station
5-2-4 Kogandori, Suwa-shi
Nagano Pref
Japan

Rémi Perelman
Ministère de l'Environnement
34, rue de la Fédération
75015 Paris
France

Duncan Poore
Commonwealth Forestry Institute
University of Oxford
South Parks Road
Oxford
United Kingdom

Oscar Ravera
Department of Physical and Natural Sciences
JRC-EURATOM
21020 Ispra (Va.)
Italy

Ignacy Sachs
Centre International de Recherche sur l'Environnement et le Développement
Ecole des Hautes Etudes en Sciences Sociales
54, boulevard Raspail
75270 Paris Cedex 06
France

Luis Sanchez de Carmona
Dirección General de Ecología Urbana
Department of Human Settlements and Public Works
Calzada al Desierto de Los Leones
Mexico 20, D.F.
Mexico

Egon Seidel
Institut für Wasserwirtschaft
DDD 119 Berlin
Schnellerstrasse 140
German Democratic Republic

Vladimir Sokolov
Institute of Evolutionary Morphology and Animal Ecology
USSR Academy of Sciences
33, Leninskiy Prospect
Moscow

Brian Spooner
Department of Anthropology
University of Pennsylvania
Philadelphia,
Pennsylvania 19104
USA

Erwin Stucki
Institut d'Economie Rurale de l'EPF
CH-1837 Château d'Oex
Switzerland

Christopher Uhl
Institute of Ecology
University of Georgia
Athens, Georgia 30602
USA

Andrew P. Vayda
Department of Human Ecology and Social Science
Cook College, Rutgers University
PO Box 231
New Brunswick, New Jersey 08903
USA

Alberto Veloso
University of Chile
Santiago
Chile

Anne Whyte
Institute of Environmental Studies
University of Toronto
Toronto, Ontario
Canada

Maurice Woodward
CSIRO
Private Bag, P.O.
Wembley, W Australia 6014
Australia

Yang Hanxi
Commission of Integrated Survey of Natural Resources
P.O. Box 767
Beijung
China

Oleg Yanitski
Institute of the International Labour Movement
Kolpachnyi per, 9A
Moscow 101831
USSR

H. A. Ødegaard
International Business Machines
PO Box 1379 Vika
Oslo 1
Norway

Nils A. Øritsland
Norwegian Polar Research Institute
PO Box 158
1330 Oslo Lufthavn
Norway

SECTION 1

# Ecological Approaches for Improving Urban Planning

CHAPTER 1

# Introduction

*Francesco di Castri, F.W.G. Baker and Malcolm Hadley*

OPINIONS differ on what will be the population of the world in the year 2000, 2050, 2100 . . ., but there is general agreement that the world's population will be increasingly concentrated in urban systems. If present trends continue, by the year 2000, about half of the world's population will be urban, with about 60 cities of over 5 million inhabitants and eight of the ten largest cities situated in developing regions. According to one estimate, urban populations in developing countries are expected to increase by an average annual rate of 3.9 per cent in the 1980s and 3.5 per cent in the 1990s, compared to equivalent rates of 1.5 and 1.1 per cent in the industrialized regions.

Observed and anticipated rises in rates of urbanization are fuelled in part by increases in total population, in part by shifts in population distribution from rural to urban settings. There is no reason to suppose that the present attraction of large cities for rural populations will turn around in the immediate future. And the irreversibility of the process of urbanization is becoming increasingly recognized by planners in national administrations and international organizations.

The expected growth in cities during the next few decades will pose enormous problems to those responsible for the planning and day-to-day management of urban systems. Increasing urbanization will have significant effects on the quality of urban life, to the extent of affecting the social and political stability of these systems. The capacity of rural hinterlands to continue to provide food, water and other supplies for urban systems will be increasingly strained.

In short, new approaches to urbanization are required — approaches that are technically feasible, ecologically sound, economically and socially viable, and practicable by relatively weak systems of governmental organization. Time will tell whether such a goal is feasible or is just wishful thinking. But one possibility is through the testing of an ecological approach to problem-solving, with its emphasis on understanding of interactions within and between systems and on long-term perspectives rather than short-term expediency.

Ecological approaches to urban systems cover a range of research methods, including social and perception surveys, assessment of health and nutritional status, energy balances, inventories of city flora and fauna, vulnerability analysis and modelling activities of various kinds. Within the MAB Programme, ecological approaches to urbanization give particular emphasis to uses and flows of energy and materials, to interrelations between urban systems and their rural hinterlands, and to integrating biological and physical analysis with more subjective, socio-cultural considerations of the quality of urban life.

One question left open in the contributions that follow is whether or not a city should be considered as an ecosystem. Some argue that a city is an entirely artificial creation, that it is an artifact of culture or technology, and that as such models derived from the natural or biological sciences are inappropriate. Others argue that, from an evolutionary perspective, man's culture and technology are no less "natural" than any other of his species characteristics, and that as such biological models of natural ecosystems can be justly used in relation to urban systems. While the use of the term "ecosystem" for urban systems might be disputed, there is less controversy that ecological thinking has a role to play in seeking a better understanding of how urban systems work and function.

In his overview, Stephen Boyden (Chapter 2) places integrated studies of cities within the overall context of ecological phases in human history. Today, three quarters of extra-somatic ("non-food") energy that flows through human society passes through cities. It is cities and the industrial processes within them that pour out a large part of the potentially damaging wastes into the atmosphere and oceans, and which are at the root, directly or indirectly, of much rural deterioration. And in more general terms, it is the urban conglomerations that are responsible for many of the ecological changes taking place in the biosphere, changes considered by some to constitute a serious threat to its life-supporting qualities.

Within this overall context, Boyden traces the evolution of MAB activities on human settlements of varying sizes and complexities, through analysis of the recommendations of successive international meetings and the scope and findings of individual field projects. Though methodological progress has been made, much remains to be done in developing overall conceptual frameworks which take full account of intangible and unquantifiable aspects of human settlements (which both affect and are affected by more quantifiable aspects). It is still premature to make an objective evaluation of most field projects. But a major difficulty identified by Boyden is the lack of "human ecologists" to play an essential full-time integrative role within each field project. Boyden concludes with the need to strengthen efforts for fostering an ecological approach to the study of human settlements, and for encouraging increased participation in such studies from the social sciences and the humanities.

In Chapter 3, Oleg Yanitsky addresses some methodological problems in planning the human settlements of the future. Five consecutive stages in an integrated approach to urban planning are recognized, passing from basic and applied research through urban design to the stages of construction and creation of a functional urban system. Three qualitative levels for integrating knowledge within this process are proposed: spatial-temporal, socio-functional and historico-cultural. The architect and urban planner have a key role in mediating between science and the social life of the city, between the scientist and the urban dweller. Another point underlined by Yanitsky is that the presentation of visual designs and plans is important as a means of communication and feedback between those that study, plan and live in urban systems.

Yona Friedman's concern in Chapter 4 is also with the shape of cities to come. But his emphasis is on the need for reducing the dependency of urban systems on their rural environments. In arguing that such a reduction would improve the quality of life in urban as well as rural settings, Friedman traces some of the main steps in the evolution of cities as objects of consumption, and suggests ways through which urban systems could become more self-sufficient in food, water, crafts and services. The promotion of increased self-sufficiency would require new approaches to the allocation of space and time within the city, and would call for well-focussed, innovative research in such fields as biotechnology, agronomy and architecture.

The first of seven case studies, presented by Mirilia Bonnes, describes how a group of scientists, town planners, management agencies and the general public are collaborating in an ecological study of the historical city of Rome, a city whose population has increased by a factor of 15 in a little over a hundred years. The Rome project covers areas as diverse as transport, energy, urban spread, pollution, flora and fauna, environmental perception. The study has aroused such interest that the city council has decided to set up a permanent centre for the study of urban ecology, with a view to promoting research in this field.

Mexico City is one of the world's fast-growing, large cities: 14 million inhabitants now, 30 million in the year 2000 if present trends continue. It is a city where accelerated growth has led to a whole series of problems, such as water supply, waste disposal and air pollution. It is also a city where an urban ecology unit has been established within the federal department of human settlements and public works, whose mandate includes the development of a firmer scientific basis for urban and regional planning in the Valley of Mexico. The case study presented by Luis Sanchez de Carmona gives an overview of the work of this unit in such fields as land allocation, provision of basic services, waste recycling, environmental quality, restoration of degraded environments, assessment of impact of engineering works, development of suburban food plots, and so on — in sum, fields of concern to urban planners and managers in many parts of the world.

The city and its hinterland considered as a biocybernetic system is a central concern of Alexander von Hesler's contribution (Chapter 7), which describes efforts in the Frankfurt region of the Federal Republic of Germany to develop tools useful for decision-making. Steps in the development of a sensitivity model for the Frankfurt region are outlined, including definition of the system, choice of variables, testing the relevance of the variables, identification of patterns of interactions within the system, refinement of a simulation model. The model is useful in elaborating planning alternatives, but also provides a tool for better understanding the functioning of the system, for education and for communication.

Lae is the major industrial city in the newly emerging nation of Papua New Guinea. Like many other cities of the tropical world, Lae has grown rapidly in the last ten years or so, with considerable migration from rural areas and increasingly acute problems linked with the use of energy. Ken Newcombe describes how an ecological analysis of the sources, flows and end uses of energy in Lae has had a marked practical impact, within one to three years of the research project starting. A series of steps have been taken for conserving energy, including use of "waste" energy for compost. Measures for diversifying energy sources have included replanting of denuded hillsides with fast growing trees for firewood and the development of subsistence level agro-forestry systems in town allotment gardens. Another practical follow-up is that the research project has triggered the formulation of the government's national energy policy, and a strategy for its implementation.

The effects of urban systems on their surrounding rural environments provides the context of the study in Chapter 9. Tokio Okino describes the impact of urban wastes and other nutrient inputs on an aquatic ecosystem, Lake Suwa in Japan. Long-term intensive studies on Lake Suwa have produced a detailed biological picture of the lake. The contribution of ten different sources to nutrient loading of the lake has been determined: the major source of phosphorus loading is wastewater from permanent residents and a sanitary plant, while rice and vegetable fields where fertilizers have been applied are the main source of nitrogen inputs. A simulation model of the effects of sewage treatment on the eutrophic state of the lake has been developed, and it is hoped that completion of the sewerage system in 1985 will enable 50 and 75 per cent of nitrogen and phosphorus loads respectively to be collected.

In contrast to the biological approach at Lake Suwa, the case study presented by Howard Andrews has entailed a systematic and detailed compilation of services and settings relevant to children in Toronto. Among the conclusions of the Child and the City programme, is that the cultural values, social structures and institutions that determine the characteristics of urban space and social organization are fundamentally more significant in influencing children's lives than the locational attributes and characteristics of urban spaces and facilities *per se*. A final conclusion is that genuine concern for the interests of children in urban settings implies a far greater sensibility on behalf of planners and policy-makers to the changing circumstances in which families function and manage their lives on a day-to-day basis within urban-industrial society.

CHAPTER 2

# Integrated Studies of Cities Considered as Ecological Systems

*Stephen Boyden*

Abstract

The need for an integrative ecological approach to the study of the problems of human society, and to planning for the future is the central theme. In particular, attention is drawn to the need for integrative concepts and methods which reflect the ecological significance of cities and other human settlements (in terms of their impact on the natural environment and its food-producing systems), and which take account of the relationships between societal development and the environments, life style, health and well-being of people. The MAB Programme, through its Project Area 11 on ecological studies of urban systems, has played a unique and very interesting role in clarifying the conceptual and methodological issues and in promoting and encouraging a number of pioneering integrative ecological studies of human settlements. The strengths and weaknesses of this aspect of the MAB Programme and suggestions for future work are examined.

*Résumé*

*La nécessité d'une approche écologique intégrée pour étudier les problèmes de la société humaine et la planification du futur constitue le thème central. L'attention se porte en particulier sur la nécessité d'élaborer des concepts et des méthodes intégrées exprimant l'importance écologique des villes et d'autres établissements humains (du point de vue de leur impact sur l'environnement naturel et ses systèmes de production alimentaire) et qui tiennent compte des rapports entre le développement de la société et l'environnement, le style de vie, l'état de santé et le bien-être des populations. Le Programme MAB, et plus spécialement les activités relevant du projet No. 11 sur les études écologiques des systems urbains, a joué un rôle exceptionnellement intéressant pour clarifier les questions conceptuelles et méthodologiques et pour promouvoir et encourager un certain nombre d'études écologiques intégrées d'avant-garde concernant les établissements humains. Les points forts et les points faibles de cet aspect du Programme MAB ainsi que des directives pour le travail à entreprendre sont examinés.*

## INTRODUCTION

IN addressing the question of "Ecological approaches for improving urban planning", it is suggested that "improving urban planning" has two fundamental objectives: first, to make cities and other human settlements[1] better places to live in; second, to ensure that what goes on in and around cities does not damage, and is indeed protective of the natural environment and its life supporting ecosystems — on which human beings and civilization are dependent for health and survival.

These two objectives are reflected in what can be called the "triangle of human ecology"[2](Figure 2—1). This triangle summarizes the concerns of human ecology,

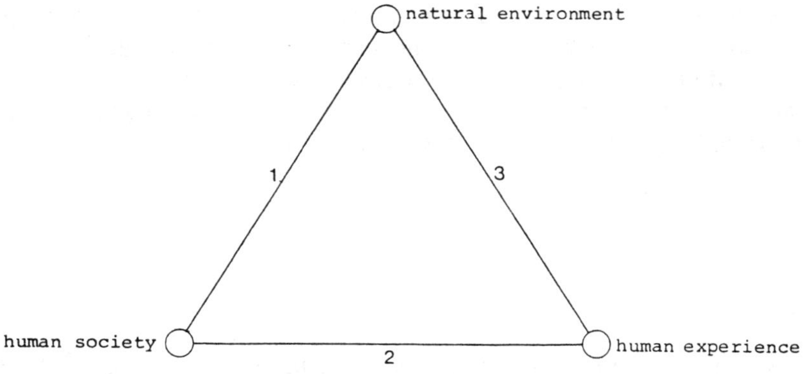


**Figure 2—1.** The triangle of human ecology

[1]The title of this contribution refers specifically to "cities". However, most of what is written here applies not only to cities, but to human settlements in general.

[2]The term "human ecology" has been used for a number of different approaches to the study of human situations. The sense in which the term is used is explained in the first part of this contribution.

which is the study of dynamic interrelationships and interdependencies in human ecosystems. The three corners of the triangle are designated human society, natural environment and human experience. It reminds us that we are interested in what goes on in society (e.g. patterns of machine and energy use) in relation both to their impact on the life-supporting processes of the natural environment and to their implications for the quality of human experience, especially that aspect of human experience known as health and well-being. It is imperative that plans for change in urban ecosystems take account of these dual concerns.

The theme addressed here must be considered against a background of certain human ecological principles and perspectives. There is not space however to deal with these in any depth. However, a few brief comments are necessary. Thus in human history four distinct "ecological phases" can be recognized (Boyden, 1979). These four phases, which coexist, are listed below in the order in which they first made their appearance:

Phase 1. The primeval (or pre-domestic, hunter-gatherer) phase. This was by far the longest.

Phase 2. The early (subsistence) farming phase. This began about twelve thousand years ago in some regions of the world.

Phase 3. The early urban phase. This started about five thousand years ago in Mesopotamia and not long afterwards in China and India.

Phase 4. The modern, technological phase. This began with the industrial transition, starting in Europe and North America about 170 years ago. The transition to the fourth phase is still taking place in many parts of the world.

Each of the four phases has quite distinct characteristics with respect both to the ecological interrelationships between human society and the natural environment and to the life conditions and health and disease patterns of people. The characteristics of the phases will not be described here. However, the point that the characteristics of Phase 4 (e.g. rapidly accelerating use of extrasomatic energy and of non-renewable resources, increasing outpouring of gaseous and chemical wastes and consequent disequilibrium between human society and the biosphere) are such that it cannot possibly last indefinitely is simply emphasized. This is an ecological truth, about which there can be no argument. The only thing which is debatable is the question of how much longer the phase can go on before there occurs irreparable damage to the biosphere. In other words, there lies ahead a *fifth ecological phase* of human existence.[3] The great task facing humankind today is to work out a way of making a smooth transition to a humanly acceptable ecological Phase 5.

[3]This is taking the optimistic view. There are many who believe that mankind is incapable of bringing under control the accelerating processes which threaten the biosphere and civilization, and that the collapse of the system will be total and will involve the extinction of the human species.

A brief analogy helps to illustrate the scale of the problem and the magnitude of the impact of human society on the biosphere. For this analogy, the time-scale must first be changed, and it imagined that the domestic or neolithic transition to ecological Phase 2 occurred some sixteen hours ago, and that at that time a section of the human species jumped into a new vehicle that it had invented (that is, agriculture and the domestication of animals and plants for food production). Let us also suppose that the speed of this vehicle at a given point of time is proportional to the actual amount of energy utilized by human society — this being a good indicator of the overall impact of civilization on the biosphere. Sixteen hours ago, then, our ancestors stepped into their new invention, and set off at a speed of 1 km $hr^{-1}$. Eight hours later they had gathered a good deal of speed, and were travelling at 50 km $hr^{-1}$ — that is, the human species was having a 50 times greater impact on the biosphere. Then the vehicle continued to accelerate so that by twenty minutes ago it had reached a speed of 800 km $hr^{-1}$. And then, quite recently its rate of acceleration increased dramatically, and now, at this moment of time, it travels at 10,000 km $hr^{-1}$.

If certain widely accepted predictions about future increases in the level of energy use turn out to be correct, by the year 2050 humankind would be using about as much energy as is used by all other animals and plants put together. One does not have to be an ecologist to be able to see that this kind of growth in energy use, associated as it is with a parallel increase in use of resources and out-pouring of wastes, cannot go on in the biosphere for ever.

It is, of course, the industrial nations of the West which are mainly responsible for the accelerating use of resources and energy and discharge of wastes and for the perceived threats to the biosphere. Nevertheless, this ecological issue is also of immense relevance to non-industrial countries in many ways, and it is particularly pertinent to their policies for future development. We are, after all, in the same boat — we are all supported by the same biosphere. As far as the non-industrial countries are concerned, it is possible that, through the wise use of both accumulated knowledge from all over the world, and through the wise application of local traditions and experience, they may be able to enter the fifth phase — the phase of ecological equilibrium and human equity and well-being — before this change is achieved in the West, and perhaps with much less social disruption.

The ephemeral nature of the fourth ecological phase of human existence is clearly of overwhelming relevance to the MAB Programme as a whole. Its special pertinence to considerations of ecological approaches to improved urban planning lies in the following facts:

1. The characteristics of Phase 4 which relate to the ecological interrelationships between human society and the natural environment and which are not sustainable in the long run, are mainly due to the activities or "metabolism" of cities.

2. The cities of the fifth ecological phase of human existence will necessarily be different in many ways from those of the fourth phase — a fact with important implications for urban planning.

Thus, with respect to side 1 of the triangle of human ecology, planning for societal and economic change in human settlements must take full account of the interrelationships between settlements and the natural environment — in terms of both the capacity of the immediate hinterland to support life and to produce food, and of the inevitable changes which must take place on the global level if an ecological Phase 5 is to be successfully attained. At the same time, with respect to side 2 of the triangle, urban planning must also take account of the implications of different societal options for the health and well-being of people.

In the following paragraphs, the key ecological problems involved in the planning and management of urban systems are discussed. The role of scientists (and other professional groups) in this area is then referred to, before turning to discussion of the extent to which MAB has successfully addressed these problems and the extent to which it has failed to do so. Finally, the author's views on the role that Unesco, through its MAB Programme, might play in the future are presented.

The comments in this paper are based solely on the author's association with Unesco's MAB Programme, MAB Project Area 11, and on ecological research in Hong Kong (Boyden *et al.*, 1981) and that of the author's colleagues in Papua New Guinea (Christie, 1981; Dalton, 1979; Jeffries, 1978; Newcombe *et al.*, 1980). It is stressed that he himself has no experience in urban planning.

## ECOLOGICAL PROBLEMS IN THE MANAGEMENT AND PLANNING OF HUMAN SETTLEMENTS

It is self evident that there are very big differences in the nature and the scale of problems of societal planning in different regions and places. The challenges facing planning authorities are very different, for example, in the following places, all of which are sites of MAB field studies on human settlements: the settlements of Kuru Kururu in tropical Guyana; the city of Rome in Italy; the city of Lae and its hinterland in Papua New Guinea; the city of Mexico (with its population of 14 million); the administrative city of Jalapa, capital of the State of Veracruz, also in Mexico; the island of Götland in Sweden; the settlements in arid areas in the northern Sahara in Egypt; and the city of Toronto in Canada. Indeed, the problems facing planners in these various places are so disparate and diverse that critics might be excused for questioning whether it makes any sense to try to deal with them all together under a single heading.

It is the theme of this contribution, however, that there is one fundamental problem of enormous significance which is common to all these human settlements — industrialized or not, urban or rural, tropical or arctic, wet or arid. It is a problem which has been recognized at all meetings on MAB Project Area 11, which is referred to later as being of critical relevance to wise constructive planning for development in the future.

In essence it is the problem of fragmentation, compartmentalization and narrowness which characterizes the conventional approach of decision-makers and their advisers, and which leads again and again to ill-conceived, piecemeal, short-term and frequently disastrous results. To put it the other way around, the problem lies in the failure of governmental authorities and policial groups to take a comprehensive, integrative and interrelational approach to the analysis of urban problems or to the planning of changes for the future.

There is no shortage of evidence in the Western urban conglomerations of the detrimental side-effects of narrowly conceived developments of different kinds. In non-industrial countries the same principle applies. Development programmes, often involving the transfer of technologies from the industrial countries, have again and again ended in failure (numerous such cases are reported, for example, in the fifty chapters of the volume *The Careless Technology* edited by Farver and Milton, 1973). Some of the undesirable consequences have been in the form of severe damage to local ecosystems and others have involved harmful effects on the physical health of human populations. In other instances, development has interfered with aspects of the social experience of people.

Clearly, something is seriously wrong with the approach which has been taken up to now. Plans for development have been drawn up, approved and implemented without any proper understanding of the total situation and without due regard to all the likely consequences of the proposed changes.

A little more should be said about an integrative ecological approach, emphasizing that it is not only academic research or educational programmes being referred to, but also policy formulation. Take the energy question as an example. Everything that happens in a human settlement or a nation involves the expenditure (and conversion) of energy. Moreover, the amount and form of energy used, and the way in which it is used have profound effects both on ecosystems and on human experience. Policies with respect to energy use relate, for instance, to all the following issues: programmes of industrialization and farming; alternatives with respect to labour intensive versus energy intensive industry; problems of air and noise pollution; threats of ionizing radiations; transport systems; recreation patterns; the impact of society on the natural environment; the spatial distribution of place of residence in relation to place of work for the work force; the amount of time individuals spend commuting.

It also is self-evident that the way energy and machines are used is relevant to the daily experience of individual people — not only in relation to goods and commodities available for them in the consumer society — but also because it affects life style — the kind of jobs people have and their level of job enjoyment and satisfaction, the extent to which they practice learned manual skills, the degree of variety they experience in daily life, and the extent to which individuals are offered incentives and opportunities for personal creative activity. Indeed, there are good arguments for the view that the pattern of energy and machine use can ultimately influence such relatively intangible aspects of life experience as family interaction, community experience, sense of personal involvement and sense of belonging.

The pattern of energy use is, of course, a reflection of the kinds, number and size of machines used, as well as of the organization of society and of the behavioural characteristics of corporations (as distinct from the behavioural characteristics of individuals). In turn, the pattern of machine use is the resultant of various cultural factors, including, in particular: the technology of the society (that is the state of knowledge relating to the design, manufacture and use of machines); the dominant value system, the economic system and the power structure of a society as these affect decisions and actions influencing the applications of technology.

Thus, when the energy problem is considered, an interrelational as opposed to a piecemeal approach is absolutely essential, because options for the future with respect to energy use must be considered in relation to all these different aspects of the situation. But the same principle applies to all other societal issues, whether they concern the design of the built environment, overall economic policy, educational programmes, international relations, technological options for the future, problems of unemployment, drug abuse or increasing crime rates. And all these issues relate to each other.

The very nature of the threats and unresolved problems which face our societies today is such that they will persist and worsen, until such time as we apply to them a truly interrelational integrative holistic approach. These generalizations are unlikely to arouse much controversy. It would appear that most people agree that the fragmentation and compartmentalization so characteristic of our societies and their institutions are undesirable and that a more holistic and integrative approach to planning is necessary. The disturbing thing is that, despite this general consensus, little or no improvement in the situation is discernible.

There are several reasons for this, not least of which is the fact that the integrative approach, essential as it is, is also very difficult. It is not sufficient merely to arrange interdepartmental conferences with specialists from different areas sitting around a table, each throwing in his penny's worth of expertise, only to return after the meeting to the security and comfort of his own area of specialization — leaving no one at the table to continue the integrative process. As will be discussed later, more people and groups of people are urgently needed whose full-time effort is devoted to the study of the dynamic interrelationships in human situations as they relate to the problems of the present and to the options for the future.

The contribution of "scientists" (by which is meant natural scientists, social scientists and students of the humanities) towards the solution to this problem will now be considered.

## THE ROLE OF SCIENTISTS

Recognizing the urgent need for a better understanding of interrelationships in human situations between different kinds of variables (e.g. biotic, physical, chemical, cultural, economic, social, etc.), other contributors to this publication urge more effective cooperation and interaction between natural scientists and social scientists. Such interaction and cooperation is not sufficient to overcome the problem. The difficulty lies deeper than this.

The greatest impediment to progress in this area is the almost complete lack of a profession of full-time *human ecologists.* This point can be illustrated by reference to the situation which has developed in biological science. Participating in the study of nature (as distinct from the study of human society), many groups of specialists can be recognized — zoologists, botanists, physiologists, geneticists, microbiologists, biochemists, physicists, chemists, molecular biologists, ethnologists, geologists and others. And within each area of specialization there exist numerous subspecializations. However, for the study of the interrelationship and interdependencies between different aspects or components in natural ecosystems, there has necessarily developed another group of specialists — the ecologists. Clearly, the knowledge gained by the zoologists, botanists and other specialists is of great relevance to these interrelationships. But the ecologists have had to develop, and are continuing to develop a conceptual basis and methodology for studying, understanding and communicating about the dynamic interrelationships between biotic, chemical and physical processes. In so doing, they have come to recognize a series of important ecological principles relating, for example, to biogeochemical cycles and patterns of energy flow. There has also been a good deal of controversy, as in the case of the debate which exists today concerning the relationship in natural ecosystems between diversity and stability.

In principle, the situation in human ecosystems[4] is similar, the main difference being that there are further sets or kinds of variables which have to be taken into consideration. Thus, while the work of the zoologists and other specialists listed above is still relevant, so is the work of medical scientists, political scientists, sociologists, economists, psychologists, historians and students of the arts. But, as in the case of natural ecosystems, understanding of the interrelationships and interdependencies in human situations demands a great deal of hard integrative work by individuals who are prepared to devote their efforts on a full-time basis to developing concepts and methods aimed at improving holistic understanding of human situations. These "specialists" are referred to as human ecologists.

One of the tasks of human ecologists is to interact with specialists who have sympathy for the integrative approach and to seek from them information relevant to the situation which is the subject of an integrative analysis or description. By communicating the essentials from their respective areas of expertise, the specialists help the integrators to improve understanding of important interrelationships and to identify key issues (Figure 2—2).

Another important task of the human ecologists is to interact with planners and interested community groups who are involved in the formulation of societal

[4]The term human ecosystem is used to mean an ecosystem of which human beings are a part.

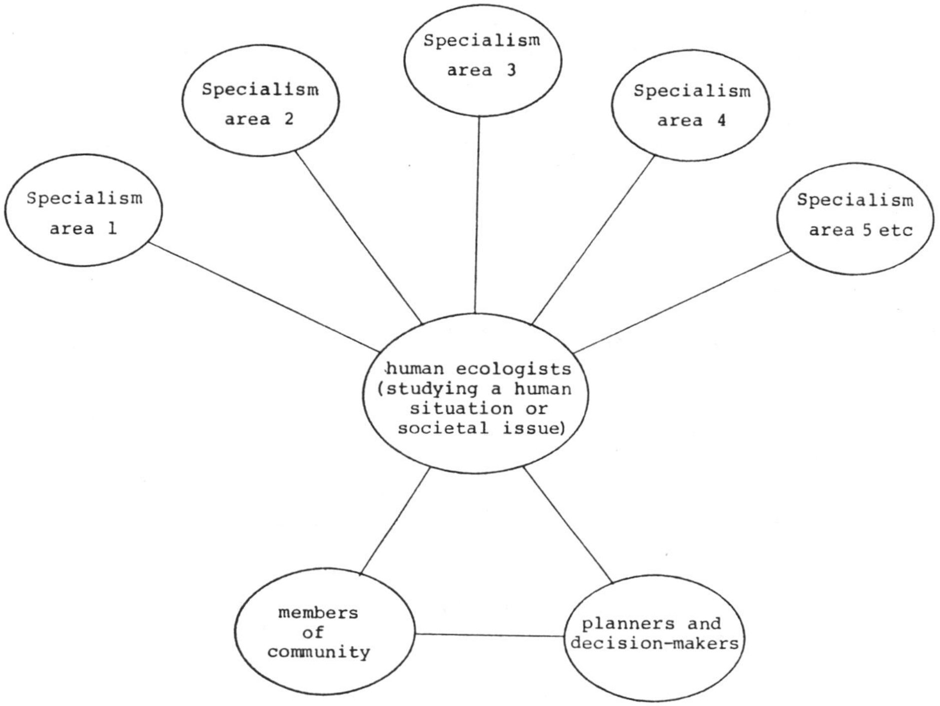


**Figure 2—2.** Interaction between human ecologists and specialists, members of community and planners

policies. The purpose of this interaction is partly to identify the problems as seen by planners and members of the community, and partly to draw their attention to the important interrelationships between different aspects of the situation. Furthermore, human ecologists can assist by introducing planners and concerned members of the community to methods which will ensure that they give consideration to the full spectrum of variables relevant to human well-being and relevant to the effects of societal processes on the natural environment. Human ecologists thus act as a facilitating and integrative link between specialists, planners and members of the community.

To summarize the situation, there is an urgent need for a great deal more intellectual effort to be devoted on a full-time basis, to the study of the dynamic interrelationships in human situations between different aspects of these situations, both natural and cultural. As a consequence of such intellectual effort, there will emerge a body of theory and methodology which will be of immense benefit to human society. To put it another way, the extraordinary imbalance which exists at the present between specialism and generalism in academic institutions must be corrected. Every encouragement must be given to individuals, whatever their academic background, who are prepared to step outside their initial area of specialism and to concentrate on the study of human situations and societal issues in a truly integrative way, thus contributing to the development of the infant discipline of human ecology. Unesco, through its MAB Programme, can make a useful contribution to this process.

## THE ROLE OF THE MAB PROGRAMME

### International and regional meetings

First, to quote from the final report of the first session of the International Coordinating Council of the Programme on Man and the Biosphere, 1971:

The Council recalled that the Man and the Biosphere Programme is an interdisciplinary programme of research which emphasizes an ecological approach to the study of interrelationships between man and the environment . . . It will focus on the general study of the structure and functioning of the biosphere and its ecological divisions, on the systematic observation of, and research on, the changes brought about by man in the biosphere and its resources, on the overall effects of these changes upon the human species itself, and on the education and information to be provided on these matters (Unesco, 1972).

With respect to urban problems, the report had this to say:

The extremely rapid growth of cities and the development of industrialization, in both developing and developed countries, is associated with problems of ecosystem modification and of human adjustment to changing environments. These problems range from those of environmental imbalance to those of nutrition and public health, to those concerned with the broadest aspects of national and regional economy and society.

In order to obtain a proper understanding of these problems, it appears, on the face of it, that it is necessary to approach the study of these systems from multiple points of view, in terms of their influence on other parts of the biosphere and by using the community knowledge of both the social and natural sciences. These systems are in a state of continuing disequilibrium, and their transformation from one sort of complex system to another requires investigation of factors relating to their disequilibrium state. Urban environments are aggregates of human beings intensively using primary and secondary products produced elsewhere, accumulating wastes which are disposed of rather than recycled and thus radically changing the natural order of nutrient distribution and availability.

An holistic comparison of the dynamics of urban and industrial ecosystems can best be approached by focussing attention on certain particular characteristics of urbanization and industrialization.

A useful entry into the problem follows from the recognition that the interactions between man and the rest of the biosphere increase with the product of the number of men and the energy expenditure per man. Experience has shown that these interactions are, all too often, deleterious. Thus it would seem appropriate to promote studies on the per-capita needs of different human societies and settlements, directed particularly toward the research and development of technologies which will permit human life patterns of high quality but with more rational use of energy.

Member States, with the assistance of competent international organizations, are invited to participate in an interdisciplinary programme of studies on the functioning of urban and industrial systems, with emphasis placed on the unifying and fundamental role of energy in man-environment interactions in these systems (Unesco, 1972).

This statement referred in particular to MAB Project Area 11, which was then known as "Ecological aspects of energy utilization in urban and industrial systems". It provided the starting point for a unique and interesting series of meetings on this theme at international and regional levels.

The first of the meetings was that of an *ad hoc* panel of experts held at Bad Nauheim, in the Federal Republic of Germany in October 1973 (Unesco, 1974). This was followed by a meeting of an interregional task force in Paris in June 1975, the report of which sets out in more detail, guidelines for MAB field projects on human settlements (Unesco, 1976). Since that time, regional meetings have been held in Amsterdam (Netherlands), Poznan (Poland), Caracas (Venezuela), Mexico City (Mexico), and Khartoum (Sudan). These meetings have been attended by individuals of diverse backgrounds, including architects, town planners, physicians, economists, biologists, psychologists, nutritionists, systems analysts, historians, and sociologists. The most striking aspect of the meetings has been the consistency with respect to the general "philosophy" behind the approach of MAB Project Area 11, and the extraordinary degree of consensus about what needs to be done. In general, the views expressed and the recommendations made agree with, or are at least consistent with, the assessment of the key problems outlined in the first part of this contribution. This is illustrated by some quotations below.

During the course of these meetings, the scope of MAB Project Area 11 has been widened, and the project area now deals with integrated ecological studies on human settlements. This change reflects the point made above, namely that the fundamental principles of the human ecological approach which is advocated apply to all kinds of settlements — and not only to cities.

## The triangle of human ecology

The meeting of the *ad hoc* panel held at Bad Nauheim in 1973 (Unesco, 1974) recognized that there are fundamentally two reasons for interest in the ecology of urban systems and other human settlements: (1) the various effects of societal change and development on the ecosystems of the biosphere which sustain the human population; (2) the interrelationships between, for example, changing patterns of use of resources and of energy, and the health and well-being of people.

The panel recognized that the ecological problems which can be studied in the city are multiple. However, it was appreciated that for the purposes of this report, the discussion and recommendations should be restricted to sets of problems of particular relevance to the Man and the Biosphere Programme. In essence, these problems are related to the complex relationships between man, his urban environment, and the biosphere. The urban environment, which is largely determined by various facets of the social and economic structure (technology, value systems, social organization etc.), significantly affects both the well-being of man himself and the state of the biosphere on which he ultimately depends. These relationships need to be examined so that the city can develop in the best possible way. There are problems in both understanding and controlling these relationships and it is to a purposeful examination of them that this report is directed (Unesco, 1974).

This aspect was developed further at later meetings and was particularly clearly stated in the report of the Poznan meeting:

The meeting strongly reaffirmed its support for the integrative ecological approach to the study of urban ecosystems and to the application of the results of such studies in physical, economic and social planning. This support is based on appreciation of the fact that in the long-term urban ecosystems must: (1) exist in a state of ecological equilibrium with the ecosystems of the biosphere on which they depend, and (2) satisfy the material, social and experimental needs of their inhabitants. That is to say, the future physical, technical, social and economic development must ensure that the integrity of the biosphere is protected, providing at the same time conditions which are consistent with an enjoyable life for a healthy human population. The development must be in harmony with nature and in harmony with the human spirit (MAB National Committee of Poland and Unesco, 1977).

**The need for a more integrative approach to urban planning:** This theme has dominated all the MAB Project Area 11 meetings. One quotation from the Poznan meeting, will suffice.

Certain impediments hindering the integrated ecological approach to the improvement and development of cities were noted, and suggestions within the framework of MAB Project 11 are expected to contribute to the removal of these impediments. For example, in most countries the emphasis on specialism (e.g. in governmental departments and academic institutions) is very strong, and this is reflected in a very compartmentalized and fragmented approach to urban problems. Clumsy bureaucratic machinery also impedes the progress towards enlightened ecological planning (MAB National Committee of Poland and Unesco, 1977).

With respect to MAB Project Area 11 field activities, several of the meetings have emphasized the need for the involvement of full-time integrators (or "human ecologists" in the sense described above) at the centre of each field activity. To quote from the Paris (1975) meeting:

Each project would involve a local integrating group and a number of specialist groups. One member of the local integrating group should be designated to act as coordinator of the project as a member of the regional integrating group.

*Local integrating groups:* Each local integrating group should consist of a central core of members interested primarily in the integrative process, including the coordinator of the local project and at least two other professional staff. The local integrating group might also include the leaders of the major specialist groups, planners, decision-makers and, in some cases, representatives of the community.

The functions of the local integrating groups might be as follows:

1. To interpret the conceptual framework and hypotheses of the programme in terms of the local circumstances;
2. To take responsibility for the overall organization of the local study;
3. To organize local specialist groups;
4. To take responsibility for all funding of the local study, including work of the specialist groups where necessary, by direct communication with the funding organizations;
5. Ultimately to prepare an integrated and comprehensive description of the local system in terms of the conceptual framework of the programme. This description would be based on the data from numerous sources including the following: reports of government departments and the academic literature, new investigations or assessments of existing information by the specialist groups and new integrative and multidisciplinary studies carried out by the integrating group and aimed, for example, at providing information on interrelationships between environmental and life style variables on the one hand, and human health and well-being on the other (Unesco, 1976).

**Concepts and methods:** The fact that there has not yet emerged a widely accepted integrative conceptual framework for the ecological study of human settlements has been recognized throughout successive meetings as a serious deficiency.

The report of the Paris (1975) meeting had this to say on the subject:

A detailed conceptual framework was acknowledged by the task force to be essential for conducting integrated ecological studies on human settlements. Based upon background materials presented at the meeting and a review of several projects currently underway, a general statement of concerns and guiding principles in matters of substance and organization was drafted and approved by the task force. It is intended to serve as an initial step toward evolution of a conceptual framework for the programme. Elaboration of these concerns should be a continuing task in Phase II and Phase III.

While recognizing the inadequacies of present methodologies, the task force considered the development of an ecological approach to the analysis of human settlements an objective of the programme which will require conceptual model building, testing and continual reformulation throughout the programme. Similarly it will be necessary to build up a methodology for integrated studies of the type proposed, drawing upon the more promising approaches found within informational and organizational theory as well as research and analytical techniques developed in the natural and social sciences (Unesco, 1976).

In other words, the meeting was suggesting that the MAB Programme itself should contribute to the development of a scientifically valid and useful conceptual framework and an integrative methodology for the study and planning of human settlements.

It was recognized that systems analysis, involving mathematical modelling, has a most important contribution to make in this area.

The limitations of mathematical modelling have, however, been recognized, and considerable emphasis has been placed on the need for a conceptual or methodological approach which takes full account of intangible and unquantifiable aspects of human situations (which both affect and are affected by the more quantifiable aspects). This viewpoint is reflected in the following quotation from the Amsterdam meeting in 1976:

> Furthermore, easy quantifiability should not be the criterion for selection of variables for consideration. There is no law of nature or of society to the effect that there is likely to be any relationship between quantifiability and importance. Variables for analysis and description should be selected on the basis of what is judged to be important in the light of the concepts and principles of human ecology and concern for human well-being. Obviously it is easier to deal with figures than with vague notions or ideas and, wherever possible, efforts should be made to collect data in quantifiable form. Nevertheless, very great caution is called for in this area, as well as a constant awareness of the danger of attaching numerical values to aspects, for example, of human behaviour or experience on the basis of possibly false assumptions (Unesco, 1978).

**The involvement of planners and decision-makers:** It has been implicit in all the meetings that MAB projects should be useful to society, and not mere academic exercises. Although most of the projects have been inspired and initiated by members of academic institutions, the advantages of involving planners and decision-makers in the planning and implementation of these projects have frequently been stressed. This aim is clearly spelled out in the main policy statement drawn up at the Paris task force meeting in 1975:

> . . . emphasis will be given to involving decision-makers in all stages of the project in order to closely link research with development planning. Constant interaction and feedback between decision-makers and research scientists would help clarify the methodological and managerial approaches necessary for problem solving (Unesco, 1976).

The extent to which this policy has, in fact, been put into effect varies considerably from project to project. It has certainly been a feature in particular of the Polish and the Papua New Guinea contributions to MAB Project Area 11.

**Education:** Another recurring theme has been the contribution that MAB projects on human settlements might make to education in the principles of human ecology — principles of great relevance to the problems of human settlements and to options for future change. This aim is spelled out in the following quotation from the report of the 1975 Paris meeting:

> Educational activities related to the programme could be of crucial importance in building public understanding. An informed public is more likely to be supportive of the effort, receptive to its results and responsive to the decisions made in accordance with those results, than an uninformed public. Changes in management policies of human settlements related to the results of scientific research work are more likely to be accepted by the general public if the process and principles used are understood as well as the research results (Unesco, 1976).

Again, there is tremendous variability in the extent to which MAB projects include educational themes. There is a strong emphasis on education, for example, in the Lower Main Project in the Federal Republic of Germany, where a series of exhibitions for the general public has been prepared to encourage holistic and ecological considerations in planning. A fifty-minute documentary film entitled *Hong Kong Project* was prepared by the Australian Broadcasting Commission and Rediffusion Hong Kong about the integrative study of the ecology of Hong Kong.

**Involvement of the public:** There has been increasing discussion in MAB meetings on human settlements of the involvement of communities in planning and decision-making and, indeed, in MAB field projects themselves. This concern is reflected in the following paragraph from the report of the Poznan meeting:

> Many participants stressed the importance of involving the community in planning and decision-making and in the evaluation of alternative strategies relating to physical, technical, economic and social planning and of options relating to applications of technology. However, as in the case of professional planners and politicians, such involvement ought to be based on an understanding of the integrative ideas and principles of ecology, and consequently there is a need for improved communication between human ecologists and communities — at all levels of education (MAB National Committee of Poland and Unesco, 1977).

## SUMMARY

The following statement from the Paris meeting in 1975 summarizes well the approach of MAB Project Area 11 and at the same time explains in a nutshell how the integrative ecological approach is fundamentally different from that of other scientific disciplines:

> Settlements have previously been modelled and measured as systems in terms of flows of energy, people, material and information, etc, The importance of an ecological approach is not in the elaboration of any one of these models as much as in the attempt to:
>
> 1. Integrate these various approaches within holistic studies located in defined geographical areas or settlement systems, while maintaining a concern for the analysis of the multiple interrelationships;
> 2. Focus attention on the interactions between settlements and their life support systems;
> 3. Emphasize ecosystem limits and capacities as a necessary basis for settlements planning, to complement more generally accepted economic, technological and socio-political criteria (Unesco, 1976).

Attention should be drawn in particular to the second item, which recognizes the vital importance of studying human settlements as part of larger ecosystems rather than in isolation, each interacting with, and ecologically dependent upon the local natural environment, other human settlements and ultimately, of course, the biosphere as a whole. This simple but essential concept has been emphasized throughout the MAB meetings and is depicted in Figure 2—3.

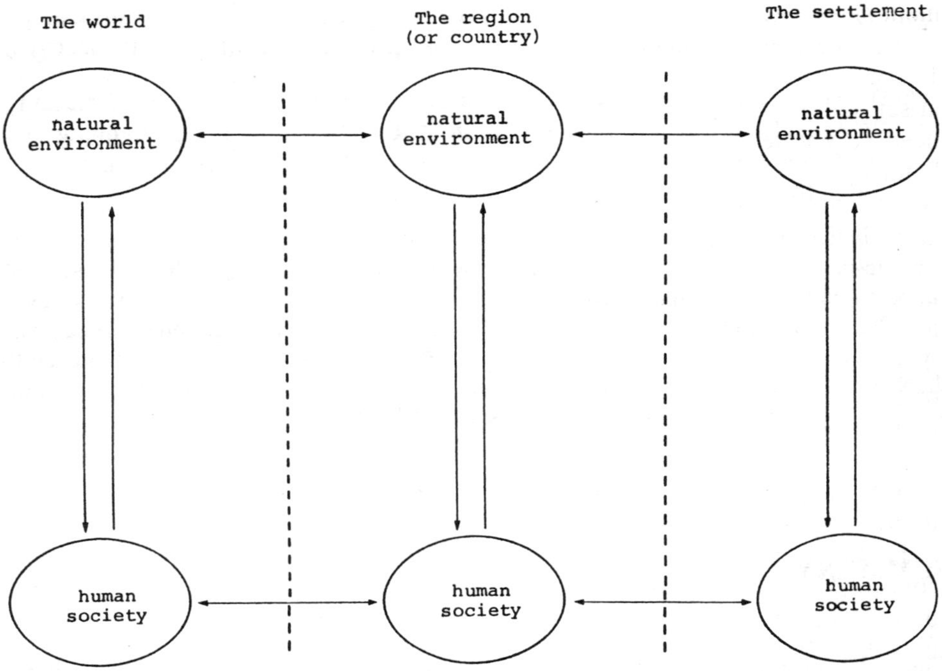


**Figure 2—3.** Ecological interrelationships between the settlement, the region (or country) and the rest of the biosphere

## Field projects

To attempt a thorough evaluation of the various studies which are listed as contributions to MAB Project Area 11 would, for several reasons, be an extremely difficult, and not very rewarding task. There are about 70 studies, very diverse in purpose, design and conceptual basis. Indeed, it is suspected that the only thing common to them all is a sympathy shared by the project leaders for the philosophy of the Man and the Biosphere Programme and of MAB Project Area 11 as described in the various reports. These projects include, for example, studies on the patterns of flow of extrasomatic energy in an island society (e.g. in Götland), the social and economic problems of children in a large city (e.g. Toronto), the use of plants as indicators of chemical changes in the urban environment (e.g. in Vienna and in the Federal Republic of Germany), the development and testing of environmental education materials (e.g. in Indonesia), and the application of sophisticated techniques of systems analysis to the problems of urban settlements (e.g. the development of so-called "sensitivity models" in the Federal Republic of Germany).

Broadly, two classes of study in the list of contributions to MAB Project Area 11 can be recognized. On the one hand there is a large group of projects which have arisen independently of the MAB Programme, and which would have probably been very little different had MAB 11 Project Area never existed. They form a heterogeneous group with respect to objectives, subject matter and general conceptual and methodological approach.

The second, very much smaller group is made up of those projects which either owe their very existence to MAB 11 or which at least appear to have been much influenced by the international meetings which are referred to above. They can all be truly described as "integrative ecological studies on human settlements", and they are referred to here as the "special MAB 11 projects". They are situation-oriented and represent a deliberate effort to put into effect the principles outlined in the reports of the MAB meetings. They reflect concern about the disadvantages of narrow, compartmentalized approaches to the problems of human settlements, and they involve the application of the interrelational concepts of human ecology to the study of the problems of these settlements.

These projects are concerned with the development and implementation of concepts and methods which can be used for research which is aimed at improving understanding of the dynamic interrelationships within human settlements and between them and their hinterlands. It is intended that these concepts and methods will also be useful in integrative and comprehensive planning and policy formulation. Their purpose in the latter context would be to make sure that, when options are considered for the future (e.g. the introduction of a new technology, or the establishment and design of a new settlement), all likely consequences for the natural environment, for the economy, and for people, will be taken properly into account.

Included in this group of special MAB 11 projects are a series of projects in which the main emphasis is on research — such as the projects in Egypt, Guyana, Indonesia, Italy, Japan, Pakistan, Papua New Guinea, Thailand and Venezuela. Also included are certain other MAB 11 projects, such as those in Poland and Mexico in which the main emphasis is on planning, but which, to some extent at least, reflect the guidelines and principles set out in the reports of the MAB meetings.[5] More interaction between individuals involved in these planning projects and those working in the more research-oriented projects would be very useful.

Most of these projects share the following characteristics to a greater or lesser degree:

[5]There are a few projects associated with other areas of MAB which have developed independently of MAB Project Area 11, but which share enough of its general philosophy and objectives to be regarded as belonging to this group (e.g. the human settlement component of the Turan programme in Iran).

1. They have taken the reports of the MAB meetings on human settlements as a starting-point for planning.

2. They are concentrating on a given settlement or region.

3. They are concerned with variables ranging from patterns of energy use and the patterns of flow of materials (e.g. water, foodstuffs, polluting substances) to descriptions of the quality of life and the health and well-being of people.

4. They are based on integrative conceptual frameworks.

5. They involve the active participation of individuals from a variety of different academic disciplines, and in some cases representatives of decision-making bodies are participating in the work.

These pioneer studies are, with the exception of the Papua New Guinea project, in their early stages, and it is too soon to assess their usefulness and success in terms of results. But the very fact that in each of these countries multidisciplinary teams of individuals are working together with the aim of improving understanding of the ecological interrelationships in human ecosystems (e.g patterns of flow of energy and materials) and their implications for the natural environment and for the life experience of individual human beings is a significant and exciting development.

Most of these projects have distinct relevance to the formulation of policies for the future. In the case of Guyana, for instance, the integrative ecological approach is being applied to the study of some newly planned rural settlements in the interior. It is hoped that the analysis and description of these settlements, in terms of the integrative concepts of human ecology, will assist authorities in planning further new settlements in such a way as to ensure their ecological, economic and social success. Another aim of the Guyana study is to demonstrate to the government the usefulness of a holistic integrative ecological approach to societal problems.

In the Venezuelan project, a comparative study is being carried out on two contrasting cities — Ciudad Bolivar and Ciudad Guayana. The former is a relatively traditional, colonial-type town, the latter a modern, planned, industrial city. It is hoped that a comparison of these two settlements, in terms, for instance, of patterns of use of energy and machines, the design of the built environment, the use of material resources, the services offered to the community, and patterns of health and well-being and of behaviour in the human populations will yield insights which will be useful in the formulation of plans for the future, and will encourage governments to take a more integrative approach to planning.

In Papua New Guinea, the study had been centred on Lae, but extended to its hinterland, including the Highlands, and also took account of the interrelationships between this region and the wider global ecosystem. The project has dealt with patterns of energy and land use and has also been concerned with the real implications — for the physical health of people and for the way they spend their time and their sources of enjoyment — of the transition from a rural to an urban way of life. It has resulted in recommendations and in some policy changes.

Another good example is the project in Thailand called "Integrated ecological studies of changes in the suburban area north of metropolitan Bangkok." This study is centred in the Institute of Environmental Research, Chulalongkorn University. The multidisciplinary nature of the project is indicated by the following list of themes under investigation: demographic and socio-economic characteristics; land use and settlement of human beings; health in relation to the settlement of human beings; natural ecosystems in conjunction with background environmental information; environmental quality in connection with human settlements; the balance of energy as a result of human settlements and its implications for future planning.

These projects are all breaking new ground and are consequently facing many difficulties; it would be surprising if we could not find in them some weaknesses and deficiencies. But each of them is likely to be useful both in improving understanding of the problems of human settlements, and in influencing planning in the future.

It has been possible to examine some of the working papers and early progress reports produced by individuals involved in planning and implementing the various projects which are underway. Even at this early stage, some comment can be made on the extent to which these projects achieve the aims and objectives of MAB Project Area 11, as set out in the various reports of MAB meetings, and to make note of certain difficulties which have been experienced.

## Integration

The projects in this group aim to make an integrative approach to the study of the ecology of certain human settlements. They are all concerned, to a greater or lesser extent, with such aspects of the study area as patterns of energy flow, patterns of flow of certain materials, the impact of societal processes on the natural environment and on the life-style and patterns of health and disease of the human population. They all involve the participation of a considerable number of experts from different areas of academic specialism.

However, almost without exception, these projects all suffer from one major difficulty. They lack a group of human ecologists to play the essential integrative role on a full-time basis at the centre of the project. Of course, since human ecology is not as yet a recognized academic discipline, professional human ecologists who could take on this role hardly exist. On the other hand, this part could be played by individuals with backgrounds in anthropology, biology or human geography, provided they are willing to step outside the paradigms of their original disciplines and to put intellectual effort into developing a truly integrative and comprehensive conceptual framework for the research. The trouble is that even when such people do exist, institutional constraints, reflecting the conservative attitudes and the general inertia of academia, interfere with their ability to take on this role.

**Involvement of planners and decision-makers:** The task force which met in Paris in 1975 and all subsequent regional meetings have recommended that planners and decision-makers be invited to participate in the planning and implementation of MAB field projects. This ideal has not been achieved in all the special MAB 11 projects.

A good example of a MAB project which has achieved this aim, is the Papua New Guinea study based on the city of Lae. Members of the Lae City Council and members of the staff of the council were actively involved in the project, which clearly contributed in various ways to new policies and action (Newcombe and Pohai, 1981). This interaction between scientists and planners has also been a feature of the MAB 11 programme in Poland.

**Involvement of community groups and dissemination of information:** Although the various MAB meetings have advocated community participation in MAB projects and the dissemination of information arising from the projects to the public, this has not been a feature of most of the MAB research programmes on human settlements. According to the proposals for the Rome project, however, this may well prove to be an exception. In the Polish MAB programme there is also much emphasis on public participation in the process of planning.

Work is also in progress at the Australian National University which aims to generate discussion at the community level based on the philosophy and conceptual framework of MAB Project Area 11.

## CONCLUSIONS AND COMMENTS ON THE FUTURE

In summary it is suggested that the most important contribution that the MAB Programme has made to an ecological approach to urban planning comprises two aspects. First, a series of unique documents arising from meetings of an *ad hoc* panel of experts, an international task force and various regional meetings to discuss MAB Project Area 11 has been produced. These reports reflect a real consensus on the urgent need for an integrative ecological approach to research, planning and education and they are seen as an early groping towards the development of a scientifically valid and practically useful conceptual framework on which such an approach might be based.

The second important aspect of the MAB contribution in this area consists of a small and, it is believed, unique series of pioneering projects aimed at studying certain human settlements in a comprehensive integrative ecological way. These projects follow to a greater or lesser degree the recommendations and guidelines described in the various MAB reports. They are aimed at improving understanding of the dynamic ecological interrelationships in human society and those which exist between human society and the natural environment and between human society and the environment, life style, health and well-being of the people who live in it.

Most of these special MAB 11 projects have received financial support from Unesco and UNEP.

In conclusion three points should be emphasized. First, the author strongly agrees with the fundamental philosophy and the broad recommendations of the MAB 11 reports. These recommendations can be summarized and depicted as in Figure 2—4 which represents a combination of Figures 2—1 and 2—2.

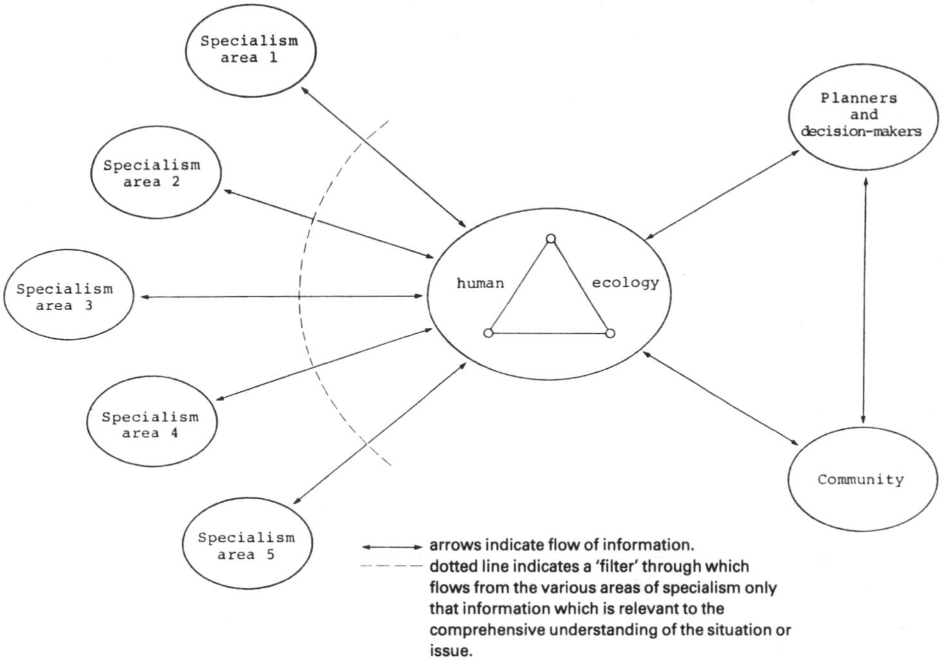


**Figure 2—4.** The academic and social role of human ecology

In essence, it is clear that an area of study and activity must be developed, which has been referred to here as "human ecology", the function of which is to bring together facts (often derived in different areas of academic specialism) in the context of given situations or issues, and to study and improve understanding of the interrelationships between these different aspects of reality, especially as such understanding can contribute to wise policy formulation. The approach must be one that takes account of the developments of human society and options for the future, in so far as they affect the natural environment and as they affect the quality of human experience (hence the inclusion of the triangle of human ecology in the centre of Figure 2—4). The other essential function of human ecology is to maintain lively interaction with communities and decision-makers. This field of study and activity should thus act as an effective bridge between specialism and the realities of the everyday world, especially in relation to the ecological sustainability of human settlements and the environments, life style and health and well-being of their inhabitants.

In the author's view, Unesco, through its MAB Programme, has an extraordinary opportunity, indeed an obligation to nurture the infant to which it has given birth in the form of MAB Project Area 11. It is hoped that new ways will be sought for encouraging those who are actively working in this area in different countries, and who are trying to develop an effective integrative theoretical framework, as well as practical methods for applying this theory to the very serious problems which face human society in both industrialized and non-industrialized countries. It is suggested that consideration be given to the possibility of establishing international task forces with this purpose in mind, to further the exchange of ideas in this area. There is also tremendous scope for the establishment of international training programmes.

The second point has already been implied in the first. However, it cannot be over-emphasized: all integrative work (e.g. on patterns of energy use, land use, on the impacts of human society on natural ecosystems) will be wasted if human beings and their institutions are left out of analyses, descriptions and plans. For, it is self-evident that they are extremely important elements of human ecosystems, both as major influences on the ecological characteristics of such systems, and also because they are much affected by these ecological characteristics. Thus, the MAB Programme in the future should be strengthened by greater input from the social sciences and the humanities. All too often the approaches of the natural sciences, social sciences and the humanities are seen as alternative approaches, even to some extent in competition with each other. This view makes no sense at all in terms of our efforts to understand human situations in a comprehensive way.

Finally, a point should be made about the ecological importance of human settlements. The MAB Programme is concerned, as pointed out earlier, with ". . . the changes brought about by man in the biosphere and its resources . . ." (Unesco, 1972). It is abundantly clear that it is the urban conglomerations which are mainly responsible for the ecological changes which are taking place in the biosphere and which are considered by many ecologists to constitute a threat to its life-supporting qualities. Seventy-five per cent of the extrasomatic energy flowing through human society is used in cities. It is the cities and the industrial processes within them which discharge the major part of the potentially damaging wastes into the atmosphere and into the oceans. Moreover, much of the ecological deterioration in rural areas is indirectly the result of urban influences. However, as already implied, there is a real need for the integrative ecological approach to be applied to the study and planning of human settlements of all sizes, and not just cities. Indeed, there is a good deal of practical value to be learned from comparative ecological studies of settlements of different sizes and at different stages of development (bearing in mind that not all development has been to the advantage of people in the long run).

In conclusion, it is believed that Unesco has made a very promising and worthwhile start in promoting an integrative ecological approach to the study and planning of human settlements. It is also believed that it can make a substantial contribution, by continuing to foster such work, to a successful transition of society from the ephemeral fourth ecological phase to a lasting fifth phase of human existence.

**References**

Boyden, S. (1979), *An Integrated Ecological Approach to the Study of Human Settlements*, MAB Technical Notes 12, Unesco, Paris.

Boyden, S., S. Millar, K. Newcombe and B. O'Neill (1981), *The Ecology of a City and its People: the Case of Hong Kong*, Australian National University Press, Canberra.

Christie, M. (1981), *Changing Consumer Behaviour in Papua New Guinea: its Social and Ecological Implications*, Report No. 3, Unesco/UNEP/CRES Papua New Guinea Human Ecology Programme, Centre for Resource and Environmental Studies, Australian National University, Canberra.

Dalton, K. J. (1979), *Chimbu People under Pressure: the Social Impact of Urbanisation*, Report No. 2, Unesco/UNEP/CRES Papua New Guinea Human Ecology Programme, Centre for Resource and Environmental Studies, Australian National University, Canberra.

Farver, F. M. and J. P. Milton (Editors) (1973), *The Careless Technology: Ecology and International Development*, Tom Stacey, London.

Jeffries, D. J. (1978), *From Kaukau to Coke: a Study of Rural and Urban Food Habits in Papua New Guinea*, Report No. 1, Unesco/UNEP/CRES Papua New Guinea Human Ecology Programme, Centre for Resource and Environmental Studies, Australian National University, Canberra.

MAB National Committee of Poland and Unesco (1977), *Coordinating Meeting on MAB Project 11: with Emphasis on the Implementation of Ecological Research in Integrated Planning in Industrial Countries*, Institute of Architecture and Physical Planning of the Technical University of Szczecin, Poland.

Newcombe, K., K. Bowman, M. Christie and J. Pokris (1980), *Energy for Development: the Energy Policy Papers of the Lae Project*, Report No. 4, Unesco/UNEP/CRES Papua New Guinea Human Ecology Programme, Centre for Resource and Environmental Studies, Australian National University, Canberra.

Newcombe, K. and T. Pohai (1981), The Lae Project: an Ecological Approach to Third World Urbanization, *Ambio*, 10(2-3), 73-8.

Unesco (1972), *International Coordinating Council of the Programme on Man and the Biosphere (MAB)*, First Session, Final Report, MAB Report Series No. 1, Unesco, Paris.

Unesco (1974), *Expert Panel on MAB Project 11, Ecological Effects of Energy Utilization in Urban and Industrial Systems*, MAB Report Series No. 13, Unesco, Paris.

Unesco (1976), *Task Force on Integrated Ecological Studies on Human Settlements, within the Framework of MAB Project 11*, MAB Report Series No. 31, Unesco, Paris.

Unesco (1978), *Planning Meeting for MAB Project 11, with Emphasis on Industrialized Settlements*, MAB Report Series No. 42, Unesco, Paris.

CHAPTER 3

# Integration of Social and Natural Sciences for Urban Planning

*Oleg Yanitsky*

Abstract

Urban planning is an important stage in the development and implementation of an integrative concept for human settlements of the future ("eco-city"). Five consecutive stages in this process are identified (fundamental research, applied research and development, urban design, construction and development of a functional urban system), as well as three qualitative levels in the integration of knowledge (spatio-temporal, socio-functional and historico-cultural). Among problems discussed are: the integration of knowledge and know-how at each stage of the urban planning process; applying the integrated knowledge and know-how when passing from one stage to another; social interpretation of ecological knowledge; the methodology for defining the object of research and management; correlation of technological and socio-cultural dimensions of urban design; correlation between different levels of integrated knowledge and the forms of social activity.

*Résumé*

*Une étape importante du développement et de la mise en oeuvre du concept intégré des peuplements humains du futur ("éco-ville") est la planification urbaine. Dans le processus sont sélectionnées cinq étapes successives (recherche fondamentale, recherche appliquée et développement, planification proprement dite, réalisation technique, mise en place du système urbain), ainsi que trois niveaux qualitatifs d'intégration des connaissances (spatio-temporel, socio-fonctionnel et historico-culturel). Les problèmes suivants qui sont parmi les plus importants, sont considérés:*

*l'intégration des connaissances théoriques et appliquées à chaque étape, et durant la transition entre ces étapes; traduction sociale des connaissances écologiques; méthodologie pour la définition de l'objet des recherches et de la gestion; corrélation entre les dimensions techniques et socio-culturelles de la planification urbaine proprement dite; et corrélation entre les niveaux de connaissance intégrée et les formes d'activités sociales.*

## INTRODUCTION

THE relationship between integrated knowledge and urban planning, which itself serves as a necessary link between science and urban life is considered here. During the last ten years, thanks mainly to research within the MAB Programme, significant success has been achieved in the development of a scientific (sociological and ecological) basis for urban planning. The following achievements seem to be among the most significant:

1. Development of an interdisciplinary approach to the city, making it possible to reveal important links between its economic, social and ecological substructures and its spatial organization (Boyden, 1977, 1979; Chistiyakova, 1975; Glazychev, 1978; Kostrowicki, 1979; Stearns and Montag, 1979; Yanitsky, 1970, 1980a, 1982); development of methods for integrated, ecologically oriented planning of urban regions (Zaremba, 1977, 1980); creation of mathematical models, making it possible to forecast the "behaviour" of urban systems (Vester and von Hesler, 1980).

2. Strengthening links between a complex of social and natural sciences and urban design and policy. In socialist countries, the compliance with economic, social and ecological criteria is envisaged at all levels of spatial planning: beginning with a general plan of settlement on a national scale extending to plans for the complex development of cities and their districts (Borschevsky *et al.*, 1975; Khodzaev *et al.*, 1977).

3. Formation of an independent branch of applied knowledge — ecology of town planning — with the task of translating ecological, social and other knowledge into the language of urban planning and design (Chistiyakova, 1975; Glazychev, 1975; Kogan, 1976; Vladimirov, 1980). Normative bases for these projects and model urban planning schemes corresponding to social and ecological requirements have been developed from this research.

4. Development of theoretical and empirical studies of ecological aspects of urban and rural ways of living, of the influence of people's attitudes towards

the natural environment upon the way which the environment is modified and its resources used, of the ecological needs of various groups in a population and their participation in improving their environment (Baranov, 1980; Yanitsky, 1978, 1980b). Such studies, in addition to their cognitive value, also represent a form of feedback between the life of the city and applied urban science and urban design.

5. Elaboration of scientific bases for ecological education and training of town planners and urban managers (e.g. Unesco, 1977) also constitutes a necessary link between science and practice.

Thus a certain "critical mass" of knowledge and know-how aimed at the solution of urgent tasks of designing and building cities has been accumulated. At the same time, when projecting into the future, when considering the future of human settlements, a much closer interaction of social, natural and technical sciences is required; a much deeper analysis of the entire system "science-urban design-urban environment" is called for.

In order to define more sharply the emerging problems, these problems can usefully be considered in respect to the prospect of creating the "eco-city", i.e. a human settlement of the future in which social and ecological processes are combined in the best possible way. Such an approach seems to be justified insofar as it helps to reveal fundamental problems of socio-ecological synthesis. Despite the fact that subsequent analysis in this contribution is largely theoretical, practical experience has been gained in all the problems that at present characterize the real interdependence of sciences and urban planning, or are expected to emerge in the foreseeable future.

The initial methodological prerequisite is the fact that urban design should be considered in a broad scientific and social context. In other words, design is one of the stages in proceeding from fundamental research to the formation of an urban environment. There are five such stages (Figure 3—1): (i) fundamental research; (ii) applied research and development; (iii) urban design proper; (iv) construction processes; and (v) formation, or maturing of a city "organism". The importance of the last two stages for urban design should be particularly emphasized, since they tend to have been neglected in the past. Though primary interest is in moving from the first stage to the last, there is also an important feedback, for example of the influence of construction techniques and methods on the design process.

## LEVELS OF INTEGRATION

Given the interest in the degree of integration of knowledge which reflects the degree of integration of social and ecological concerns in the development and implementation of the concept of "eco-city", Figure 3—1 attempts to portray

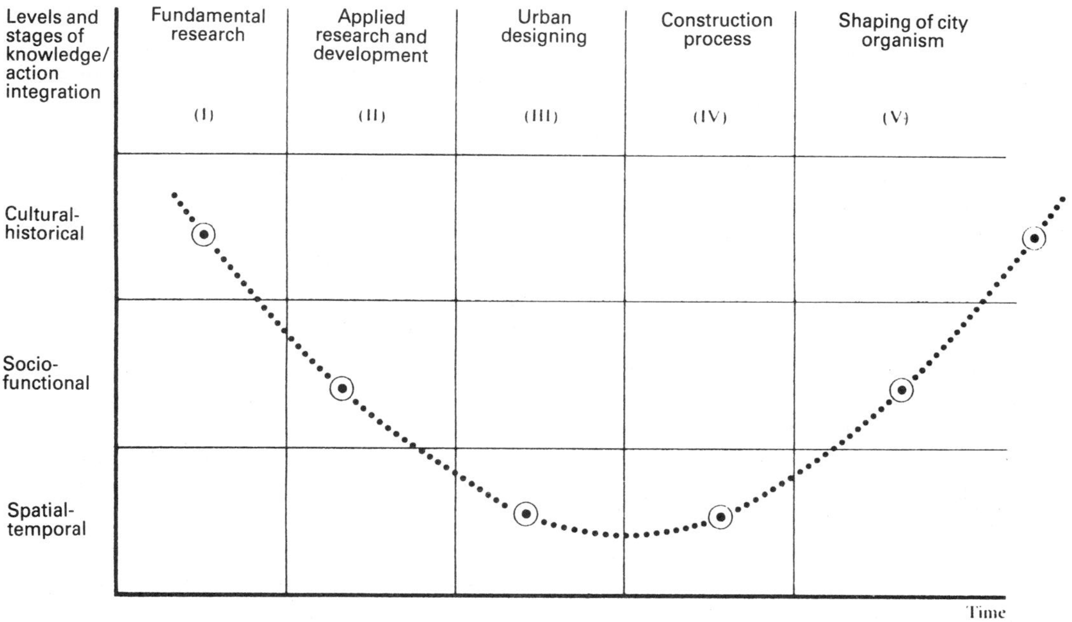


**Figure 3—1.** A matrix scheme illustrating the development and implementation of the "eco-city" concept. The abscissa represents the development of the process in terms of time, the ordinate the levels of integration of knowledge of the reality it represents

relations between the five stages mentioned above. The abscissa represents the development of the process in terms of time, while the ordinate shows the level of integration of knowledge or the reality it reflects. The concept of level of integration requires some explanation.

As applied to the object under consideration, three qualitative levels of integration of knowledge can be recognized: spatial-temporal, socio-functional (organizational) and historico-cultural. The spatial-temporal level is very important because both social and ecological processes of the city have a clear-cut organization in terms of space and time. In addition, the entire man-made technical environment which modifies the interaction between social and biological systems also has clear-cut spatial boundaries, but only at the first level. Any attempt to consider it as the only possible level or to try to reduce all social relations to spatial ones, as was done by the founders of the Chicago school of urban ecology, will mean the adoption of a position of social positivism.

The second level is that of urban organization and of the totality of knowledge that reflects such organization at the socio-functional level. It does not annul the validity of the previous level, but limits it to a condensed form. Complex systems of organization exist behind man's "direct" interactions with nature. On the one hand, there are social systems (production systems, social infrastructure, etc.), on the other hand biological ones (natural and man-made biocenoses). Though different qualitatively, both of them have one common link — that of exchange processes carried out by man in urban areas through technological systems. It is this second level of organization of urban ecosystems that has undergone the most thorough analysis within the framework of the MAB Programme.

The third level of interdisciplinary analysis is a cultural-historical one. Its object is an analysis of deep socio-economic causes that result in conflict between city and nature, of the ways of achievement of those goals set by society, and of the actions required of each individual to overcome such conflicts. The leading role in this process is played by socio-philosophical and historico-cultural analyses. The classic tenets of Marxism-Leninism have invariably emphasized the fact that changes in the conditions of human activities result, in the final instance, in change of man himself (Marx and Engels, 1968). Today, at the same time as enjoying the advantages of urban civilization, man must take care to conserve and protect the environment and nature, at least in the state that we find them at present. Thus, new social and cultural goals emerge, goals guiding interdisciplinary interactions in a special way.

It goes without saying that the above ensemble of stages and levels in the development of an integration process is no more than a bare outline. But it represents in a graphic form one view of the "universe" within which efforts should be placed for the most complete and effective integration of knowledge in urban design.

It is clear that two types of problems exist: "vertical" (or synchronous) problems, relating to the integration of knowledge or practical activities within each of the above mentioned stages, and "horizontal" (or diachronous) problems, that emerge during the transition from one stage to another or that are associated with the entire cycle of "knowledge-practice". Connections between these two types of problems depend on the general character of the transition from integrated knowledge to action (i.e. from the development of the "eco-city" concept to its practical implementation). But given a gradual process of stage-by-stage implementation, the development of the concept may be represented by a U-shaped curve (Figure 3—1). Although this is only a model, it does reflect changes in emphasis in the sequence from knowledge to practice. Several points are of particular importance here.

First, the shape of the curve indicates that, in passing from the integrative concept of "eco-city" to its implementation, there is a discernable lowering in the degree of integration of knowledge and action, because it is not possible as yet to bypass Stage (IV) in the process, namely the phase of creating the technical (construction) basis of the urban settlement. The role of technology in mediating between the social and biological processes of the city explain the necessity of "lowering" theoretical thinking and practical activities to a spatial-temporal level. Whether it will always be necessary to pass through this bottleneck of spatial-temporal reduction, or whether there exists other ways of realizing the integrative concept of "eco-city", is a matter for debate.

Second, note should be made of the similarity between the right and left hand parts of the curve, i.e. between the development of the concept and its practical implementation. In accordance with V.I. Lenin's theory of reflection, this can be considered as a correspondence between the levels of theoretical analysis of an object and the levels of organization of its vital activities.

Third, direct informative links exist between the focal points of the right and the left hand parts of the curve, at each of the three levels. For the left hand branch, i.e. for various levels of knowledge and skill, the corresponding points on the right hand side serve as goal reference-points. They represent meaningful social reference-points in the creation of the problem-solving organization of knowledge and skill at each level in the development of the "eco-city" idea. On the other hand, in the process of translating ideas into practice shown on the right hand side, each succeeding stage is controlled and regulated by the corresponding notions developed during theoretical analysis and design.

## SOME METHODOLOGICAL PROBLEMS

Different categories of problems are encountered at different stages in the task of providing integrated knowledge for urban design. At the first stage, i.e. during the development of an integrative concept of the "eco-city", a central methodological problem is that of interdisciplinary transition, of mutual interpretation of data on the subject at hand from both social and natural sciences points of view. The objective basis for this interdisciplinary transition is the interaction between the various components and processes that combine to create the integrity of the urban ecosystem. Ecological problems have today become social ones due to the emergence of socially meaningful links between human needs and the quality of life.

In solving simple problems, such mutual transitions from one branch of knowledge to another come about relatively simply, as a result of human experience. But knowledge based on experiment has proved inadequate for the study and subsequent creation of such complex systems as large urban areas. There is therefore a need to develop special methods for social interpretation of ecological knowledge, as well as for ecological interpretation of social knowledge. Such methods can only be elaborated as a result of systematic cooperation among representatives of the social and natural sciences in the course of solving a common problem. The development of the "eco-city" concept is an excellent field for such cooperation. In this respect, the idea of "full-time integrators" put forward by Boyden (1979) has acquired the character of an organizational imperative.

Naturally this does not mean that ecological research and knowledge should be absorbed and engulfed by social studies. Ecological knowledge is included in social research only in proportion to its social significance. Social sciences utilize only a certain socially meaningful part of ecological knowledge, but this nevertheless does create problems of interdisciplinary interaction. Such communication across disciplines is a two-way process, as the ecological significance of social phenomena are gradually introduced in ecological research.

The problem of interdisciplinary transition has been evident for some time, as instanced by the well-known works of Dubos (1970) and Shvartz (1974) on the social and biological significance of the levels of diversity of urban environments. This concept has also acquired demographic and genetic meaning, as manifest by discussions on the integral genetic characteristic of the "adaptive norm" in a human population (Dubinin and Altuchov, 1976). The development of an integrating notion of "human health" (Sidorenko and Prockopenko, 1976) follows a similar direction. Long-term research in the field of comprehensive socio-economic and ecological assessments of urban areas — carried out by scientists from Czechoslovakia, the German Democratic Republic, the Federal Republic of Germany, Hungary, the Soviet Union, and other countries within the MAB Programme — also has a direct bearing on the question at hand. The development of a multi-level research strategy makes it possible to study the interactions between the component parts of an intricately organized system. Such a methodological approach is particularly important for studies dealing with the problems of spatial planning. In this respect, an interdisciplinary approach should be implemented within the framework of general historico-materialist theory, and particularly in respect to those concepts which in sociology are called "theories of middle level". In the present case, the concept of urbanization (Yanitsky, 1980a) provides the link between the historical process and the development of the urban system.

## The socio-functional concept of the urban ecosystem

One of the central problems of the following second stage is the socio-functional concept of the object, that is the urban ecosystem. The urban ecosystem has social, ecological and technological dimensions, and thus calls for interdisciplinary study. At the same time the system must be represented spatially, as a basis for the next stage of spatial (urban) design. Analysis of the urban system is therefore a complex one, requiring simultaneous synthesis and disaggregation of knowledge, as well as its technical embodiment and spatial representation. In this context, it should be emphasized that technical solutions are not simply a means of implementing a functional concept, but they also have an important influence on shaping the concept.

Consideration of the city as an ecosystem is far from clearcut. Thus there are diametrically opposed views about the possibility of existence of such an object. Some authors, while resorting to biological interpretations of the concept of the ecosystem, maintain that a city is not an ecosystem because it builds itself by destroying nature. Others, also on the basis of biological interpretations of the concept, feel that a city could become an ecosystem in the future if it becomes completely included in natural cycles. In my view, if this standpoint is taken to its logical conclusion, it amounts to the denigration of the very specificity of the city as a social organism. Finally, there is a third group of authors who maintain that the social history of the transition from hunting and gathering to economic activity, from the use of natural resources to their reproduction, serves as a basis for the development of the "eco-city" concept. This third point of view, being of a historical character, appears to be the best founded, and it has its confirmation in the idea of "ecosocial development". But there is still need for thorough study in concrete ways of the amalgamation of production and reproduction functions in relation to society and nature within the framework of an urban ecosystem.

For instance, despite the importance of studies that reveal the structure of materials and energy flows in contemporary cities, such studies are not sufficient to allow us to comprehend the integrity of an urban ecosystem. If one aim is to control the vital activities of the urban ecosystem, the inadequacy of these kinds of data becomes apparent. At present there is need for detailed studies on energy and materials flows in qualitatively different subsystems that make up the overall system, analyses that would be based on joint socio-historical and bio-evolutionary research. To date, such studies have developed independently. The separate strands need to be brought together in the interest of understanding the interconnected development of the social and ecological subsystems of the city.

There is a big research challenge in making the methods of an interdisciplinary approach operative. The notion of "filter" (Boyden, 1979), widely used at present, is not completely adequate in that it does not go beyond the framework of a quantitative, materials-energy understanding of man's interaction with the urban environment. Thus, the concept of the diversity of urban environment, which is of such great interdisciplinary significance, is not embraced by the notion of "filter". It is also insufficient for characterizing the great variety of types of interactions between man and this environment — direct and indirect, unilateral and multilateral, stimulating or obstructive. Social sciences have a leading role to play in the development of methods for study of these diversities and interactions.

In the course of socio-economic and urban-ecological analysis, the need of making more accurate assessment of the dimensions of the object of study, for design and management purposes, has become more apparent. At present, the urbanized area (Pchelintsev, 1980; Vladimirov, 1980) is considered to be synonymous with such an object. This fact greatly increases the scope of disciplines involved in interactions and complicates the tasks of the knowledge synthesis. While building the concept of such an object the researcher, guided by the goals of ecosocial development, must at the same time draw upon an established method of historical analysis. Regional systems represent very stable natural-historical entities, sometimes preserving over a period of centuries a number of their special features. Contemporary research has revealed the significance of entire territorial systems for social and economic development; their importance to the cultural development of each individual is far from declining, in spite of the growing mobility of production and population.

The structure of these regional systems has been, and is being, profoundly modified by the scientific-technological revolution and resulting alterations in the requirements and way of life of people as well as changes in the ecological situation. Thus the problem of object has both prognostic and historical dimensions, whose interrelations require an interdisciplinary analysis. The relationship between spatial and sectoral systems — a traditional concern in regional analysis but one that is acquiring new dimensions — is also linked with this question.

### Objects of study and of management

A related aspect concerns the relationship between the object of study and the object of management. To date, we have mainly studied the influence of the urbanized environment upon the natural environment. But it is clear that the city cannot be created without urban habitats and other ecosystems artifically created and maintained by man. Yet another functional aspect is the role of the city in supporting such translocal processes as the migration of birds. These various interrelations underline the inadequacy of existing monodisciplinary knowledge for developing a functional concept of "eco-city". For instance, questions relating to the creation of and technology for maintaining urban and regional habitats have not been sufficiently studied.

In a corresponding way, the question of a sociological model of the region — and first and foremost that of the organizational level of such a model — is becoming more and more urgent in sociology. In spite of the fact that the major tasks of regional sociology were formulated over a decade ago (Pahl, 1970), concrete sociological studies often consider city and nature as different, isolated "milieux" of human activities. Thus, the very question of a single object of study for a number of sciences has revealed the need for reassessment of research boundaries in some individual disciplines.

The complexity and qualitative aspects of the urban ecosystem throw into relief the inadequacies of traditional forms of representation of this object, which has developed a three-dimensional structure subject to temporal changes. Therefore spatial representation can only be considered as one of the initial cross-sections of the object.

## THE STAGE OF URBAN DESIGN

The chief problem of the third stage is that of designing itself. It may sound trivial but the tasks facing the urban planner are far from being trivial. This stage in many respects is the most sensitive, the most tense link in the transition from science to practice. In this stage, urban planning tasks (in the sense of proper compositional tasks) are inseparably linked both with the goal of interdisciplinary synthesis and with the task of technological and organizational preparations for the next stage, that of implementation.

## Functions of the design process

At least five main functions of designing urban systems can be recognized. The first function is to design a spatial representation of the "eco-city" model, developed through drawing on descriptions developed through separate disciplines such as sociology, ecology and urban planning. Matrices can help in synthesis of knowledge from different fields, and in incorporating within the design process complex dependencies between social, ecological and technological systems. They can also help in heightening the appreciation of the effects of the urban environment on nature and on man, effects which are often insufficiently accounted for in the designing activity. Here one can see a real contradiction in the process of implementation of synthesized knowledge. On the one hand, this stage is necessary. But on the other, as follows from the experience in the development of "ekistics" theory, approaches to the synthesis of knowledge based on spatial criteria are fraught with the danger of losing sight of the very entirety of the object for which we are striving.

Second, a design is always aimed at implementing a certain goal, a concrete social order, which signifies an additional number of requirements and limitations.

Third, until now the development of a design has been the principal means for representing an urban concept and facilitating its translation into action. But it must be questioned whether the static form of representation is the best way of approaching the creation of a qualitatively new, higher level entity, that of "eco-city". Also is the state or form of representation adequate for the task of guiding the development of the future urban environment?

Fourth, the process of urban design is today exerting an increasing influence upon decision-makers as well as on public opinion. In contemporary cultures, urban design is one means by which the future shape of the local environment is represented. It is a form open to the general public, and as such is an important means of communication between science and various social groups in the community. The present-day architect is correspondingly an important mediator between science and the social life of the city, between science and public consciousness. For these reasons, and paradoxical as it might sound, it is important for the "eco-city" concept to have its own visual image.

Fifth, but in reality of first importance, designing is an art. How does the task of interdisciplinary analysis relate to its laws, its canons? This is a very important question, because the design is the basis from which the "evergreen tree of life" of the future city will sprout. Over two decades have passed since the publication of Lynch's classical "The Image of the City" (Lynch, 1960), which showed how the large city is perceived by the city-dweller. Even earlier, in the 1920s, one of the first methodologies for utilizing historico-cultural and sociological knowledge of the city for the purpose of vocational training and general education was developed in the Soviet Union (Antsyferov and Antsyferov, 1926). Notwithstanding these and other studies, the forms in which people perceive the urban environment have not so far exerted any significant influence on the way urban complexes are constructed.

### Urban planning and urban design

In spite of its shortcomings, the professional subculture of urban planning remains the basis for seeking new approaches to the design of our cities and towns. It represents an amalgamation of science, art, technical skills, professional excellence, and in addition an ability for multilateral communication. These qualities provide a solid basis for making significant progress in refining and implementing the "eco-city" concept.

Thus, the design of any city is both the means and the goal, a phenomenon resulting from technology and creative art, a product of synthesis of knowledge and culture. Further analysis of the diverse links between these different aspects is needed, and there is no question that sociologists and ecologists must collaborate with urban planners in the development of "eco-city" models. These contacts cannot be postponed until the time that all major scientific problems have been resolved. And, in fact, urbanists, in spite of the practical orientation of their professional activities, are extremely perceptive to new ideas.

## PUTTING THE DESIGN INTO PRACTICE

Stages IV and V of Figure 3—1 incorporate various phases of a single process of implementing the concept of the future city. This process has an inner logic of its own, its own laws and regularities connected with the interactions between natural, socio-economic, technological and other systems. These interactions are complex, which explains the need to consider certain problems specific to each of the above mentioned stages.

Construction of a new city inevitably results in a negative impact upon nature. This impact, as demonstrated by an analysis of capitalist urbanization (Yanitsky, 1979), consists of several characteristic stages. During the last few decades, the efforts of scientists, urban planners and the general public have resulted in a reduction of the scale of these negative consequences. At the same time, the requirements of urban industries and populations in relation to nature and its resources are constantly growing. This makes a comprehensive approach to urban construction and thrifty use of land the most important criteria in the implementation of any design.

## Scenarios for implementation

In order to implement adequately the entire complex of requirements for an "eco-city" it is necessary to develop, along with the urban design, a scenario for implementation. Programming methods long used in contemporary large-scale goal-oriented research and developments should be more widely adopted, in promoting a scenario of city formation in time, with an emphasis not on description of spatial structure but on the task of comprehensive control of system development (Kochetkov and Pchelintsev, 1980). The development of scenarios like this make it possible to link together all the desired stages in the formation of the city and help to reduce the undesirable effects which are virtually inevitable when traditional methods of representation of the system are used. The scenario will significantly influence the urban design istelf. As shown by the experience of urban planning in different parts of the world — for example, in the development of new towns around London — the architectural-spatial composition of the city depends to a large extent on whether its construction started with the formation of the city's central part or whether it finished with it.

## Maintaining the integrity of the system

A central idea in the "eco-city" concept is that of the integrity of the system, of its coordinated biosocial development. As such, one of the most important criteria in scenario elaboration is that of the time needed for achieving this state of integrity, that organic integrity which will be capable of being sustained and at the same time will fulfil a number of functions in relation to a larger whole.

From the viewpoint of creating the scenario, the problem of time has another important aspect: the question of which components of the future settlement should be built, and when, by the efforts of society, and which should be the result of the city's self-organization and self-development. This is in practice a vitally important problem. Both man and nature must be at home in the city, making the city an integral part of biological and social spheres. "Eco-city" must be, at the same time, both an organism and an environment, regenerating and supporting the lives of man and nature.

These various questions are directly linked to the problem of the scenario. Take, for instance, the problem of forming scientific centres. The importance of the movement of knowledge, and that of the entire system of the production and dissemination of such knowledge, is not in question. Nevertheless this extremely mobile system of human activity requires for its development the existence of a spatially localized, stable, well-developed socio-cultural environment. The atmosphere of scientific creativity evolves gradually, sometimes over many decades, and is extremely sensitive to all controlling influences from without. The question arises, where are we to begin in order to create these conditions rapidly enough? Experience has shown that the effect of a "lived-in" environment is achieved in various ways, depending on circumstances: through retaining links with the previous living environment, through reconstruction of the previous environment's most significant characteristics in a new locality, through promoting especially favourable conditions for creative work, etc.

The task of creating the scenario for implementing the "eco-city" concept becomes much more complicated when account is taken of the fact that nature must also be able to "make itself at home" in the city. Newly created urban habitats must enter into relationships with agricultural areas surrounding the city, with natural landscapes, with different types of protected areas and other types of natural and man-modified system.

## Communication and participation

The stages and levels of "eco-city" creation outlined above involve considerable public participation, much greater than hitherto. It is understandable enough. Until now, public involvement was mostly aimed at exposing and countering the negative social and ecological consequences resulting from rapid growth of settlements. In other words, these efforts were aimed at solving the current problems of the city. But a broader involvement of the public in efforts for the protection of nature[1] is an urgent requirement of contemporary life. In the case of urban development, it means, first and foremost, the participation of the public at all stages in the city formation — spatial, functional and socio-cultural. Actually, these three levels of integration of the urban organism signify three interconnected but different spheres of social activity of city dwellers. These spheres of activity generally correspond to the three levels of knowledge integration shown in Figure 3—1.

This social involvement of city dwellers can best be promoted within a programme developed on a scientific basis, in the present case within the scenario of implementation. At present, in the socialist countries, public activity aimed at promoting nature conservation is planned in the course of the elaboration of comprehensive plans for social development of cities and their districts (Baranov, 1980). The socio-cultural aspect of such activity is of particular interest. During the process of formation of new cities, the question of their cultural nuclei has always been a critical one. Recent experience in urban planning shows that natural areas protected during construction, particularly those of cognitive and historic significance, can easily fulfil a socially integrating function, particularly in relation to a newly established city.

Consideration of the future should take an increasingly larger place in the social activities of the population, and this explains the importance of promoting such involvement at as early a stage as possible. Such reasoning is based on two

[1]See for example, the documents of the 26th Congress of the Communist Party of the Soviet Union, 1981.

considerations. First, the foundation of the future integrity of a city is laid during its construction. Specialists are well aware of the difficulties involved in later corrections to the master city plan. Second, time is needed to obtain the required social reference-points as a result of socio-ecological research. This raises the question of communicating the results from one stage to another, and representing adequately the results of the work at each stage of the process. Observations show that, even if the level of ecological consciousness is high, long-term anticipatory planning for public participation in nature conservation and other activities is required, otherwise such activities remain within the framework of the city's current requirements. Systematic discussions of future construction projects with city dwellers have almost invariably resulted in a long-term integrating effect. Similar considerations also relate to the vocational education of urban managers. They need an image of their future city not only to enable them to make correct decisions but also to help them establish contacts with the population and to achieve mutual understanding in problem-solving efforts.

Another important aspect of the social activities of the population is how to adapt the concept and its implementation to local requirements and conditions, and at the same time take advantage of local opportunities and resources which can contribute to the implementation process. It goes without saying that the real scope and possibility of the population's social activities are invariably greater and more diverse than those described or shown in Figure 3—1. The aim of this figure is merely to outline the sphere of social activity corresponding to urgent social requirements and consequently bearing a socially significant content (Volkov, 1981). Finally, even a schematic consideration of the problem shows how relative are the boundaries between professional (institutional) and public activities. The latter are acquiring an increasingly organized scientific character, and call for new, expanded initiatives for interprofessional interaction and the integration of science, culture and art.

## CONCLUDING REMARKS

In viewing the overall challenge of putting "ecology into practice", the study of ways and means of implementing a scientific idea is as important as its development, and this is particularly true in dealing with concepts like the "eco-city" which entail important considerations of perspective. Examination of the "eco-city" concept and its implementation, even at theoretical and model levels, has revealed a number of research tasks requiring thorough study in the near future.

The scheme proposed in this contribution holds significance in at least four different fields. First, in the field of cognition, it represents a process of transition from science to practical urban planning. Second, in the social field, it provides a context for implementation of a certain interdisciplinary idea. Third, in the cultural field, it advocates a process of interaction between science and culture. Fourth, in the organizational field, it proposes insights for the composition of interdisciplinary teams. It is important to emphasize once again the necessity of further study of the socio-cultural context of the process of urban planning and design. Only after understanding the entire system and its interactions at various stages and levels in the process of research and implementation can one be confident that the recommendations of scientists will be taken into account.

Finally, it appears that one of the interdisciplinary tasks of the MAB Programme could be to sponsor an experimental verification of different variants of the idea of "eco-city". One example of an activity that could contribute to such a venture is the "Ecopolis" programme developed in the Soviet Union by a group of scientists with the participation of urban dwellers and incorporating cognitive, socio-organizational and cultural-educational dimensions (Anon., 1981). Such experimental approaches should unquestionably be tackled on a democratic basis, entailing broad-ranging sociological and historico-cultural analysis and the participation of the general public in their development and implementation. This would be another important direction of social science research within the framework of the MAB Programme which would respond to the dual needs for the protection of nature and for social progress.

## References

Anon. (1981), "Ecopolis", *Literaturnaja Gazeta* (Literary newspaper), January 1, 1981.

Antsyferov, N. and T. Antsyferov (1926), *Gorod Kak Vyrazitel Smeniajuzchihsia Kultur*, (A city as a mouthpice of changing cultures), Brockhause and Efron Publishing House, Leningrad.

Baranov, A. (1980), Urban Population and Environmental Pollution, *Bulletin of Commission of the USSR for Unesco*, 3-4 (41-42), 31-7.

Borschevsky, M. V., S. V. Uspensky and O'I. Shkaratan (1975), *Gorod: Metodologicheskie Problemy Complexnogo Economicheskogo i Sotsialnogo Planirovania*, (City: Methodological Problems of Composite Economical and Social Planning), "Nauka", Moscow.

Boyden, S. (1977), Integrated Ecological Studies of Human Settlements, *Impact of Science on Society*, 27, 159-69.

Boyden, S. (1979), *An Integrative Ecological Approach to the Study of Human Settlements*, MAB Technical Notes 12, Unesco, Paris.

Chistiyakova, S. B. (1975), *Ozdorovlenie Okruzajuschej Sredy Gorodov* (Sanitation of Urban Environment), Town Planning Institute, Moscow.

Dubinin, N. P. and Y. P. Altuchov (1976), Integralnaja Geneticheskaja Characteristica "Adaptivnoj Normy" v Populiatsii Cheloveka (Integral Genetic Characteristic of "Adaptive Norm" in Human Populations), *Doclady Academii Nauk SSSR*, 4(230), 957-60.

Dubos, R. (1970), The Human Environment, *Ekistics*, 30(178), 172-3.

Glazychev, V. L. (1975), Problema Gradostroitelnogo Znania (The Problem of Urban Planning Knowledge), In *Problemy Teorii Sovietskoi Architectury* (The Problems of Soviet Architecture Theory), Moscow, 5-7.

Glazychev, V. L. (1977), *Organizatsyia Architecturnogo Projektirovania: Voprosy Teorii*, (Architecture Designing Organization: Theoretical Problems), Strojizdat, Moscow, 131-6.

Glazychev, V. L. (1978), Development of an Ecosystem Town. In *Sociology and Problems of Social Development*, Nauka Publishing House, Moscow, 199-209.

Khodzaev, D. G., A. V. Kochetkov and F. M. Listengurt (1977), *Sistema Rasselenia v SSSR*, (Settlement Systems in the USSR), Ekonomica Publishing House, Moscow.

Kochetkov, A. V. (1980), Regional Systems as the Object of Planning and Management, In *Regionalnye Systemy: Strategii Razvitija i Upravlenija* (Regional Systems: Strategies of Development and Management), Collected Papers, Issue 10, Institute for System Studies, Moscow, 8-28.

Kochetkov, A. V. and O. S. Pchelintsev (1980), Sistemy Rasselenia: Strategii Razvitia i Upravlenia (Settlement Systems: Development and Management Strategies), In *Metodologia Sistemnogo Analiza Regionalnogo Razvitija* (Regional Development Systems Analysis Methodology), Collected Papers, Issue 7, Institute for System Studies, Moscow, 6-15.

Kogan, L. B. (Editor) (1967), *Razvitie Gorodskoi Kultury i Formirovanie Prostranstvennoj Sredy* (Urban Culture Development and Building up the Urban Space), Town Planning Institute, Moscow.

Kostrowicki, A. (1979), Voprosy Sotsioecologii Gorodov i Character Etoj Discipliny v Polshe (Urban Socio-Ecological Problems and Features of this Science in Poland), In *Novye Idei v Geografii* (New Ideas in Geography), Progress Publishing House, Moscow, 160-75.

Leontiyev, A. N. (1977), *Dejatelnost, Soznanie, Lichnost* (Activity, Consciousness, Personality), Second edition, Politizdat, Moscow.

Lynch, K. (1960), *The Image of the City*, MIT Press, Cambridge, Mass.

Marx, K. and F. Engels (1968), *Sotchineniya* (Collected Works), Second edition, Moscow, Vol. 46, part I, 483-4.

Pahl, R. E. (1970), *Whose City? and Other Essays on Sociology and Planning*, Longman, London.

Pchelintsev, O. S. (1980), Kontseptsija Regiona i Razvitie Funkcij Territorialnogo i Programmno-Tselevogo Planirovanija (Regional Concept and Development of Function in Territorial and Programme-aimed Planning), In *Metodologia Complexnogo Issledovania Sotsialno-Economicheskih Sistem* (Methodology of Complex Socio-Economical Systems' Research), Collected Papers, Issue 1, Institute for System Studies, Moscow, 5-12.

Shvartz, S. S. (1974), Problemy Ecologii Cheloveka (Human Ecology Problems), *Voprosy Filosofii*, 9, 104-9.

Sidorenko, G. I. and Y. I. Prockopenko (1976), In *Vzaimosvijz Nauk pri Reshenii Ecologicheskich Problem*, (Intercommunication of Sciences in Ecological Problem Solving), Moscow-Obninsk, 33-8.

Stearns, F. W. and T. Montag (Editors) (1974), *The Urban Ecosystem: a Holistic Approach*, Dowden, Hutchinson and Ross, Stroudsburg, Penn.

Unesco (1977), *Human Settlement Managers Training Programme*, Two volumes, Unesco, Paris.

Vester, F. and A. von Hesler (1980), *Sensitivitäts-Modell*, Regionale Planungsgemeinschaft Untermain, Frankfurt am Main.

Vladimirov, V. V. (1980), Aktualnost i Predposylki Ekologicheskogo Programmirovanija v Raionnoj Planirovke (Prerequisites of Ecological Programming in Regional Planning), *Voprosy Geografii* (Problems of Geography). Mysl Publishing House, Moscow, 113-6.

Volkov, Y. E. (1981), Public Activities: Their Essence and Some Problems of Development, In *Voprosy Filosofii* (Problems of Philosophy), 4, 39-51.

Yanitsky, O. N. (1970), *Socio-Informational Aspects of Urbanization*, Paper presented to Seventh World Congress of Sociology, Moscow, Soviet Sociology Association.

Yanitsky, O.N. (1978), Urbanization and Social Development, In *Sociology and Problems of Social Development*, Nauka Publishing House, Moscow, 187-98.

Yantisky, O.N. (1979), Fazy Kapitalisticheskoj Urbanizatsii i ich Vzaimodejstvie s Okruzajuschej Sredoj (Capitalist urbanization phases and their interrelationships with nature), In *Sotsialno-Ecologicheskie Problemy Kapitalisticheskogo Goroda* (Socio-Ecological Problems of Capitalist City), Moscow, 18-37.

Yanitsky, O.N. (1980a), K Probleme Integracii Ecologicheskogo Znania v Teorii Urbanizacii (Integrating Ecological Knowledge in Urbanization Theory), In *Nekotorye Problemy Urbanizacii i Gradostroitelstva* (Some Urbanization and Town Planning Problems), Moscow, 72-9.

Yanitsky, O.N. (1980b), Contribution of Social Sciences to the Solution of Ecological Problems, *Bulletin of the Commission of the USSR for Unesco,* 3-4, (41-42), 8-18.

Yanitsky, O.N. (1982), Towards an Eco-city: Problems of Integrating Knowledge with Practice, *International Social Science Journal,* 34(3), 469-80.

Zaremba, P. (1977), Methodology of Integrated Environmental Planning of Urbanized Areas, *Technical University of Szczecin Special Series,* 4, 2-42.

Zaremba, P. (1980), Principles of the Ecological Oriented Integrated Urban Planning, In *The Development and Application of Ecological Models in Urban and Regional Planning,* Bonn, 82-94.

CHAPTER 4

# Towards a Policy of Urban Survival

*Yona Friedman*

Abstract

Cities should be able to assure survival of their inhabitants without total dependence on means imported into the city from the outside. Reduction of the present levels of dependency of cities on their hinterlands would make both the city and the countryside more liveable. The self-contained city could be a realistic concept even in our industrial civilization, once appropriate space and time allocations were made. A policy in favour of urban autarchy should suggest new orientations in physical and social planning, orientations corresponding to an appropriate use of both space and time. The promotion of such policies would require innovative research in biotechnology, agronomy, architecture and other fields.

*Résumé*

*Les villes devraient être capables d'assurer la survie de leurs habitants sans être totalement dépendantes des biens importés dans la ville en provenance de l'extérieur. Ainsi à la fois la ville et la campagne deviendraient plus vivables. L'auto-suffisance de la ville pourrait répondre à un concept réaliste, même dans notre civilisation industrialisée, si des affectations appropriées, aussi bien de l'espace que du temps, étaient faites. Une politique en faveur de l'autarcie urbaine devrait suggérer de nouvelles orientations dans la planification des milieux physiques et sociaux (ces orientations correspondant à un usage approprié aussi bien de l'espace que du temps). Pour une telle politique, la recherche scientifique dans des domaines comme la biotechnologie, l'agronomie et l'architecture, devrait être revue.*

## URBAN SELF-SUFFICIENCY

IF one tried to give a "moralist's" definition of the concept "city" it could be: the city is a tool conceived in order to assure a more agreeable and more effortless way to survival for all those who live in it. This definition is that of a "moralist", because it defines the city's goal, without determining the social and physical means necessary for reaching this goal, and because it is non-selective (it assumes that all people who live in the city should profit from the existence of this goal.)

There is at the same time a lacuna in this definition, as people who do not live in the city (the "others") are not mentioned, and we do not know whether they profit or not by the city's existence. This lacuna is very important. It could be neglected, if the city were a self-contained survival tool (as it was in most historical periods and in most regions), and if the effect of this survival tool on the "others" would not be particularly harmful: but that is not the case with a modern city. It is a characteristic phenomenon of cities in industrial, and particularly "neo-industrial" civilizations, that these cities destroy larger and larger regions around them.

Already, the "paleo-industrial" city (that of the middle ages to the nineteenth century inclusive) was harmful, but only for its immediate geographical vicinity, by exploiting economically and dominating the countryside near it. Paleo-industrial cities produced a "labour-drain" as a result of an artificial impoverishment of the countryside: thus started a process of colonization of the countryside, by systematic under-evaluation of its products and by the over-evaluation of the merchandise produced in the city. Thus the paleo-industrial city started to lure people away from rural occupations in order to furnish cheap labour for the city. But, in spite of this process, the paleo-industrial city was still largely self-sufficient: it still produced a part of its food in addition to industrial products, and did this in order to survive sieges, epidemics and economic crises. Even cities like Paris in 1870 produced some food; if not sufficient for indefinite survival, at least enough to keep alive its population for many months without help from outside.

The neo-industrial city continued further the development away from self-sufficiency, and became not only completely dependent upon supplies coming often from faraway regions (even water supply is assured, in many cases, by "importing" it from far away, often spoiling areas far from the city). Economists and management people, instead of denouncing this tendency, made out of dependence an ideology of development. The city's predominance is today openly based on its parasitic quality. This is similar, in a way, to the parable of Menenius Agrippa about the stomach and the members: the members revolted against the stomach, which enjoys all food procured for the organism by the members, but when the members refused to procure food for the stomach, the organism died. It is evident that this parable is not true in the case of urban parasitism, on the contrary: if the city "dies", the countryside continues to exist, but if the countryside dies, the city dies with it.

The difficulty with the ideology of urban predominance based on the city's parasitic existence could be better illustrated by a different parable with the city as a cancer, an overgrown organ which takes all the food, so much food that it can no longer perform its proper function: and cancer is a lethal illness. Cities at present have a population performing mainly unproductive and socially less useful activities. For example, in a city like Paris, about 90 per cent of the active population performs "tertiary" activities and this in a country where a majority of the population is involved in the tertiary sector (in France about 65 per cent). This indicates that cancer is not anymore an image, but a reality.

The results of cancer are disfunction and decay. Our global economic crisis is largely a result of "over-tertialization" — of the colonization of the productive sectors by the unproductive ones. Neo-industrial civilization can less and less guarantee survival for people, and particularly that of people residing in urban areas. Unemployment, in neo-industrial civilization, causes the unemployed to be alienated from that civilization. Unemployed people can not survive if unemployment reaches a large scale and becomes durable. The only way for survival for the unemployed is to develop a parallel economy, by starting "quaternary" activities, i.e. activities producing goods and services necessary for the producer's own consumption. Quaternary activities and quaternary production is not included in the statistics of mainstream economy. The external aspects of "quaternary" survival in the neo-industrial civilization will often be different in industrialized and in non-industrialized countries, but the base is common to all: the emergence of a self-help subsistence economy.

Thus cities grow all over the world, exploiting and destroying the countryside: economic mismanagement produces global crisis. The only way out of this situation is local self-sufficiency, particularly that of urban areas. A new type of urban ecology, that of very small systems as self-contained as possible, could become a topic of first-rank importance, if we want cities to become again "tools" assuring more agreeable and more sure survival for those who live in them, and do this without harm for those who live outside them.

## POLICIES FOR URBAN AUTARCY

The basic means for urban autarcy are available space and available time (time for self-help labour). Indeed, all activities necessary for subsistence, self-help economy need first of all space: urban food production, subsistence crafts, repair, etc., all require a certain amount of space. For example, FAO estimates that 100 $m^2$ is necessary to feed a person for a year: this might be true for vegetables and cereals. As for "quaternary" animal husbandry, there is no necessity to allocate space as private property: chickens, goats pigs, cows and so on, can subsist, like all pedestrians, on communal surfaces (streets, etc.); similarly trees. Obviously, we need to indicate to whom these animals and trees belong. Many small crafts can be practised in the street. But public or private, the total surface necessary for subsistence for a number of people is surely not less than the total surface under a roof necessary for the same people.

There are many possibilities for assuring sufficient surface for subsistence activities: private areas adjoining the houses, kitchen gardens, concentrated areas for private subsistence farming ("jardins ouvriers"), communal areas allowing transient use (roadsides, etc.). But it would be an error to consider a city as liveable without assuring for it the reserves necessary for subsistence activities. The following proportions give an idea of a possible allocation of land for different purposes.

| | |
|---|---|
| Housing (under-roof area) | 20 per cent of total area |
| Communal area | 15 per cent of total area |
| Circulation area | 5 per cent of total area |
| In-town subsistence area | 60 per cent of total area |

Given that all special areas (such as housing, circulation and some communal areas) could be situated in multilevel infrastructures, a rough estimate can be made of the population density that a city could support if assuring its own subsistence. Using Indian standards for poor populations, of 6 $m^2$ covered area per person, 1 $km^2$ of city could feed and house up to 50,000 persons if developed at ground level alone, and up to 150,000 persons if developed on three floors. If the settlement is organized appropriately for water collection (all roofs and street areas used as water collectors, i.e. 80 per cent of the total surface), 1 $km^2$ of city tissue could collect, for every 100 mm of rainfall per year, about 100,000 $m^3$ of water. Taking the Indian average annual rainfall of 600 mm as an example, this quantity would represent about 6 $m^3$ water a year for an inhabitant. Such an amount is not sufficient in itself for all uses (industry, etc.), but is enough to ensure basic needs like cooking and cleaning which can be estimated at 15 litres a day per capita.

These figures are given here for example only, in order to indicate that a settlement largely self-sufficient in food, in water, and assuring all necessary crafts and services, is not contradictory to relatively high density land use. It is obvious that these figures are cited only to give an idea of a threshold density, and not of an optimal density. But these figures do show that it is possible to work out guidelines for space allocation adapted to complete or partial urban autarcy.

As regards the time allocation, the total time required for subsistence self-help labour is not a limiting factor: unemployment liberates such time for "quaternary" activities. As illustration of the time investment required for subsistence, compared to other, activities, the following can be cited:

| | |
|---|---|
| Subsistence food production | about 10 hours per week |
| Crafts and services, for self-help or barter | about 12 hours per week |
| Intellectual, social and cultural activities | about 28 hours per week |

A week with 50 hours of activity (compared to the current activity week consisting of about 40 hours paid work, 5 hours transport and about 20 hours household activities a week) is a better balanced time plan than that of people "employed", and surely better remunerated, even if its remuneration is not "in cash". A city which would work on the basis of such a 50-hour week could survive by itself, if, in addition to the time allocation, the necessary space allocation could be guaranteed.

Urban self-sufficiency could not be promoted without prior scientific research, for developing appropriate methods and tools. Present forms of urban autarcy, inspiring as they can be, are not always adapted to present-day conditions, and it is unlikely that the general context will change much in this respect. Thus, it is necessary to find solutions which are compatible with the status quo and to the development of urban autarcy at the same time. Some fields where a new, survival-oriented research effort could help bring about such solutions can be identified as follows:

1. Engineering and architectural research oriented to improvement and adoption of techniques for water-collection and recycling: the city as a water-collecting apparatus could lead to new designs for roofs, road surfaces and urban layout.

2. Urban food production involves not only new concepts for planning (how to assure available surfaces on one or more levels), but a new orientation of agronomic research might also emerge. This research would look for plants demanding less care and plants which would produce a continuous small harvest instead of a large harvest at a single time. A continuous small harvest would provide food for long periods without the necessity of storing the crop. The labour for continuous harvesting is easier than for subsistence gardening. If, for example, a tomato plant produces two tomatoes every week, this plant is more useful for urban subsistence than a plant which yields 10 kg of tomatoes once a year.

3. Design research should look for ways to recycle waste. Packaging crates and industrial waste could be looked upon as raw materials for small crafts. Tins, bottles, boxes, etc., can be transformed into various useful objects, and design research could be oriented to making them appropriate for such recycling.

4. Cities are generators of micro-climates, and their designers should be made conscious of the micro-climates that would be created in an urban autarcy. Present design processes do not use the necessary scientific methods, simply because such methods are lacking. Thus, a design-oriented, microclimate-research programme would be very helpful.

5. The relations of present-day urban societies to nature and habitat is often manifest in large-scale, radical modifications of ecosystems, through such means as deforestation, massive movements of earth, etc. The search for new, less-impacting techniques for shaping our landscapes and habitats should be intensified.

6. The use of microorganisms for subsistence production, for water cleaning, for protein production, for modifying air temperature, etc., is a research theme which up to the present has been mostly investigated for industrial

purposes alone. The use of microorganisms in a subsistence economy would also be a suitable theme for new research.

7. The city as an organization has already served as a topic for many researchers and for various research programmes. Such programmes might valuably be broadened from the perspective of urban survival in crisis (urban autarcy).

Research programmes in these fields could yield coherent and much needed knowledge about the capacity for urban survival. Research in most of these fields could yield results quite rapidly, as experience in many of them is not lacking. A five-year programme on urban survival would be, in my view a realistic proposition.

CHAPTER 5

# Mobilizing Scientists, Planners and the Local Community in a Large-Scale Urban Situation: the Rome Case Study

*Mirilia Bonnes*

Abstract

Scientists, planners and urban dwellers are working together to evaluate the characteristics of the city of Rome from an environmental viewpoint. Modifications to the city's environment from 1870 to the present have been examined using various techniques including remote sensing. The needs for energy, water and housing have been assessed and are being used in the preparation of plans for future expansion and for the improvement of the quality of life in the city. Negative aspects, including allergens and animal pests are also taken into account. The study endeavours to provide an integrated examination of the needs and problems of a large-scale urban complex composed of a wide variety of environments which range from examples of classical architecture to urban shanty dwellings.

*Résumé*

*Savants, planificateurs et citadins travaillent ensemble pour évaluer les caractéristiques de Rome d'un point de vue environnemental. Les modifications subies par l'environnement de la ville depuis 1870 jusqu'à nos jours sont étudiées selon différentes techniques, y compris la télédétection. Les besoins en énergie, en eau et en habitat ont été évalués et servent à la préparation de plans pour l'expansion future de la ville et l'amélioration de sa qualité de vie. Les aspects négatifs, y compris les allergies et les nuisances animales sont également considérés. L'étude vise à établir un examen intégré des besoins et des problèmes que pose un complexe urbain de grande dimension composé d'une large variété d'environnements qui vont d'exemples d'architecture classique aux "bidonvilles".*

## OVERVIEW OF THE ROME STUDY

> The city has to be treated primarily as a biological system . . . ; all biological organizations constitute open systems and those bonds of solidarity link each living system to all the rest.
>
> It may seem extremely difficult, if not impossible, to apply this approach to a city; but the fact is that we have no alternative. Only through a new spirit of solidarity among all the sciences, among men of learning, citizens, and all those responsible for decision-making, can we hope to find acceptable solutions and come to grips with problems of behaviour, custom and human solidarity calling for an understanding of life.

THESE words were those of Valerio Giacomini, who founded and directed the MAB study on Rome until his death in January 1981. They reflect his belief that it is possible for people from all walks of life and all professions to work together towards a common goal — in this case the research for planning on how to make Rome a more humane, more emotionally satisfying and more functional city (Giacomini, 1981; Giacomini and Hinrichsen, 1981). And it is in this spirit that the Rome project has advanced and developed.

In its early years, the main effort has been devoted to construction of a multidisciplinary team which can study Rome as an ecological system. At the same time the project is constantly evolving because, while it includes a range of disciplinary interests, its main focus is on problem-solving. This contribution provides a general description of the project team and methodology and an overview of the main problems under study. Figure 5—1 shows an organizational chart of the Rome project, indicating the topics and fields of inquiry of the various working groups and the institutions taking part in the project. The solid lines represent established groups, the dotted lines represent recently formed or tentative groups.

### Botanical survey

In the middle of the organization chart appears the name of the Botany group which, under the direction of Valerio Giacomini, represents the starting point of the MAB Rome project. This group has been engaged in an extensive and detailed mapping of the green spaces in the city. The group has been involved in the study of the different green areas in the city and has prepared detailed maps, which

**Figure 5—1.** *(Opposite)* The institutional and cooperative framework of the Rome project. Solid lines represent established groups, dotted lines recently formed or tentative groups. Source: Italian MAB National Committee

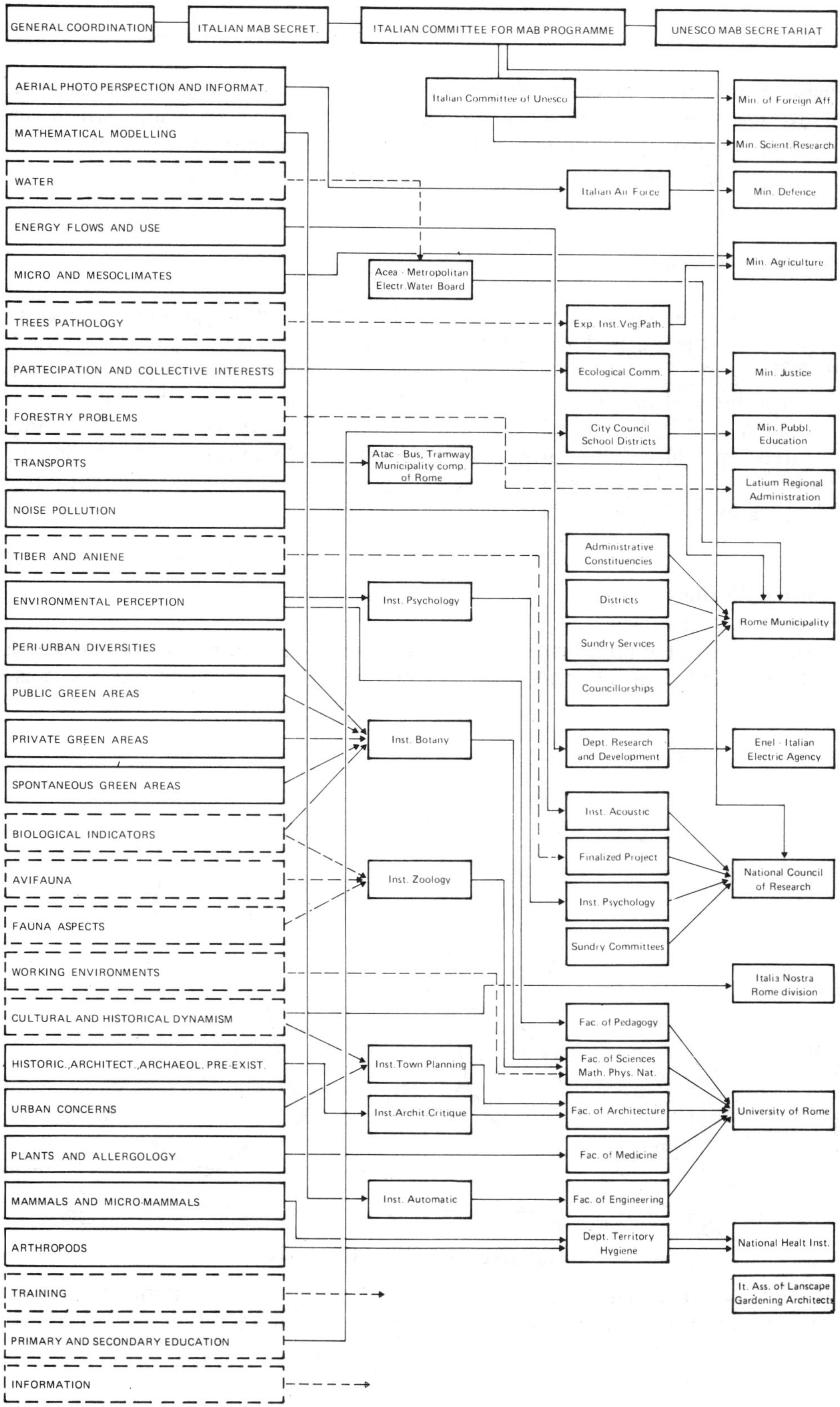
GENERAL COORDINATION
ITALIAN MAB SECRET.
ITALIAN COMMITTEE FOR MAB PROGRAMME
UNESCO MAB SECRETARIAT
AERIAL PHOTO PERSPECTION AND INFORMAT.
MATHEMATICAL MODELLING
WATER
ENERGY FLOWS AND USE
MICRO AND MESOCLIMATES
TREES PATHOLOGY
PARTECIPATION AND COLLECTIVE INTERESTS
FORESTRY PROBLEMS
TRANSPORTS
NOISE POLLUTION
TIBER AND ANIENE
ENVIRONMENTAL PERCEPTION
PERI-URBAN DIVERSITIES
PUBLIC GREEN AREAS
PRIVATE GREEN AREAS
SPONTANEOUS GREEN AREAS
BIOLOGICAL INDICATORS
AVIFAUNA
FAUNA ASPECTS
WORKING ENVIRONMENTS
CULTURAL AND HISTORICAL DYNAMISM
HISTORIC.,ARCHITECT.,ARCHAEOL. PRE-EXIST.
URBAN CONCERNS
PLANTS AND ALLERGOLOGY
MAMMALS AND MICRO-MAMMALS
ARTHROPODS
TRAINING
PRIMARY AND SECONDARY EDUCATION
INFORMATION
Italian Committee of Unesco
Acea - Metropolitan Electr.Water Board
Atac - Bus, Tramway Municipality comp. of Rome
Inst. Psychology
Inst. Botany
Inst. Zoology
Inst.Town Planning
Inst.Archit.Critique
Inst. Automatic
Italian Air Force
Exp. Inst.Veg.Path.
Ecological Comm.
City Council School Districts
Administrative Constituencies
Districts
Sundry Services
Councillorships
Dept. Research and Development
Inst. Acoustic
Finalized Project
Inst. Psychology
Sundry Committees
Fac. of Pedagogy
Fac. of Sciences Math. Phys. Nat.
Fac. of Architecture
Fac. of Medicine
Fac. of Engineering
Dept. Territory Hygiene
Min. of Foreign Aff.
Min. Scient.Research
Min. Defence
Min. Agriculture
Min. Justice
Min. Pubbl. Education
Latium Regional Administration
Rome Municipality
Enel - Italian Electric Agency
National Council of Research
Italia Nostra Rome division
University of Rome
National Healt Inst.
It. Ass. of Lanscape Gardening Architects

distinguish between private green areas, public green areas and spontaneous green areas. The maps show the dominant vegetation of the green spaces, which can sometimes be of great variety and richness. Many interesting plants have, for example, been found growing spontaneously in the ancient ruins of the city.

## Aerial photography

The Botany group is working closely with two other working groups indicated in Figure 5—1 the Aerial Photoperspection Information group (top) and the Noise Pollution group (middle of the figure). The Aerial Photoperspection group is taking aerial pictures of the city and of its region in order to appraise the whole urban settlement and its hinterland. For more detailed descriptions, the group is using infrared film, which shows variations in the infrared radiation of trees and buildings. Through this kind of technique, it becomes possible to identify the different forms of vegetation and appraise their condition.

## Noise pollution

A survey on traffic noise has shown that the noise levels in Rome are among the highest recorded in different European cities. Figure 5—2 compares the noise

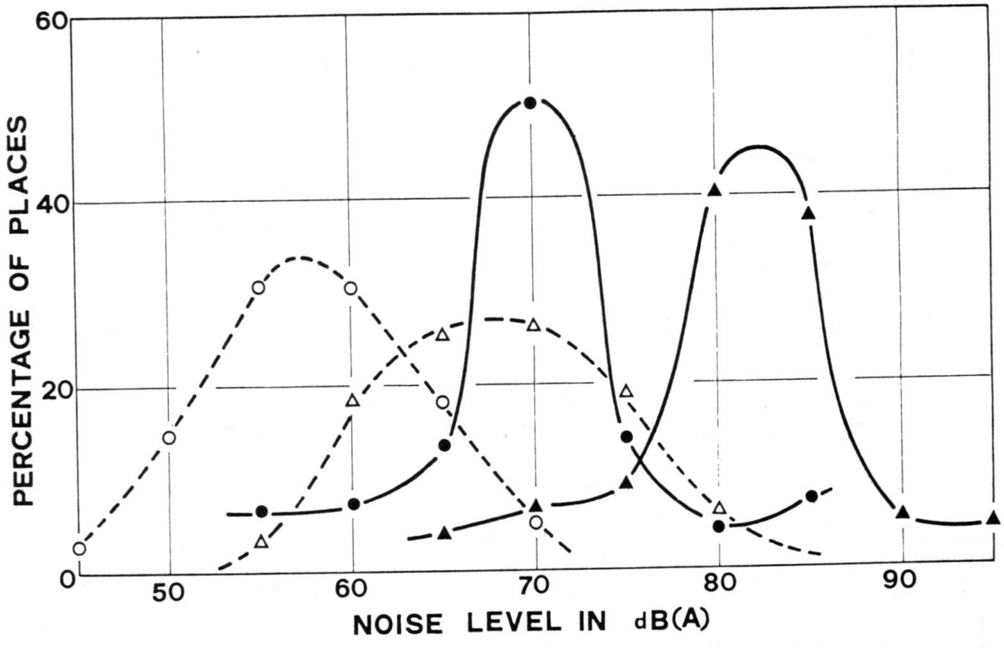


**Figure 5—2.** Comparison of noise levels in different parts of Rome (solid lines) with levels in London reported by the London Noise Survey in 1963

measured in different areas of the city with that recorded in London in 1963 by the London Noise Survey. The Noise Pollution group is compiling a noise pollution map to help in finding an effective way of reducing noise in the city. To this end the group is working with the Botany, Public Transport and Electric Energy groups indicated in the upper part of Figure 5—1. Collaboration with the Botany group is aimed at planning suitable acoustic barriers of trees and other vegetation along intense traffic roads. The collaboration with the two other groups is concerned with plans to introduce electric vehicles into the historic centre of the city.

## Traffic studies

Public and private vehicles cause many problems in Rome. Data from 1980 indicate that the city is inundated each day with four million vehicles. The city transport authorities are introducing a number of remedial actions, including: projected construction of 100 km of new subway; imposition of further restrictions on inner city traffic, in order to extend pedestrian areas; and introduction of a fleet of electric vehicles for use in the historic centre. The historic core of the city has been divided into four sectors of restricted traffic. In each sector, parking lots are proposed for the electric vehicles, where it will be possible to pick up and to drop off these vehicles.

The Noise Pollution group is tracing a hypothetical noise map of the city centre, derived from the present one, in which the noise levels of traditional vehicles will be replaced with those of electric ones. The introduction of electric vehicles in the city centre is also under wider study by the Electric Energy group, as a part of its programme on diversification of energy sources within the urban context, a programme which also includes work on solar panels for household uses.

## Energy and water supply

In the last 20 years, electric energy consumption in Rome has increased enormously at a rate much greater than population growth, to the extent that energy consumption per capita has more than doubled in 20 years (Figure 5—3 and 5—4). As Rome is not an industrial city, and most of the electric consumption is domestic, these data indicate a great change in the styles and the quality of life of the inhabitants of Rome during this period.

A MAB group within the Italian Electric Agency (ENEL) has prepared maps showing the energy supply system and flow for the whole city. Another group within Rome Municipality Services is dealing with problems of water supply. Five aqueducts (three of them dating from Roman Empire times) bring water into the city from the northeastern hills. Water quality ranks among the best in Europe, and only a small part of the supply needs to be treated. Today, Rome has over 2,000 drinking fountains and some 300 ornamental fountains, a number of them quite old. In response to increasing demands, the water authorities plan to use Lake

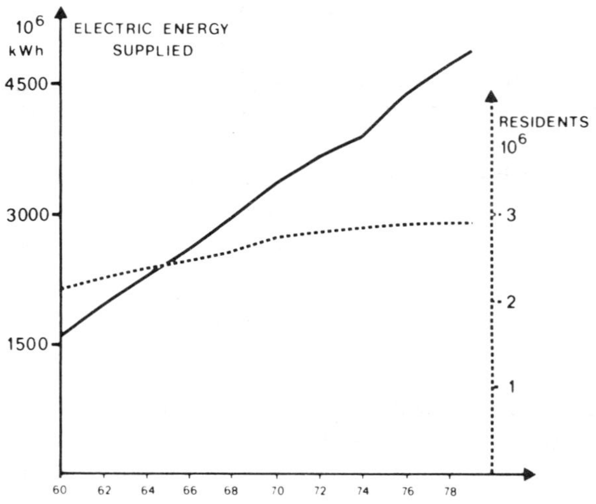


**Figure 5—3.** Trend of electricity developed in Rome during the last 20 years (solid line) and of the number of residents (dotted line)

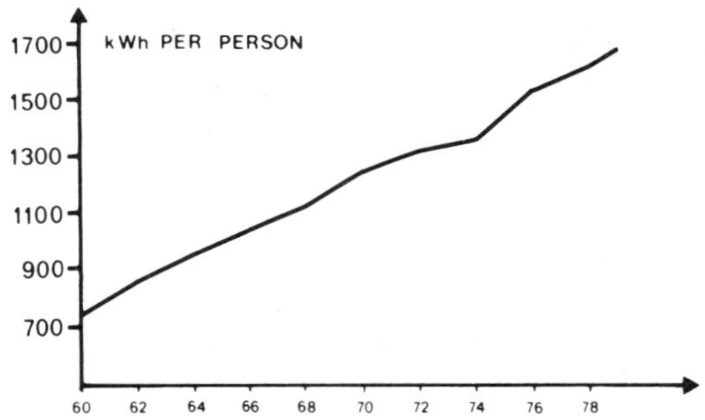


**Figure 5—4.** Energy use per inhabitant in Rome during the last 20 years

Bracciano (to the northeast of the city) as a source of drinking water and to close down a sewage treatment plant which previously made use of this lake. These groups work very closely with two other groups from the Faculty of Architecture and Urban Planning of the University of Rome, as indicated in the lower part of Figure 5—1. These groups are studying the city from the viewpoints of its historical, architectural, and archaeological aspects as well as in respect of the dynamics of urban growth.

## Dynamics of urban growth

During the last 100 years, Rome has undergone tremendous growth and change, increasing from 200,000 in 1870 before becoming the capital of a united Italy, to 520,000 in 1900, 1 million in 1930, and 2 million in 1960. An important aspect of urban growth concerns the *borgates,* illegal settlements that have sprung up in urban fringe areas, particularly during the last 30 to 40 years, as a response to urban overcrowding and rural migration. An estimated 700,000 people now live in such illegal housing settlements. The houses are built, without proper authorization, on agricultural land and are outside Rome's jurisdiction. Generally, they are separated from the legal boundaries of the city, by a green or a non built-up area. Most of the *borgates* are not true slums. The houses are generally quite big and have a lot of inside living space, but in general they suffer from a serious shortage of services. Often drains and drinking water, and electricity and gas supplies, are lacking. The roads are unpaved, and become very bad during winter.

In order to rectify this situation, the city municipality is dividing these outlaw areas into service districts and all the *borgates* will soon be given the services that they need. The municipality has identified 76 *borgates* around the Rome area which are actually involved or will be involved in this intervention plan. The MAB study, on its part, will increasingly be oriented to the peri-urban areas on the fringes of the city, where many things can be done with a small effort, so as to improve the living conditions of the whole city.

## The city fauna

The Zoology group is studying some aspects of the fauna of the city, particularly the rat and dog populations, relevant to the health and the quality of life of Rome's citizens. It is estimated that there are at least 2 rats for every Roman, with some estimates as high as 10 rats per person. In 1979, the city launched a rat extermination programme. But after one year of poisoning, the rat population started to increase again because the rats were developing strong resistance to the poison used. Researchers are trying to devise alternative control strategies and map the major concentrations of rats. The city has also a "dog problem". There are just too many dogs, with concomitant health and hygiene problems compounded by a growing population of stray dogs.

## Perception of the urban environment

The Psychology group, indicated in the middle of Figure 5—1, is studying how people perceive and evaluate their urban environment. The basic premise of this work is that individual behaviour depends not only on the physical characteristics of the environment but also on what the individual thinks and feels about it. A survey on how inhabitants perceive the Rome environment shows that the city has a quite complex and ambivalent image. In general, inhabitants are very positive about the aesthetic aspects of the city, but very negative about its functional aspects. The cognition and evaluation of the environmental quality of the city are under study, to provide better understanding of the inhabitants' perception of the city and how this is related to the physical environmental conditions. Another group, shown at the bottom of Figure 5—1, is studying the inhabitants' allergopathies, mapping the allergenic pollens in the city and developing an epidemological study of the allergopathies in the Rome area. Further information on this and other components of the Rome study is given in the second progress reports of the project (Italia-MAB, 1981).

## METHODOLOGICAL APPROACH

From the methodological viewpoint, three different levels of multidisciplinary cooperation and integration are recognized within the MAB project. Each working group can work at all three levels or only one or two. Studies are referred to as "first level" when each group studies the city from a general point of view to form a body of knowledge on the overall system common to all groups — a common body of knowledge that acts as a stimulus to communication and subsequent integration between the various groups. An example of this first level approach is that taken by the Energy group when studying the consumption of electricity, or that by the Noise Pollution group in mapping noise levels.

For "second level" studies the groups work together in the same specific, prescribed and fairly typical area of the city. One such area is the northwestern sector of the city, starting from the city-centre and extending out to the country. In this sample area, the various working groups are focussing on those aspects of the environment that affect the quality of life of the inhabitants and are using a common set of "indicators". As part of the team effort, the Environmental Perception group is exploring the subjective perception by inhabitants of the physical, objective aspects of the environment being studied by the other working groups.

Studies at the "third level" consist of specific problem-oriented assignments sponsored by individual groups and focussed on some specific aspect of the city considered as particularly interesting from the group's field of competence and inquiry. At this level, each group can work alone or with one or more of the other groups.

The overall aims of these three levels of study are to arrive at an integrated assessment of the city as an ecological system. The connections between the different groups are growing with time, as indicated by the links between the different topics studied within the project (Figure 5—5). Links with the Rome municipal authority are becoming increasingly close, and are expected to become even closer with the

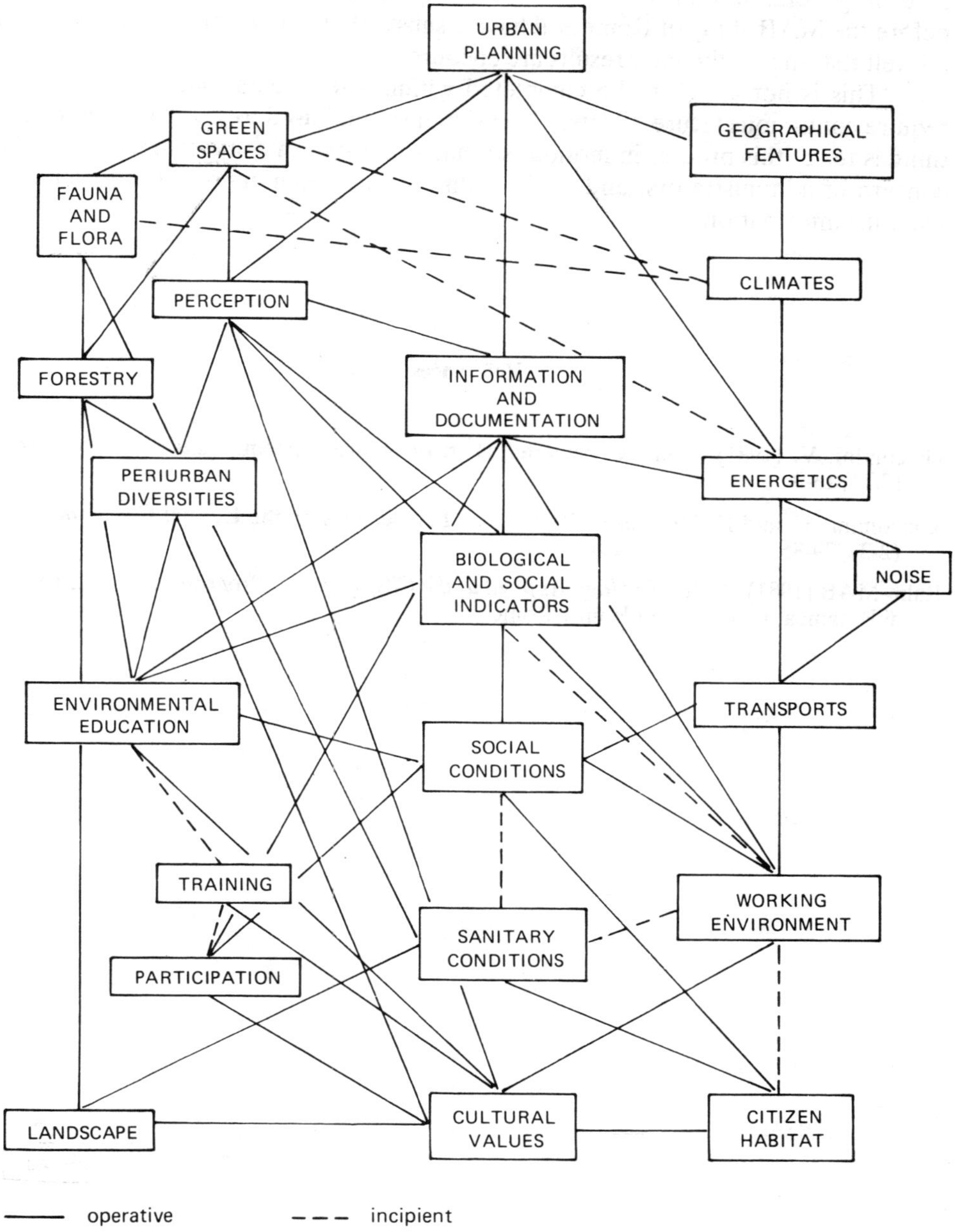


**Figure 5—5.** Interrelations between different topics and areas of study within the Rome project

setting up by the municipality of a permanent centre for interdisciplinary studies on cities in general and on the city of Rome in particular. Much remains to be done before the MAB study of Rome will have a substantial impact on city planning, but it is felt that the preliminary results are encouraging.

"This is not a task to be completed within a few years, but one which will require continuous future efforts," Valerio Giacomini used to say. "The important thing is to set this process in motion, so that scientists can come to understand the concern of administrators, and so that administrators can appreciate the value of scientific information."

**References**

Giacomini, V. (1981), Rome Considered as an Ecological System, *Nature and Resources*, 17(1), 13-9.

Giacomini, V. and D. Hinrihsen (1981), New Perspectives on the Eternal City, *Ambio,* 10 (2-3), 79-85.

Italia-MAB (1981), *Urban Ecology Applied to the City of Rome,* Progress Report 2, Istituto di Botanica, Università di Roma, Rome.

CHAPTER 6

# Ecological Studies for Regional Planning in the Valley of Mexico

*Luis Sanchez de Carmona*

Abstract

The Valley of Mexico forms the immediate hinterland for the Metropolitan Area of Mexico City, which now has 14 million inhabitants and by the year 2000 may have between 23 and 31 million. Rapid urban growth is strongly affecting the natural characteristics of the valley, resulting in resource depletion and severe environmental damage. A study of this ecological system has been carried out to identify cross effects in order to issue recommendations for the better use of the natural resources and for the improvement of environmental conditions. Preliminary findings have already influenced plans and decisions in this region.

*Résumé*

*La Vallée de Mexico forme l'entourage immédiat de l'étendue métropolitaine de la Ville de Mexico qui actuellement compte 14 millions d'habitants et pourrait en avoir 23 à 31 millions en l'an 2000. Cette croissance élevée et rapide a de fortes répercussions sur les éléments naturels de la vallée, entraînant l'épuisement des ressources et de sévères dommages au milieu. Une étude de ce système écologique a été entreprise pour identifier les influences qui se recoupent et en tirer des recommandations en vue d'une meilleure utilisation des ressources naturelles et de l'amélioration de l'état de l'environnement. Les résultats préliminaires de cette étude ont déjà eu une incidence sur les plans et les décisions dans cette zone.*

## BACKGROUND

INTERNATIONAL activities on environmental issues and human settlements have greatly influenced national action in Mexico in these fields. Following the United Nations Conference on Human Settlements held in Vancouver in 1976, the Federal Administration created a Department of State on Human Settlements and Public Works, responsible for urban development planning at the federal level. Because of the close links that exist between environmental concerns and planning of human settlements, a Direction-General of Urban Ecology was created within the Under-secretary Branch of Human Settlements.

This contribution is based on the experience of the Urban Ecology Directorate, particularly through its work on regional planning in the Valley of Mexico, which forms part of the Mexican contribution to the Man and Biosphere (MAB) Programmes. The conceptual framework of our work on Urban Ecology (Figure 6—1) includes the relationship between the natural system, the urban system and the socio-economic system, and an analysis of the mutual influences of the different elements of each of these systems. This approach allows analysis of the flow of energy and materials through the urban areas, and consideration of the inputs, throughputs and outputs of energy and materials, including identification of the feedback relations that are so important in planning remedial action.

## CHARACTERISTICS OF THE STUDY AREA

### Natural features, environmental problems

The Valley of Mexico is located in the extreme south of the central mesas (plateaux), between 19°04′ and 20°11′ N and 98°16′ and 99°30′ W and with an altitudinal range of 2,237-5,452 m above sea level. The valley covers a surface area of 9,600 km$^2$, and includes the political jurisdictions of the states of Mexico, Hidalgo, Puebla and Tlaxcala, as well as the Federal District. The land of the valley is suitable for crops, fruits, natural pastures and man-made grasslands. Fire is widely used as a tool for managing summer pastures. The main land use problems are linked to lack of nutrients, erosion, salinity, alkalinity and flooding. Overgrazing and excessive deforestation have led to extensive resource depletion in the valley.

Accelerated demographic growth has provoked serious pollution problems, due as much to the wastes resulting from daily domestic daily activities as to wastes from large industrial zones. Pollutants are emitted from three principal sources: stationary, mobile and natural. The stationary sources include a large number of industrial areas in the Federal District and Metropolitan Area. The mobile sources are mainly road vehicles, which account for 75 per cent of recorded levels of air

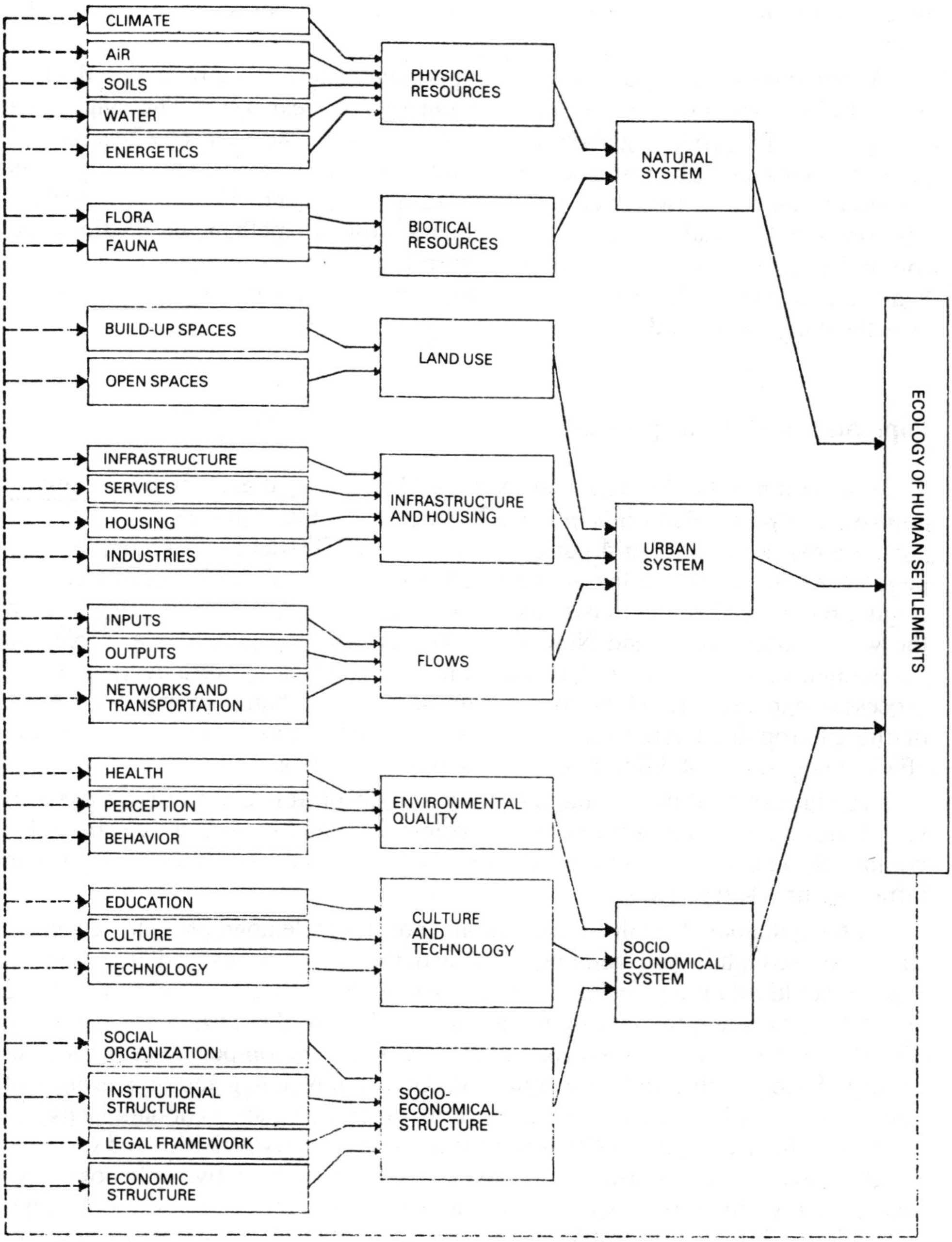


**Figure 6—1.** Interrelations between components affecting the ecology of human settlements

pollutants. "Natural" sources include the dust storms which originate in zones of high erosion, mainly those around Lake Texcoco in the northeastern part of the valley.

Waste disposal is a major ecological problem. There is a growing production of solid wastes, mainly from the metropolitan area of Mexico city. This single area is responsible for producing about 6,000 t of solid waste each day, of which only 75 per cent is collected; the rest is scattered throughout the city. The collected portion is merely poured on two open dumps, with no treatment. The result is serious environmental degradation through air pollution (due to methane production) and soil and water contamination, which in turn have important implications on the health and welfare of the urban populations, especially those in marginal settlements near the dumping grounds.

## Population and urban growth

The population of the Metropolitan Area of Mexico City is estimated currently at approximately 14 million inhabitants: 20 per cent of the country's total, with an annual growth rate of over 4 per cent. The Federal District houses most of these people, the rest of whom live in the neighbouring state of the Mexican Republic (Figure 6—2). United Nations statistics put Mexico city as the third largest city in the world, after Tokyo and New York. The growth of the city as a whole has accelerated since 1950, when the limits of the Federal District were overtaken by a process of expansion which lies outside any scheme of urban planning. Vast zones of the Metropolitan Area lack basic infrastructures, making more difficult the efficient integration of different urban spaces into planning efforts.

The lack of land use regulations has led to uncontrolled use of large areas of the Metropolitan Area, where housing, commercial and industrial uses are indiscriminately mixed. There are extensive residential areas with incomplete urban infrastructures and services.

If the anticipated results of the national policies in demography and decentralization of economic and administrative activities are not realized, the Federal District could attain a population of approximately 18 million inhabitants by the year 2000. Including the metropolitan municipalities of the neighbouring State of Mexico, the whole area would contain more than 30 million people at the turn of century. Such growth would accentuate difficulties in planning and organization of public works and services. Present patterns of predominant economic activities would also be accentuated, with further increase in tertiary activities and decrease in agriculture. The concentration of secondary administrative activities in downtown areas would be intensified which would in turn induce a shift of land use in zones currently used for housing and outdoor recreation. The deficit and costs of urban services, infrastructure and housing would be increased affecting, not only low income groups but also part of the middle class sector. Present socio-economic differences would be further emphasized, with a widening of the base of the pyramid

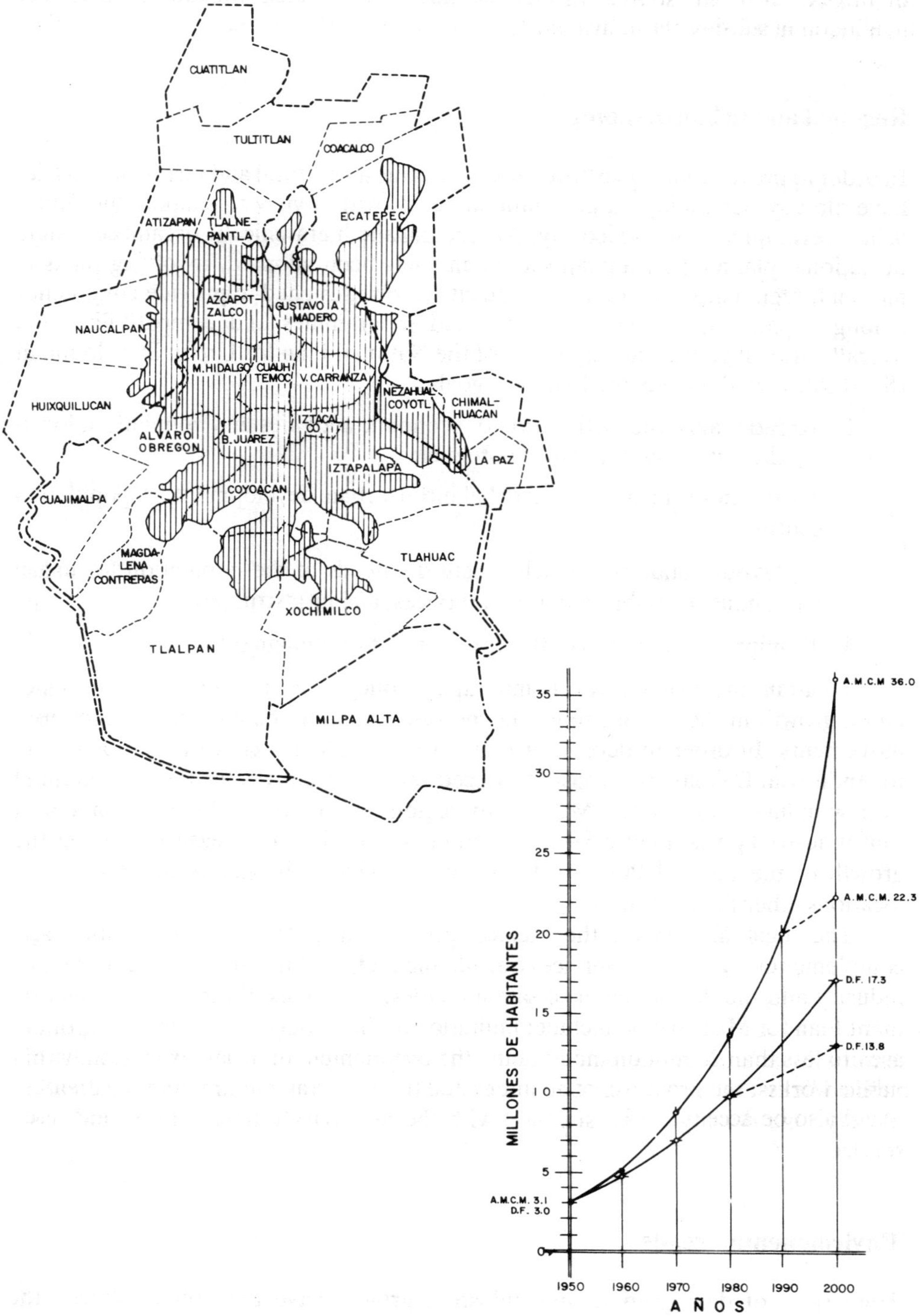


**Figure 6—2.** Population trends of the Valley of Mexico (at left), the metropolitan area of Mexico City (AMCM) and the area within the Federal District (DF)

of income distribution and with unqualified people being obliged to seek their livelihood in subsistence activities.

## Regional and urban planning

In order to prevent and solve problems such as these, sectoral and territorial policies have already been adopted at a national level, with a view to shaping the future urban development of Mexico city. An integral and hierarchically organized system of national planning on human settlements has been adopted, involving plans at national, regional, state, municipal and city levels, designed to provide congruency among objectives, policies, strategies and programmes and compatibility with overall national goals. The objectives of the National Plan for Urban Development (SAHOP, 1979a) include the following goals:

1. To rationalize the distribution of economic activities and population, locating them in areas of greatest potential;
2. To promote an integral and balanced urban development of population centres;
3. To favour conditions which satisfy the needs of the population for urban land, housing, public and urban services, and infrastructures;
4. To improve and preserve the environment for human settlements.

To attain these objectives, demographic policy advocates a decrease of population growth in the region, reducing and re-orientating the patterns of migratory movements. In order to decentralize growth from Mexico city, the National Plan for Industrial Development has introduced several incentives for the development of new industry outside the Mexico city region by promoting the development of four industrial ports (Figure 6—3). Another decentralization measure concerns the growth of the Federal Public Administration, which will mainly occur in urban locations other than Mexico city.

The regional plan for the Mexico city area includes the specialization and complementarity of urban services outside the metropolitan area, the relocation of industry and the development of satellite cities. The goals of the Urban Development Plan for Mexico city, include: limitation of the population of the metropolitan area to less than 25 million inhabitants; the confinement of urban expansion within defined limits; the provision of an integrated transit infrastructure; and the creation of very high density areas, supplied with the necessary infrastructures and basic services.

## Environmental trends

High rates of demographic and industrial growth have perceptibly altered the

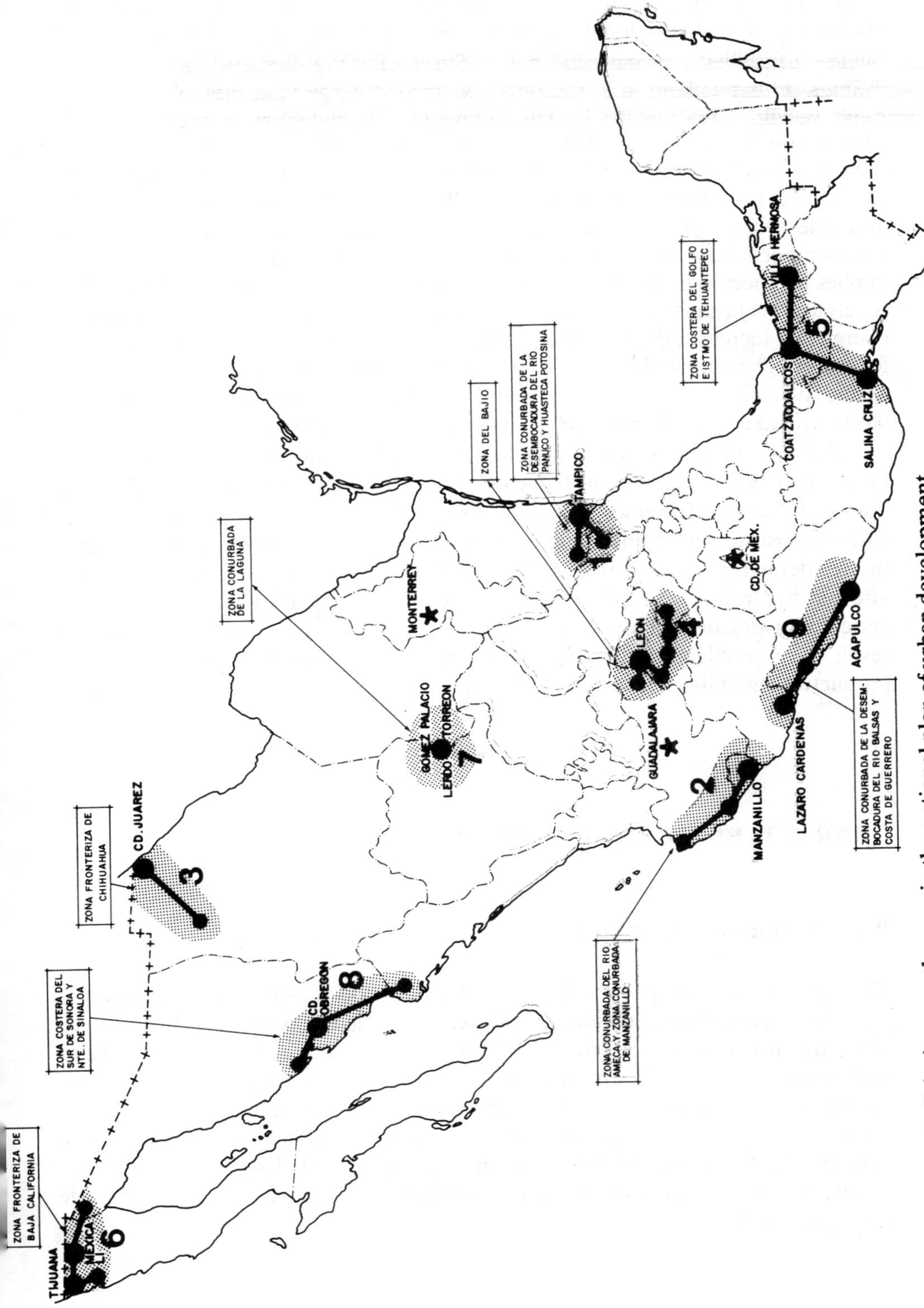


**Figure 6—3. Priority growth areas in the national plan of urban development**

metabolic functioning of the valley. Among other effects are poor processing of products for consumption and an inadequate disposal of wastes. Among the consumption products, water occupies a primordial place. Since existing demand far outweighs the possibilities of the valley, it is necessary to import large volumes of water to satisfy present needs. This results in environmental degradation in the valleys from where the water is extracted (Figure 6—4). In addition the overexploitation of groundwaters in Mexico city has led to a lowering of the water table.

In the near future, water supply will remain a priority concern, since the available volume per inhabitant will continue to be reduced. This will necessitate the continued transport of water from distant, ever more distant, valleys, which implies an enormous investment as well as serious economic, ecological and social repercussions in those places from where the water is drawn. The consumption of energy may increase enormously in the next few years, due to pumping requirements for water. The actual electrical consumption for pumping may well double in the period 1981-87, and then double again by 1990, if the waters from the Tecolutla basin have to be raised about 2,000 m over a distance of 200 km.

Pollution problems due to mobile sources are accentuated by the location of most industries to the north and east of Mexico city, from whence come the dominating winds. The gases and particles emanating from these industries, together with the dust storms originating around the Lake Texcoco, are transported towards the residential areas. At present 50 per cent of the land surface of the region is affected by problems of erosion. If deforestation, overgrazing and inappropriate agricultural practices continue as at present, an even more critical situation will result. The problem is compounded by continued urban incursion into highly productive agricultural land, which results in a lowering of agricultural production.

## CHARACTERISTICS OF THE STUDY

### Basic assumptions, goals and priorities

The regional planning project for the Valley of Mexico (SAHOP, 1980) considers interrelationships between human settlements and the environment at three levels. First, the urban metabolic process, analyzing effects such as fluctuations in water and air quality on people's health and the effects of changing patterns of land use or energy use. Second, the effects of the dominant features of the Valley of Mexico on the metropolitan area of Mexico city. Third, the limits of the maintenance systems which provide the metropolitan area with essential supplies such as air, water, nutrients, energy and manufactured products, and the capacity for waste absorption of the valley.

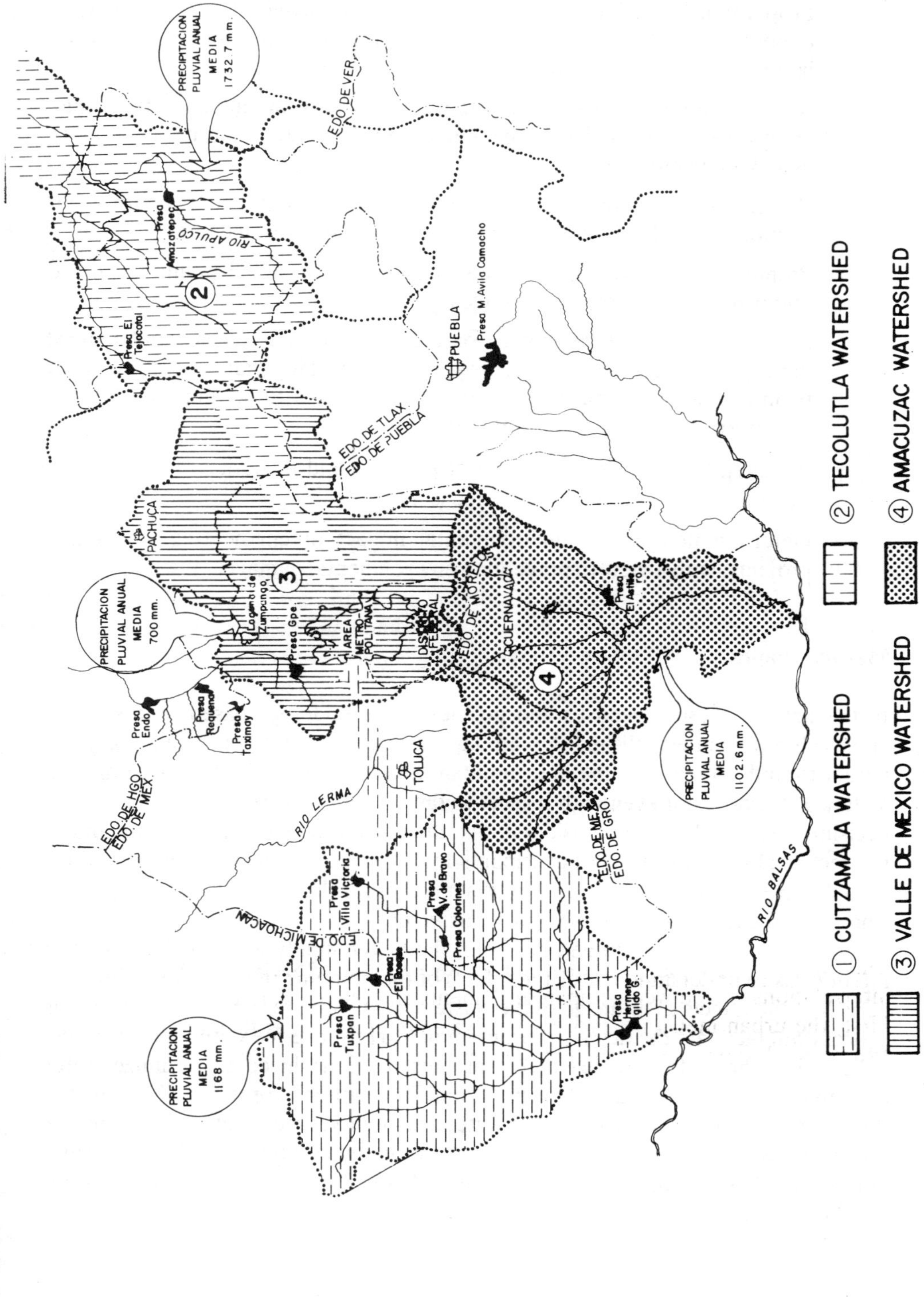


**Figure 6—4.** Annual precipitation in four watersheds in the Mexico Valley

The objectives and policies of the project can be summarized as follows:

1. Proper planning of the location of the different activities in order to attain a more favourable balance and a more equitable distribution of the population with regard to natural resources;
2. Rationalization of the use of natural resources in human settlements, applying ecologically sound techniques in the supply of basic services such as water and energy;
3. Reduction of the consumption of natural resources and energy with introduction of new habits of consumption and new sources of energy;
4. Promotion of recycling of wastes at all levels, to avoid pollution and to optimize the use of scarce resources;
5. Preservation and development of the socio-cultural, urban and natural dimensions of environment quality, thus recognizing that the well-being of people depends as much on the quality of each of these dimensions as on the interrelationship between them;
6. Regeneration and rehabilitation of polluted and degraded environments, starting with low-cost corrective measures;
7. Determination of the environmental impact of engineering works and projects, with a view to preventing environmental damage.

## Methodologies

In the light of these objectives and policies, a methodology based on a systems approach has been established. No attempt has been made to prepare a complete model to understand how the metropolitan area of Mexico city works within the Valley of Mexico, nor even with respect to the rest of the territory of Mexico. This decision was based on two considerations: the insufficiency of existing data and of the technical skills of project participants for constructing and managing complex models. In spite of this decision, the systems approach has not been abandoned, since partial analyses of subsystems are carried out.

The first subsystem studied has been the water supply used by the city. The intention is to understand better the effects of the pumping of water for urban use, as well as the effects of drainage of the waste waters. A better understanding of this process will facilitate the design of programmes to improve the water balance in the valley, and the different measures that must be introduced to optimize water availability and to minimize water pollution. The systems approach will allow the detection of mutual influences among water availability, green areas, soil conditions and land use. This may assist in selecting programmes that are different from conventional ones, and which will have a direct, beneficial influence on water conditions in the valley.

The second subsystem analysed is that of land use and land allocation, with particular reference to industrial areas. The study of land use is incorporated within an integral regional planning framework, in order to understand the combined effects of different types of land use on natural resources and on the quality of life. The study is designed to provide information for improving current land use plans and the programmes of decentralization for industry and government offices.

The third subsystem under study concerns the food of the inhabitants of the urban area. This includes analysis of food production in the Valley of Mexico, as well as the transport of food to the city for consumption therein, or for wholesale or processing for later transport to different regions of the country.

## Results

The intention is that these studies will give rise to a series of recommendations for improving the quality of life in the metropolitan area of Mexico city and for promoting environmental management through better use of existing natural resources. A set of provisional recommendations and programmes has already been identified and proposed to the authorities of the Federal District.

In relation to water balance in the valley, it has been recommended that the infiltration rates be increased in order to reduce the gap between the 40 $m^3\ sec^{-1}$ extracted from wells in the valley and the 30 $m^3\ sec^{-1}$ that normally recharges the groundwaters. In order to achieve this purpose, some small dams and reservoirs are proposed in the hills to the south and west of the valley, which may also be used for recreational purposes. Implementation of this measure will also extend the vegetation cover and forested area in the region concerned, as well as improving drainage.

It is also proposed to recycle as much water as possible, through a secondary treatment that will enable the "grey" slightly polluted, partially used water to be used in some industries and in green areas and parks. It is an absurdity to pump so much water such great distances and not make maximal use of it. With respect to industrial location, studies already carried out provide a solid basis for assessing the environmental impact of existing and proposed industries. The findings of the study support the National Programme for Decentralization of Industry, mentioned earlier.

In relation to food, it is intended to increase the production of basic food for human consumption in the city. In addition to programmes for improving the use and productivity of existing agricultural land in the valley, it is proposed to develop suburban food plots ("hortalizas") in the backyards of middle and low income houses (SAHOP, 1979b). In order to promote this concept, manuals have been published to stimulate a certain self-sufficiency in suburban and rural houses. It is further hoped that more intensive use of agricultural land in the valley will reduce air pollution induced by wind blown dust and will increase rates of water infiltration. Concerning waste disposal and management, it has been proposed to increase the collection of garbage and to dump it in landfills with sanitary measures, to prevent pollution of soils, water and air.

As a complement to these recommendations, it has been proposed that a green belt in the form of a system of parks be created around the metropolitan area, with inward extensions in the form of linear parks serving recreational purposes as well as the protection of waterways. Tourist and recreational routes have also been proposed for linking sites of importance to the natural, historical and cultural heritage of the country, which might also contribute to the leisure opportunities for the inhabitants of the city.

In conclusion, the systems approach adopted in this study, though constrained by limitations in data and trained personnel, has served to clarify certain key environmental issues in the Valley of Mexico. It has put decision-makers in a better position to take effective action to improve environmental conditions in the area. Preliminary findings have been transmitted to planners and politicians, and have had an influence on their plans and decisions. But there are still many facts to obtain. There remains an urgent need to define more precisely the type of programmes and projects that must be put into practice. The study has already contributed to the planning of the Valley of Mexico and we hope that it might provide insights useful to planners and decision-makers in other parts of the world.

## References

SAHOP (1979a), *Plan Nacional de Desarrollo Urbano,* Sintesis Informativa, Noviembre 1979, Secretaria de Asentamientos Humanos y Obras Publicas, Direccion General de Ecologia Urbana, Mexico.

SAHOP (1979b), *Cartillas de Ecotécnicas par la Vivienda Autosuficiente,* Sintesis Informativa, Diciembre 1979, Secretaria de Asentamientos Humanos y Obras Publicas, Direccion General de Ecologia Urbana, Mexico.

SAHOP (1980), *Ecoplán del Valle de Mexico,* Sintesis Informativa, Marzo 1980, Secretaria de Asentamientos Humanos y Obras Publicas, Direccion General de Ecologia Urbana, Mexico.

CHAPTER 7

# Sensitivity Models for Urban and Regional Planning: The Example of Lower Main-Frankfurt

*Alexander von Hesler*


## Abstract

A sensitivity model of Frankfurt am Main has been developed as a tool for urban and regional planning. The model facilitates understanding of the city and its hinterland as a biocybernetic system, and enables interactions between key groups of variables and sectors of the economy to be considered. Such models can play a useful role in increasing the awareness of the planner about the urban system, and can also fulfil an important function in communication of ideas between scientists and decision-makers. Steps in the development, testing and application of the model are described.



## *Résumé*

*Un modèle de simulation de Francfort sur le Main a été mis au point pour servir de moyen de planification urbaine et régionale. Le modèle facilite la compréhension de la ville et de ses environs en tant que système biocybernétique, et met en lumière les interactions entre les groupes-clés de variables et les secteurs de l'économie à prendre en considération. De tels modèles peuvent jouer un rôle utile pour la prise de conscience du planificateur vis-à-vis du système urbain; ils peuvent également remplir une importante fonction de communication des idées entre savants et décideurs. Les étapes du développement, de l'essai et de l'application du modèle sont décrites.*

## SYSTEMS PLANNING

MOST of the face of our planet has been influenced, shaped or destroyed by human hand. The earth has progressed from a natural to an artificial state. When planners concern themselves with ecosystems they are usually dealing with systems that have already been subjected to anthropogenic influences. The present-day planning task is to improve these systems subject to such influences, usually because the systems are not functioning particularly well. The intention is to facilitate, to the extent possible, their transformation into sustaining self-regulating systems. To do this, planners must first study the laws and norms of natural ecosystems, in order to apply knowledge about such characteristics as far as possible. The more successful the efforts, the greater the system's chances of being able to develop and reproduce itself.

One of the essential aims of the Frankfurt regional study within the framework of MAB Project Area 11 was to provide an instrument to help planners understand an environment as a biocybernetic system and to develop decision-making aids; instruments which are suitable for increasing the viability of the system under consideration. Experience with classical planning instruments indicates that they are useful up to a certain density and a certain diversity of utilization. But studies in many urban agglomerations have shown that these existing instruments cannot solve the acute problems of urban planning. Unlike less densely populated zones in urban areas, the different uses lie so close together with no buffering zones that a differentiation that is only spatial cannot be successful.

Such systems show a degree of interdependency (Figure 7—1) that cannot be managed with the present planning instruments. Planning legislation is such that it is difficult to identify and enumerate interdisciplinary relationships in such a way as to derive the necessary planning measures. For example, land use plans can outline residential, industrial, traffic and recreational areas to fit the forecasts for population and economic development, taking due account of the nature of soil, surface waters and climate. But it is more difficult to assess how this "system" will develop as a result of its own dynamics, how it will react to intervention and to what extent natural resources will or will not be exhausted.

Various plans (regional, land use and master plans) are constantly amended or updated: residential areas are enlarged when population growth makes it "necessary"; roads are broadened when the traffic "demands"; industry is introduced because manpower is "available". The updating of plans is mainly based on additions: more houses, more roads, more workplaces and, as a natural consequence, more exhaust fumes and dust, more noise, more refuse. To eliminate these undesirable effects, more filter systems, more embankments to screen the noise from motorways, more refuse-incineration plants are needed. This in turn calls for more laws and more regulations, which are then applied synergistically. Constant interventions induce even more interventions to eliminate or redress the inconveniences or harmful effects that have been created. Dependencies increase.

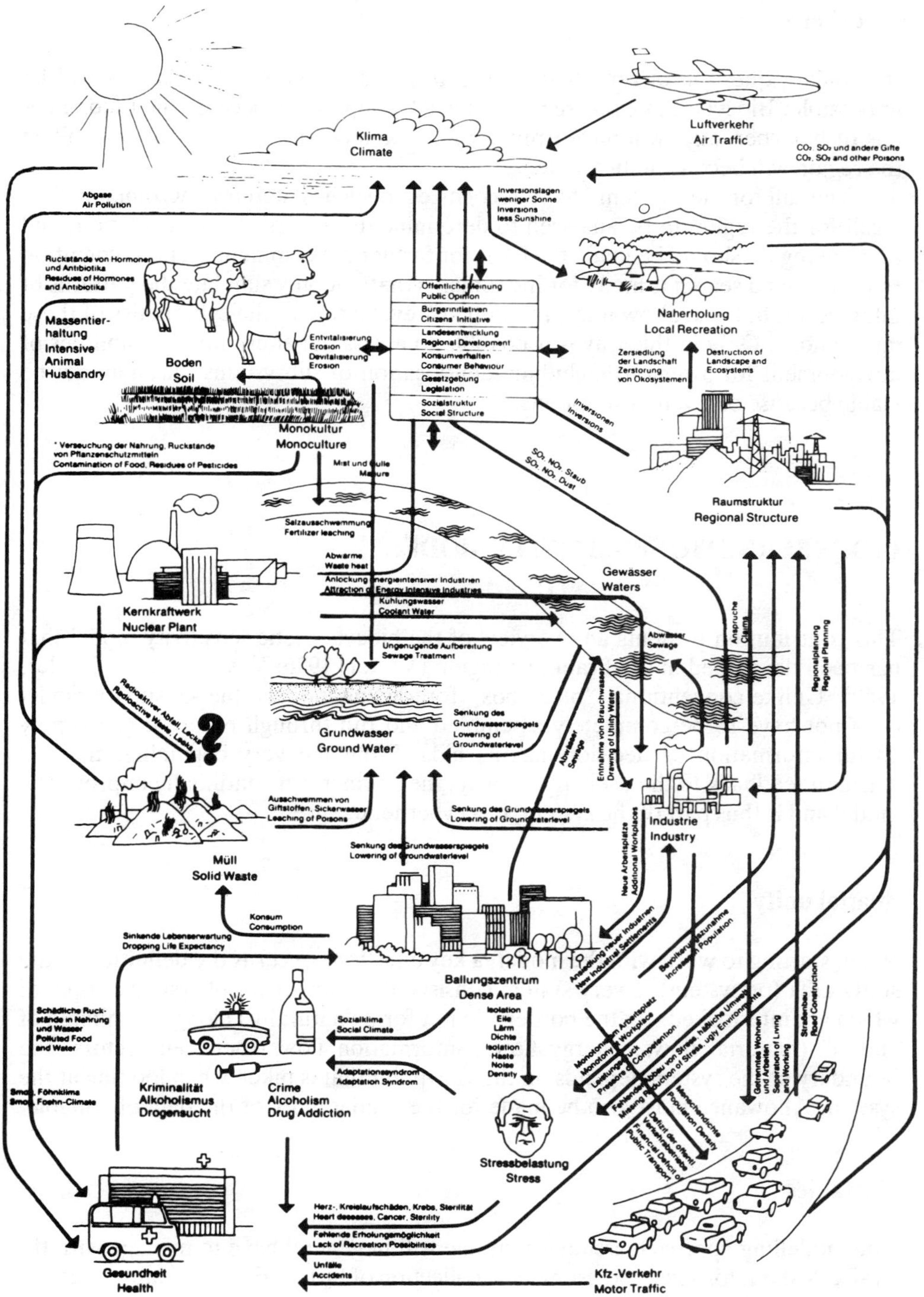


**Figure 7—1.** The urban system network

### Biocybernetics

It is only a question of time before such technical cybernetic regulations will be impossible. Biocybernetic self-regulation will be impossible because the fundamentals of biocybernetics will have irrevocably broken down. The costs of controlling the region will become unbearable.

The call for the "system" to be recognized as such is therefore accompanied by a call for the region to be analysed to determine its powers of self-regulation, and that the region should be made the basis for further developments. This recognition requires that a search is made for instruments that not only simulate the patterns of interactions but also allow interpretation and evaluation of the cybernetics of these interactions. Only in this way is it possible to assess the alternative possibilities of development for a region, including identification of subsystems that may not be viable because of given assumptions.

## BIOCYBERNETIC SENSITIVITY MODEL

This contribution presents an overview of the biocybernetic sensitivity model that has been developed in the Frankfurt region (Vester, 1976; Vester and von Hesler, 1980). Unlike conventional "black-box" forecasting models, the sensitivity model does not have to be completely assembled and run through before it can supply useful information or decision-making aids. From the very beginning, there is constant feedback to the user. In this way, the planner can continually improve the model and is thus part of the system, in a cybernetic sense.

### Spatial unity

When starting to work with the model, a key question concerns the definition of the study area (or system). Every system consists of a number of subsystems, some of which are interlocked. As the borders do not form "naturally" along the minima of "flows" (material flows, energy flows, information flows) between sectors, the boundary of the system depends on the viewpoint that is taken when looking at the system. Allowance must also be made for the spatial unity of the data obtainable.

### Variables

The modelling approach assumes that one will in general have to manage with the available data, for several reasons. The collection of further data invariably becomes increasingly difficult. Complete coverage is never achieved. Collecting the data often takes far longer than the time available. In order to demonstrate to the users the kinds of variables (data) that are necessary for the model, an exemplary set of variables is proposed as an indication of the nature of the variables that are needed. Changes in the set of variables will have to be made from case to case. The key words for the variables are generated from seven overlapping sectors: economy; population; land use; human ecology; natural balance; infrastructure; communal life.

Three main steps are involved in obtaining a set of variables tailored for each study area: First, preparation of a description of the area and its situation, as an orientational aid in the selection of variables. Second, examination of available data and factors affecting those data, and assessment of their utility for a cybernetically relevant set of variables on the basis of a "variables" record sheet and of a criteria matrix. Third, reduction of the number of variables to obtain an optimal set representative of 30 to 40 key parameters.

## Criteria matrix

The system relevance of the combination of variables is thus critical for the efficiency of the sensitivity model. The relevance can be controlled by checking the variables with a criteria matrix, which allows a quick check to be made on the right combination of system-relevant criteria for any set of variables. The criteria matrix contains six groups of criteria which must be satisfied by the totality of the variables in a balanced manner. The six criteria groups are: spheres of life; basic cybernetic criteria; basic physical categories; basic dynamic categories; system relationships; development of variables.

A card is prepared for each criteria group. Each of the holes in the card represents one criterion. If the variable covers a criterion, the hole is covered with a patch of self-adhesive paper. A set of variables, consisting of a number of cards, can then be held up against the light or laid on an appropriate surface to see which criteria are not covered. The overall coverage can be modified by removing or adding individual variable cards. The criteria matrix is always of importance when running the model. Its main functions are as follows:

1. To check the entire set of variables for coverage of all criteria;
2. To check the entire set of variables for different weightings of the criteria;
3. To check additional variables for supplementing the coverage;
4. To think out the set of variables without losing overall coverage by the removal of individual cards;
5. To compare the coverage of subgroups, for example, for an individual sphere of life;

6. To determine the scattering range of criteria by comparing the coverage of two variables.

## Pattern of interactions

The next step in developing the model is to establish a supposed pattern of interactions. The user must apply his knowledge of the data and his experience; that is, the user must first develop his own interpretation by examining the individual variables for possible relationships with all other variables. This initial examination identifies important interactions, including feedback relationships. Some variables not only affect others, but are themselves influenced through a number of other variables. As a working aid, the model illustrates some important feedback loops which appear everywhere; for example, the feedback cycle economy-population shows two interlocking feedback loops (Figure 7—2). The left-hand loop shows that manpower demands are increased with increased industrial capital. This means more job vacancies, but capital flow will decrease if the number of vacancies increases. The right-hand loop shows that the migration of population (manpower) is stimulated by job vacancies, which leads to a decrease in the availability of jobs. If the two negative feedback loops form an "eight", they supplement each other to form a positive feedback, with the danger of overburdening the region. The high degree of conformity of single feedback loops, and the overall pattern of interactions with reality, result in the following inputs which are introduced by the user:

1. Technical knowledge (geology, ecology, systems research, etc.);
2. Conventional planning (for example, regional planning and planner's observation);

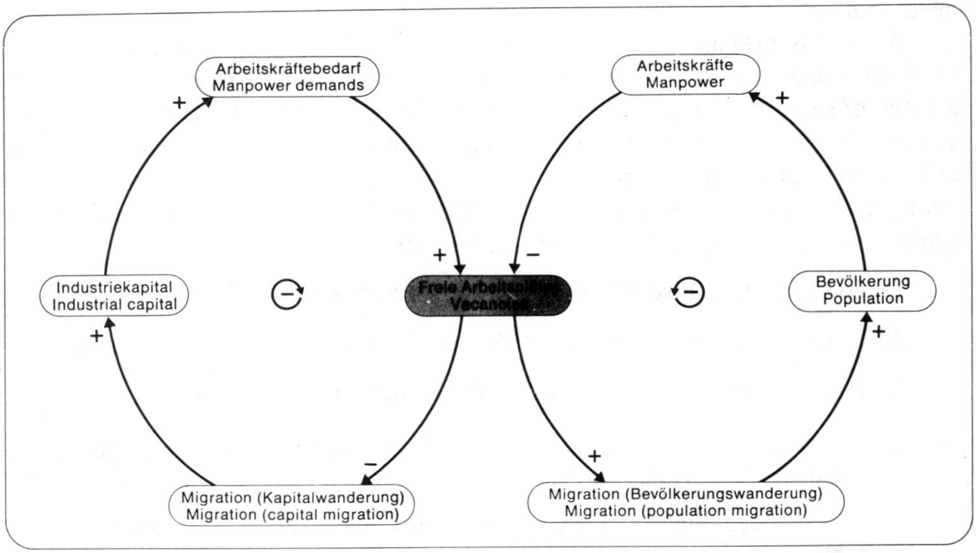


**Figure 7—2.** Use of the screen matrix

3. History of the dynamics of the variables (past developments influence factors and conclusions as to the linkages, parallelism, etc.);
4. Plain common sense in identifying obvious relationships in component sectors (e.g. a road is laid through a wood, the wooded area becomes smaller; the sewage burden grows as the population increases, etc.).

The actual behaviour of a system cannot be determined from such individual scenarios, but only when they have been linked to form an overall open model in which all spheres of life are represented. These and other chains of cause and effect (as determined from data and interviews seen against the background of the previously discerned relationships) must be combined to produce a pattern of interactions. The resultant (heuristic) pattern of interactions not only improves the user's general consciousness of the nature of the system, it also serves as a basis for the interview programme that forms part of the next step. This pattern of interactions must be modified according to the results of the survey.

## Interview programme

A catalogue of questions is used to clarify the assumed relationships in the pattern of interactions. The questions contained in the questionnaire will more or less reflect the variables themselves or the influencing factors and auxiliary variables connected with them. The answers on the other hand, because of the way in which the questions are put, reflect the relationships between the variables. Variables extracted from available statistics are checked with the screen matrix to determine their cybernetic relevancy. Much the same approach is taken in checking the empirically determined variables and their relationships. In the former case, the main concern is to check representation of the main spheres of life and the basic physical criteria. For this variable group, balanced consideration needs to be made of both the dynamic and the cybernetic criteria.

The necessary degree of differentiation in the catalogue of questions is ensured by specifying five target groups each of which is requested to answer a certain set of questions. The target groups for the interviews are as follows:

| *Group A* | *Group B* | *Group C* |
|---|---|---|
| Mayor | Doctors | Farmers |
| Opposition | Social workers | Foresters |
| Vicar | Police | |

| *Group D* | *Group E* |
|---|---|
| Young businessmen | Retail traders |
| Factory managers | Inn keepers-hotel managers |
| Workers-employees | Housewives |

## Expanded pattern of interactions

The body of data obtained — consisting of the original key variables, a number of supplementary variables (determined through the interviews) and the linkages revealed by the poll — can then be used to build up an expanded pattern of interactions. The nature of the variables, the directions and courses of linkages are recorded (Figure 7—3). Such a diagram is still relatively simple, but it does make it possible to recognize readily the relationships and influences, and to determine the general behavioural trend with the aid of suitable interpretation models. The corrected pattern of interactions represents a first coherent model of relationships between the individual factors and thus of the system as a whole.

The pattern of interactions among the 72 variables shown in the model is about twice as detailed as the first heuristic pattern of interactions from which it was developed. The number of possible dependencies has increased by the same proportion. Together with the variables, these interactions facilitate clear recognition of the system-relevant processes. A first cybernetic interpretation can now be attempted, even though this will only be roughly qualitative and still without any connection to the given data. To arrive at a model which is suitable for simulation, the variables must now be "labelled" with their roles and put into a form that will allow the input of quantitative values. The same applies for the interdependencies, which until now have only been shown as arrows. For these, a more detailed mathematical description is required of the nature and dynamics of the interdependencies.

At the same time, it is useful to build a comprehensible model which has a reduced number of elements — a model that on the one hand facilitates cybernetic interpretation and on the other still leaves sufficient leeway for action and reaction. For this, certain areas of the pattern of interactions need to be condensed once more. This time, however, allowance is made for mathematical functions. In each of these steps, the criteria matrix is used to check whether the selection is adequate, that is whether it meets the criteria of a cybernetically relevant set of variables.

## Simulation model

Aggregation is obtained by assembling either chains of variables or parallel factors. If the relationships, represented as table functions, are inserted into the connecting arrows of the aggregated pattern of interactions, the result is a visual simulation model (Figure 7—4). For this simulation model, a special computer program is developed, and is subsequently simplified slightly to allow its use with a dialogue computer. The program is so designed that at the following level of programming the interpretation and evaluation models are directly based on it. These can therefore directly link and communicate with the computer simulation program.

The following hypothetical values for the seven state variables were taken as a simulation basis for interrogating the interpretation models; these values have been taken from the Frankfurt region, the test area used in developing the sensitivity model.

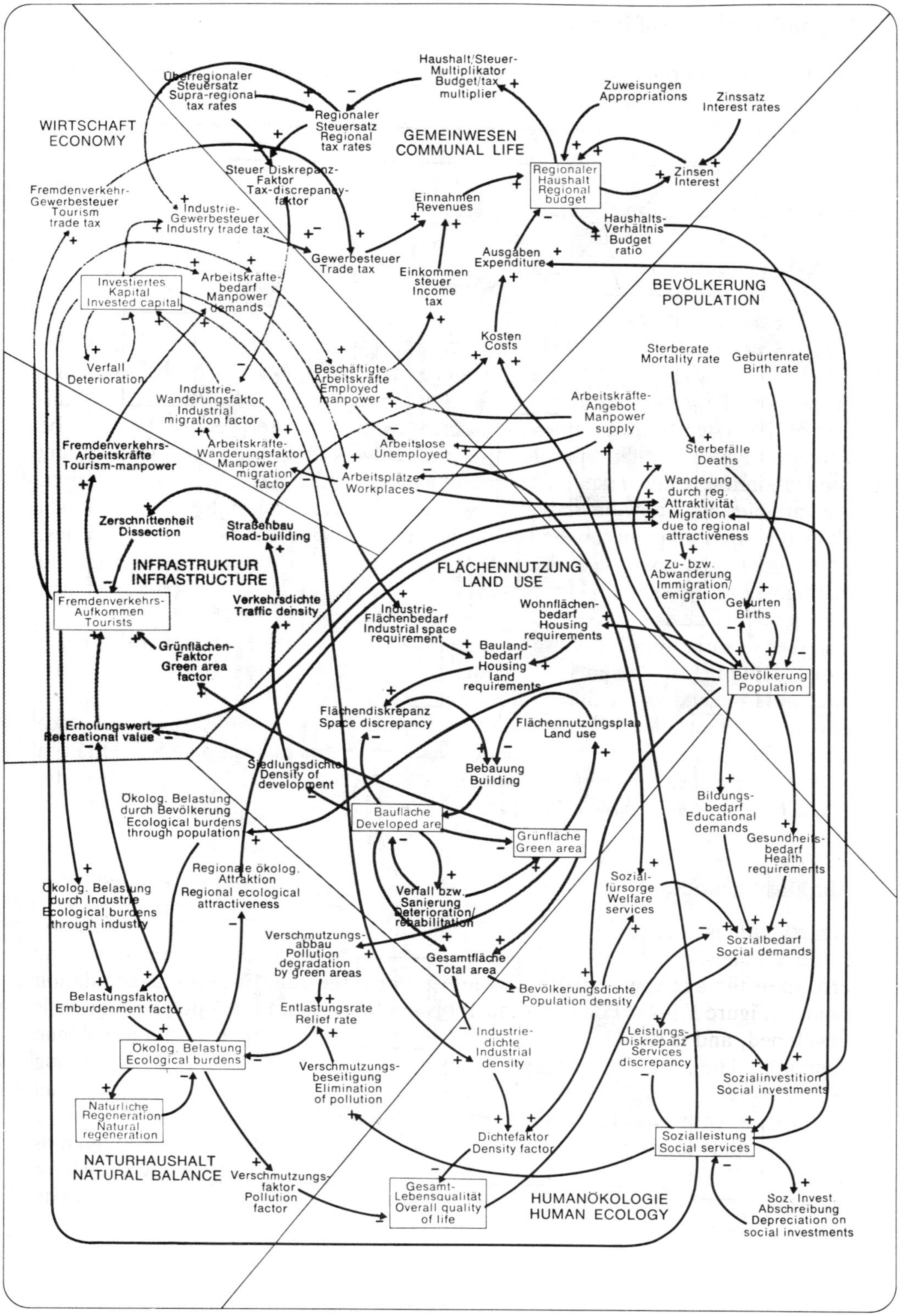


**Figure 7—3.** Economy-population feedback cycling

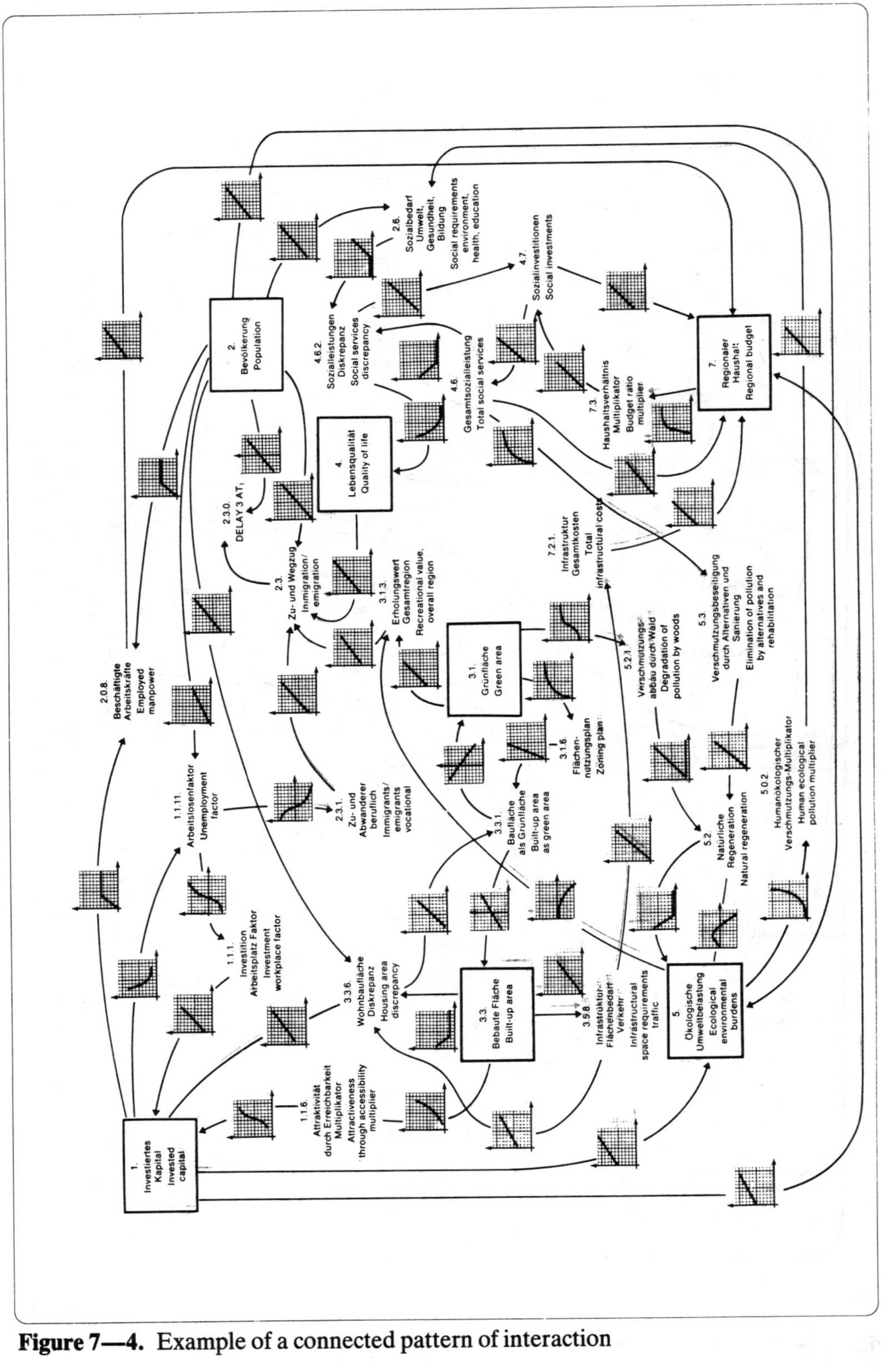


**Figure 7—4.** Example of a connected pattern of interaction

| *Variable* | *Input* |
|---|---|
| Population | 136,000 |
| Invested capital | 80 million DM |
| Regional budget | 120 million DM $yr^{-1}$ |
| Social services | 15 million DM $yr^{-1}$ |
| Built-up area | 71 $km^2$ |
| Green area | 608 $km^2$ |
| Environmental burden index | 100 Ue |

For the traffic network and density (persons per km × annual average), the corresponding values were:

| | |
|---|---|
| Expressway | 50 km (7 persons $km^{-1}$ × year) |
| Railways | 100 km (2 persons $km^{-1}$ × year) |
| Highways | 140 km (4 persons $km^{-1}$ × year) |
| Municipal roads | 150 km (10 persons $km^{-1}$ × year) |

After input of specific data of the system under investigation into the simulation program a test run is carried out to check mathematical compatibility; this consists of 10 to 20 successive operations. After the necessary corrections, the simulation program serves as a basis for the following programming levels. With the aid of standard runs the planner can improve the used variable relations and thus optimize the simulation model. Above all, the planner will develop a feeling for the dynamics of the system and thus be able to expand his own notional concept. His understanding of the interpretation of such simulations with the following interpretation program will be sharpened, and it will therefore be easier for him to progress from system dynamics to system behaviour. For example, Curve 1 (Figure 7—5) allows

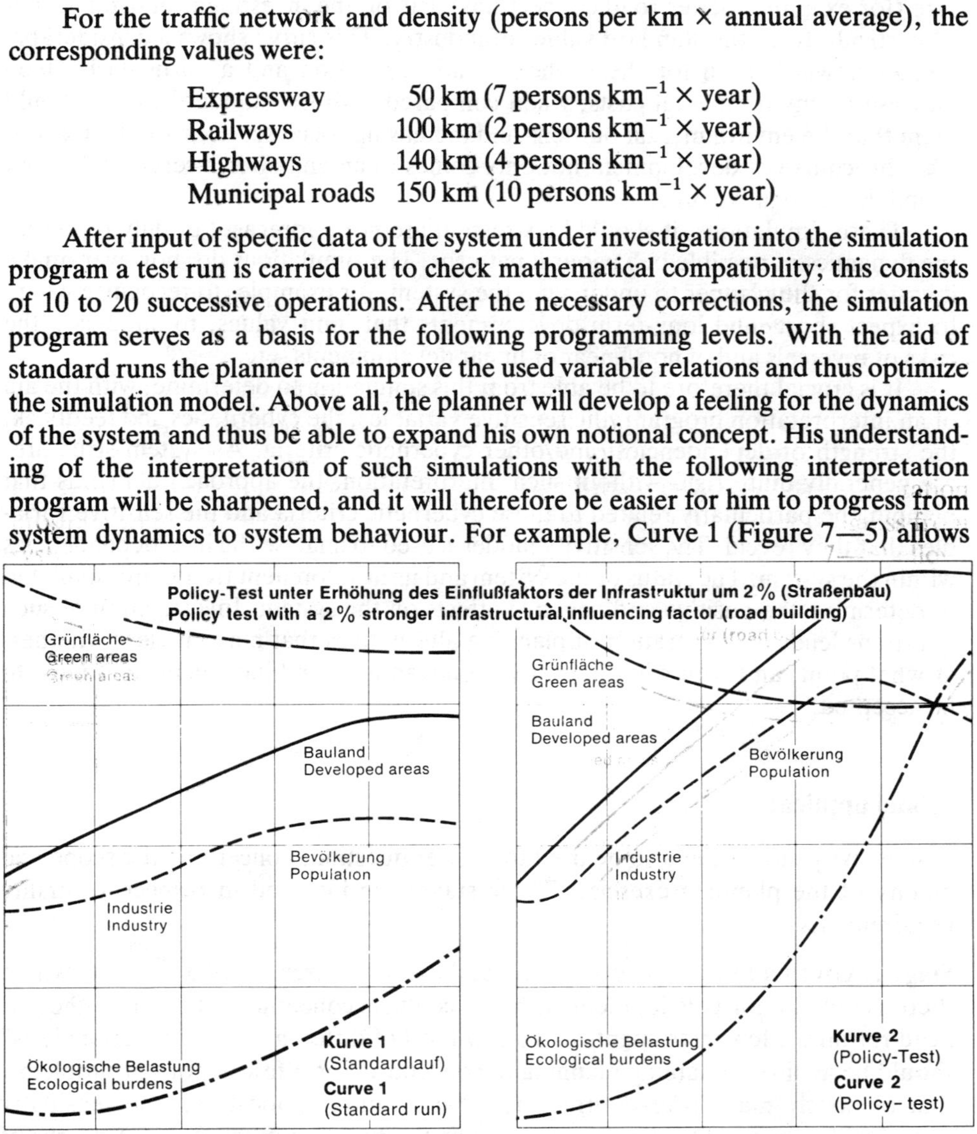


**Figure 7—5.** Visual simulation model with functions

recognition of general system dynamics in five important state variables: economic development; population; ecological burdens; building ground; and green areas. Slow and steady economic development accompanies regional population growth. The developed areas increase, as one might expect, whilst the size of the green areas decreases accordingly. Increasing burdens finally lead to a stagnant population in the region, and the amount of invested capital also ceases to grow.

In a further test (Figure 7—5, Curve 2), the standard run is verified by the specific assumption that a 2 per cent influencing factor due to improved infrastructure (for example, better road connections) acts on the investment of capital, that is, particularly on the standard values of industry. This curve shows a considerably greater upward trend for the economy and population and a particularly steep increase in environmental costs, when compared with the standard run. It would seem that the environmental burdens and increasing social services are the reasons why this course of development not only comes to an end after a certain time, but even takes a negative turn.

These simulations, it should be stressed, are not a forecast but only represent the dynamics of possible behavioural patterns. The simulations do, however, make it easier for the planner to understand the system, for example, to recognize where to expect short- and long-term developments that limit values, to anticipate the risks of reversals and of non-linear or linear developments, etc.

It is crucial therefore to be able from this simulation to determine, with the aid of an interpretation program, the sensitive variables, the cybernetics and feedback, the strength of dependencies, and other cybernetic criteria. As system structures are generally quite rigid without such interpretation, the approach demands that planning be particularly related to these cybernetic criteria and the sensitive variables that they reveal. The sensitivity model is used to analyse the interdependencies within the system. The status of the system and its development trends are simulated to determine the positive and negative effects on the system. Investigation of such questions leads to elaboration of planning alternatives that make it clear whether, at what point, and with what intensity, disadvantages and inconveniences have to be accepted.

## Model application

A sensitivity model can also be used to answer questions concerning the scope and quality of the planning exercise. Three stages are involved in respect to quality requirements.

**Stage 1:** Given a low order problem in land use, characteristic of a situation where there is only minor development with few conflicts concerning utilization, there is need for only a few instruments and a low level of differentiation. "Over-analysis" should be avoided, enabling planning to concentrate on a few important problems, that obviously may differ enormously from location to location. The sensitivity model will very quickly show — as early as the first "outputs" — what depths of study are necessary. The first steps will probably suffice for obtaining an adequate knowledge of the interrelationships in the region.

**Stage 2:** In an area with a medium density of settlement and a constant rate of development, the conflicts between claims for different land uses become readily apparent. The system interdependencies are already so diverse that careful assessment of the present situation and prediction of development trends become necessary. Because of the greater conflicts between competing claims, the plan must contain more differentiated statements on land use. A number of the above-mentioned models need to be brought into play for analysing the key questions in greater detail. In working with the sensitivity model, it becomes apparent whether it is necessary to go to the interpretation model stage to solve the problems being addressed. If feasible, it is recommended to begin to use the entire sensitivity model at this level.

**Stage 3:** In areas with high density agglomerations and in areas in which considerable changes in utilization are planned — new towns, engineering works, allocation of extensive tracts of land for agriculture and forestry — the system interdependencies and the effects of changes in utilization are no longer comprehensible. Conflicts between land use claims can occur even where they are completely unexpected. Irreversible developments can destroy the system. Planning must not only regulate land use but also consider whether the ecosystem can absorb and sustain the proposed form of usage. Here, it is absolutely essential that the entire sensitivity model be used. This is the only way of ensuring that incorrect developments are recognized at an early stage and that the necessary counter-measures are taken.

The model begins with the first, system-oriented phase of data collection, and then leads the user from the identification, collation and processing of data to the simpler model stages and, finally, to the more difficult stages of interpretation and evaluation. Working with the model is facilitated by the fact that only a very few of the computer programs are difficult to follow. The structure of the model is open, and gradually becomes comprehensible as work with the model proceeds, because of the numerous feedbacks. The benefits for users are increased by the numerous outputs from the model: the findings that result from each individual stage deepen the user's understanding and already contribute to decision-making. The structure of the model is such that the user is inevitably led to "his" variables, "his" region and thus to "his" problems via the various data and variables.

## References

Vester, F. (1976), *Ballungsgebiete in der Krise/Urban Systems in Crisis,* Deutsche Verlags-Anstalt, Stuttgart.

Vester, F. and A. von Hesler (1980), *Sensitivitäsmodell,* Regionale Planungsgemeinschaft Untermain, Frankfurt am Main.

CHAPTER 8

# Energy Conservation and Diversification of Energy Sources in and Around the City of Lae, Papua New Guinea

*Ken Newcombe*

Abstract

The social and environmental problems of the growing urban ecosystem of Lae, Papua New Guinea, were assessed by a MAB team of scientists and social planners in 1977-78. Local government and community leaders were engaged in subsequent planning and implementation from the beginning, and in parallel with ongoing analysis. Detailed energy flow analyses defined present energy usage and prospects for cheaper renewable sources in the city and from its hinterland. Waste production and disposal systems were analysed and systems of nutrient and materials cycling and energy recovery identified. The social problems of low income and subsistence dwellers in and around the city were limited to decreasing land available for gardening, and limited opportunities for work. Within a year of the start of the research programme, the City Council, assisted by the national government, had begun fuelwood production, conservation forestry, intensive food gardening, subsistence level agro-forestry systems, composting of city wastes and fruit-tree planting in and around the urban area. Within two years, charcoal production, fuelwood distribution systems, and industrial scale biogas production had begun. Major industry had begun the transition to woodfuels away from oil, and energy conservation in industry was in practice. In 1981 national government initiatives led to renewed use of solar hot-water heating and detailed design of alcohol fuel production for the city was nearing completion. Within 18 months this MAB project gave rise to the national energy policy and programme, and subsequent projects of implementation. The Lae project is a working example of the application of ecological principles and analysis towards social and economic development in a Third World city.

*Résumé*

*Les problèmes sociaux et environnementaux affectant l'écosystème urbain de Laé, Papouasie-Nouvelle Guinée, écosystème en pleine croissance, ont été évalués en 1977-78 par une équipe MAB de scientifiques et de planificateurs. Dès l'origine de l'activité, le gouvernement local et les chefs de communauté ont été associés à la mise en oeuvre des travaux, puis au déroulement des analyses et enfin à la planification ultérieure des recherches. Des analyses détaillées du flux d'énergie ont permis de définir son utilisation actuelle et les perspectives d'en obtenir de nouvelles sources renouvelables à faible prix dans la ville et ses alentours. La production de déchets et les moyens de s'en debarrasser ont été analysés ainsi que les systèmes de recyclage des nutrients et de matériaux et de récupération d'énergie. L'examen des problèmes sociaux des habitants de la ville et de ses alentours ayant de bas revenus ou simplement de quoi survivre, s'est limité à ceux de la décroissance critique des jardins et l'étroitesse du marché du travail. Dès la première année du programme de recherçhe, le Conseil Municipal, aidé par le Gouvernement Central, a enterpris de produire du bois à brûler, de conserver la forêt, d'intensifier le jardinage potager, d'établir des systèmes agro-forestiers permettant de vivre, de fabriquer du compost avec les déchets de la ville, et de planter des arbres fruitiers dans et autour de la ville. Dans les deux ans qui ont suivi, la production de charbon de bois, les systèmes de distribution de bois à brûler et la production de biogas à l'échelle industrielle ont débuté. Les plus importantes industries ont commencé de se reconvertir à l'utilisation du bois plutôt qu'à celle du pétrole, et la conservation de l'énergie a été mise en pratique dans l'industrie. En 1981 l'initiative du gouvernement a conduit à la reprise de l'utilisation du chauffage solaire de l'eau et un projet détaillé de production decombustible à base d'alcool était presque terminé. Dans les dix-huit mois suivant son début, ce projet MAB a donné naissance à un programme et à une politique nationale de l'énergie suivis de leur mise en oeuvre. Le projet de Lae est un exemple parlant de l'application des principes écologiques et de l'analyse pour le développement économique et social d'une ville du Tiers Monde.*

## INTRODUCTION

In historical terms the need for a science of city management is relatively recent, but the challenge of urbanization which demands its evolution and refinement is exploding as the world population moves inexorably towards a majority of city dwellers early next century. Conventional planning guidelines derived for the orderly expansion of European and North American cities did not explicitly recognize physical and biological limitations now being experienced, and are unable to cope with the present rate of change even in the more favourable circumstances of the industrialized economies. What then of the rapidly urbanizing Third World where the magnitude of the problem is much greater, but where the resources available for planning and for implementation are very much smaller?

Clearly there is a need for a new conceptualization of the problem; one which is economical in the time and effort required to determine the constraints, to identify the problems at their root cause, to define the potential for change, and to deliver to decision-makers realistic practical options consistent with the constraints. The complexity of the city, let alone the megalopolis, is daunting — though the functional requirements for human well-being vary little, despite the infinitely diverse overlying fabric of human culture which contributes to the uniqueness of every human settlement. In order to cope with this complexity it is useful to conceive the city as an ecosystem, and to assume that the interaction and interdependency between the flow of energy and materials and human health and well-being may best be understood by the rigorous use of systems analysis and application of the principles of ecology — even if, at best, these are only suggestive of behaviour and control (Boyden *et al.,* 1981; Meier, 1974; Meier *et al.,* 1978; Newcombe *et al.,* 1980; Stearns and Montag, 1974).

Adopting an ecosystem approach to the identification and solution of the socio-economic and biophysical problems of city growth and development is simply one means of achieving an intellectual appreciation of the interrelationships which dominate the behaviour of an urban complex, or an urbanizing region. There are parallels which may usefully be drawn between the behaviour of natural plant and animal communities and that of grossly modified man-made and dominated communities, which are, in the least, heuristic.

There should not be any confusion however that ecology displaces the paradigms and methodologies of other long standing approaches to urban planning and management, such as urban geography, sociology, and anthropology. Urban ecology is merely a unifying and integrating conceptual approach to the problem which facilitates the selection and more powerful application of existing techniques of analysis, and provides for the more efficient use of existing knowledge.

It was with this conceptual approach in mind that the Hong Kong Human Ecology Programme was born, and fostered in the early 1970s as a pilot project within the framework of the Man and the Biosphere (MAB) Programme (Boyden, 1984; Boyden *et al.,* 1981). The Hong Kong project permitted the evaluation of this human ecological approach to urbanization, revealing its strengths and its shortcomings, and it led to the delineation and testing of selected methodologies and analytical techniques. The object of any new attempt to understand better problems affecting human populations must be, in the ultimate, the betterment of the human condition.

This was no less the case with this ecological approach to the problems of human settlements. Unesco is concerned with contributing meaningful beneficial change as a result of its programmes of support for research and development.

Thus, in the light of the experience gained in the Hong Kong project, it was agreed that the second phase of activities within MAB Project Area 11 should be one which openly encouraged the formulation of practical policies and their implementation (Unesco, 1976).

This meant the involvement from the very beginning of local decision-makers, planners and community activists in a partnership with the research scientists engaged under MAB to expose the dynamics of key problems of rapidly growing urban settlements. This quite deliberate recommendation by a task force charged with charting the future development of MAB studies on human settlements (Unesco, 1976) was reflected in the operation of the Papua New Guinea Human Ecology Programme, known commonly as the "Lae Project".

## THE POLITICAL AND ECONOMIC DIMENSION

Research dedicated to understanding problems of great moment to human well-being is meaningless without subsequent action. Every significant problem of a human community has a socio-political and an economic dimension. Action and change require political decisions, and have social and economic implications. To deny this dimension in research into problems of Third World development is to deny reality. Research scientists often attempt to remain aloof from these dimensions, and attempt to deliver a synthesis of the problems free from political overtones for fear of controversy. Unfortunately this is often a behaviour necessary for personal security, but where this is not a necessity every attempt should be made to communicate the progress and objectives of research to decision-makers, and to engage the political process in the collective recognition of the nature of problems and their possible solution.

## THE POLITICAL ECOLOGY OF CITIES

Almost invariably the most significant problems of human well-being in cities are those experienced by the poor and the dispossessed minorities. The problem can be described in ecological terms as follows.

Most Third World cities resemble "immature" ecosystems as defined by Odum (1969). They grow rapidly, use the available energy resources inefficiently, have negligible reuse of vital materials such as nutrients and water, and are subsequently more highly dependant on, and vulnerable to, changes within their hinterland (or forage areas). Their infrastructure development is low, with relatively simple structures of short half-life, and with high gross productivity. As a parallel, agricultural systems with high gross productivity in energy terms, are immature ecosystems, deliberately kept simple to channel the available energy flows into single food products.

Thus highly productive agricultural ecosystems and rapidly growing cities of the Third World are not self-sustaining, and they are subject to large fluctuations in the supply of inputs essential to their maintenance as life support systems. A downturn in the supply of an essential commodity means shortages and deprivation for some, necessarily the low-income groups already experiencing "marginal" inputs. Any attempt to stabilize the urban ecosystem by augmenting and securing its supply of energy and essential materials, will reduce the rate of change, and provide more opportunities for individuals with limited resources to adapt and to cope. This in turn can be expected to reduce social and psychological stress.

The anticipated result of an ecological approach to human settlement planning and development is, then, an improvement in the conditions of life for the low-income groups and an increase in the overall level of human well-being. This is not, however, a necessary outcome. An ecological approach is a potent tool for determining options for change in the urban environment, but the direction of desirable change is a matter of political judgement.

The Lae Project was begun in 1977 with explicit recognition of the national development strategy of Papua New Guinea, and in particular of the goals of increased self-reliance and equitable distribution of the benefits of social and economic development between nationals.

## THE LAE PROJECT

Papua New Guinea is a newly independent nation in the South Pacific with 3 million people and a land area of 460,000 km$^2$. It is due north of Australia and has a common border with Indonesian Irian Jaya. Lae is the second biggest city in Papua New Guinea, with a population in 1980 of 62,000. It is located on the Huon Gulf and is the gateway to the world for the majority of the nation's primary produce. Up until the mid-1960s, it was relatively isolated from the large population of its highlands hinterland. By the late 1960s, the road connection was well established with the populous Simbu and Western Highlands, and large numbers of immigrants flowed into the peri-urban area giving an annual increase of more than 10 per cent in the urban population between 1968 and 1972 (estimates are as high as 17 per cent per year for the period).

Many of these people were unemployed, and many again voluntarily so. They created a demand for land for homes and gardens, far greater than could easily be accommodated. The heavily forested hills around the city, which previously supplied occasional timber and fuelwood, and which buffer the flow of 6,000 mm annual rainfall into the shallow rivers draining through the city to the sea, are now under intense pressure of exploitation.

Gardens were made and fertility exhausted at the expense of heavy forest depletion, and within a decade only isolated members of the former forest canopy remained. Run-off was much more rapid without either the shielding effect of the forest canopy or of the humus sponge of floor below. Erosion was severe, taking the fertile topsoil into the raging torrents of flash floods. Suburban allotments were washed away and damage to roads and bridges became more frequent.

Gardening land gradually became scarce for some settlements, aggravating the impact of unemployment in a cash economy. By the mid-1970s, petty crime was common, especially breaking and entering. The city's refuse disposal system was limited to landfill and open incineration, and was in need of relocation. Sewage was dumped at the ocean foreshore from an overburdened, conventional sewage reticulation and lagoon system. The industrial economy was entirely dependent on imported oil for transportation, and for heat and steam raising. Low-income groups turned to kerosene in the face of fuelwood shortages, even though sawmills caused severe local pollution by incinerating wood-wastes within the city boundaries. This setting is partly described in Figures 8—1 and 8—2.

## ENERGY AND WASTES

The subject of this contribution is primarily the energy component of the Lae project. The PNG Human Ecology Programme included work on the social impact of rural to urban migration, food habits and beliefs, and changing consumer behaviour (Christie, 1981; Dalton, 1979; Jeffries, 1978) as well as the energy and resource management aspects of the urban ecosystem (Newcombe *et al.*, 1980). All this research was planned together, and was conducted and analysed, in a mutually beneficial manner, indicating linkages between social behaviour, culturally determined responses to change and the growing demand for energy, food and materials.

The object of the energy analysis was to identify the present patterns of energy use, to describe their social and economic implications, and to identify alternative sources or management practices which might reduce cost, increase security, stimulate economic development and improve the well-being of, especially, the low-income Papua New Guineans in the urban and peri-urban environment. Urbanization was regarded as an inevitable and increasing trend, hence emphasis was placed on reducing the economic and social welfare burden of the city on the national and regional government by sensitive planning for increased self-sufficiency and self-fulfillment in the urban environment.

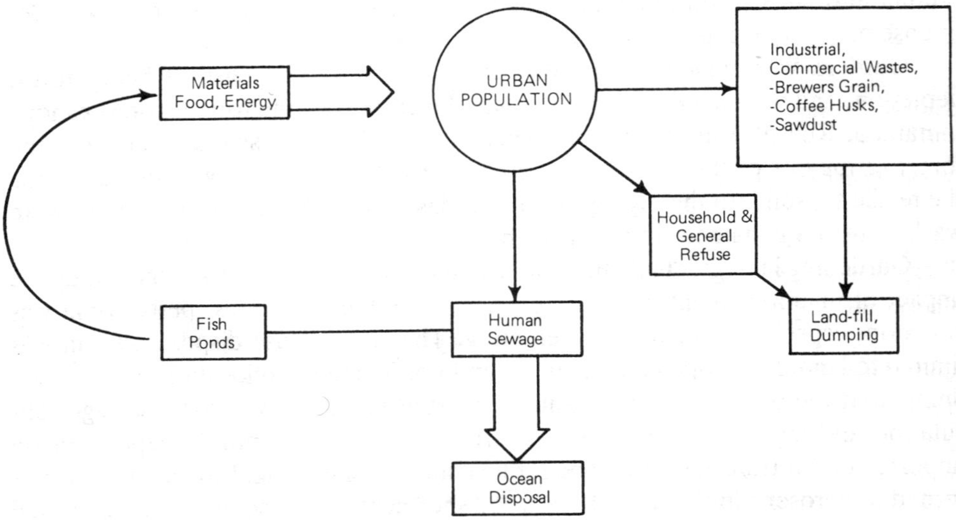


**Figure 8—1.** Lae 1977: minimal nutrient cycling

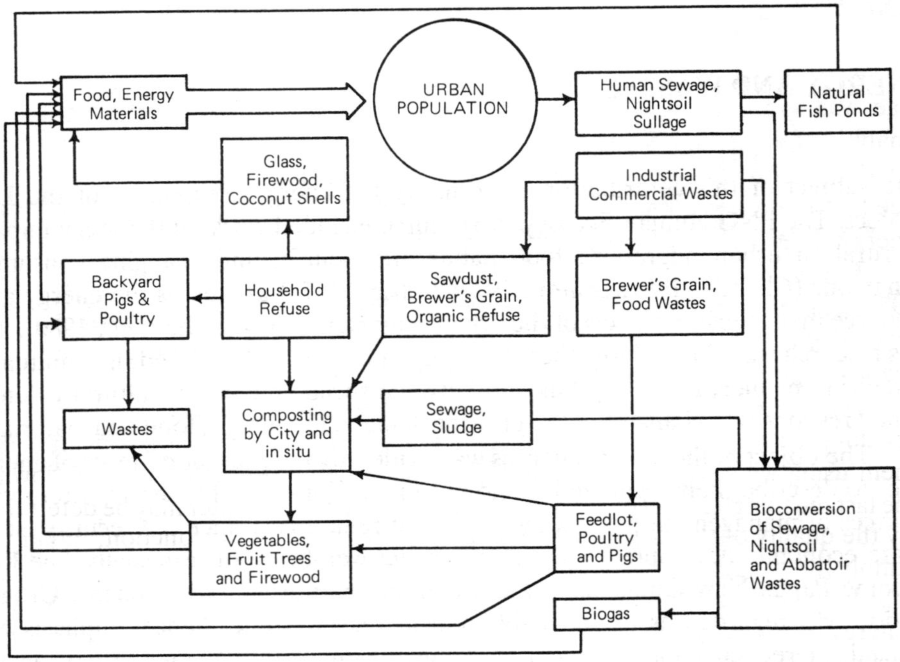


**Figure 8—2.** Lae, 1979: a developing urban-agroecosystem

**Table 8—1.** Available wastes with energy potential generated by industry in Lae urban area, 1976/77

| Waste | Origin and amounts (tonnes) | |
|---|---|---|
| Sawdust, shavings and wood offcuts less than 1 in (2.5 cm) diameter | 17,900<br>1,200 | Sawmill<br>Joineries |
| Wood offcuts larger than 1 in (2.5 cm) diameter | 33,400<br>1,800 | Sawmill<br>Joineries<br>Industrial sites |
| Paper trimmings | 70 | 2 Printers |
| Cardboard | 1,000 | Package makers<br>Bottle recycler |
| Coffee husks | 175 | 2 Coffee mills |
| Tyres | 430 | 2 Tyre companies<br>Service stations |
| Waste oil | 720 | Service stations |

## ENERGY FLOW ANALYSIS — A MICRO-SECTORAL APPROACH

Lae was subjected to a detailed micro-sectoral analysis of energy use, and an analysis of all available alternative sources of energy. Future needs were extrapolated, and potential new and renewable sources evaluated, and matched with these needs, to produce a progressive displacement of imported oil by locally produced energy forms within the ensuing 20 years.

Energy flow analysis derives from, and is a fundamental tool of, ecology (Cook, 1977). Charting the flow of energy through an ecosystem step by step, conversion by conversion, reveals the relations between all the components and activities, and readily exposes the local economy, its frugality and its prodigality.

As the term suggests a micro-sectoral approach to energy analysis means dividing the local economy into significant sectors of activity, and probing within them using whatever methodologies prove satisfactory to define end-use down to the last meaningful level of economic or social activity. The latter may be determined by the quality of the energy in use, and the importance of the function, rather than simply the magnitude of energy use.

In the industrial sector every single industrial activity was logged, and inspected. Energy use, including end-use efficiencies, and thermodynamic quality, were recorded and the production of waste heat, or waste residues of manufacturing with an energy potential, was identified. Simultaneously, known and potentially viable, locally available, alternative energy sources, and technologies of conversion and end-use were compiled, and matched with the needs serviced by imported distillates and fuel oil. The same exercise was completed for the commercial sector. The simple pie charts of end-use and fuel-use in each sector are displayed in Figures 8—3 to 8—6. Table 8—1 indicates the alternative sources of energy available to industry and commerce in the city as at 1979.

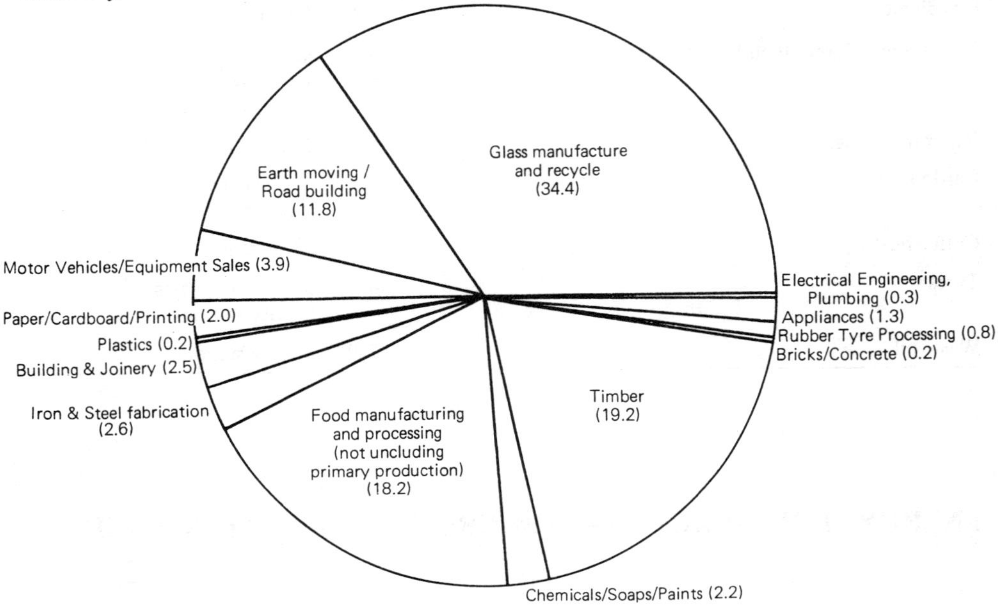


**Figure 8—3.** Industrial sector: energy use (per cent) by activity

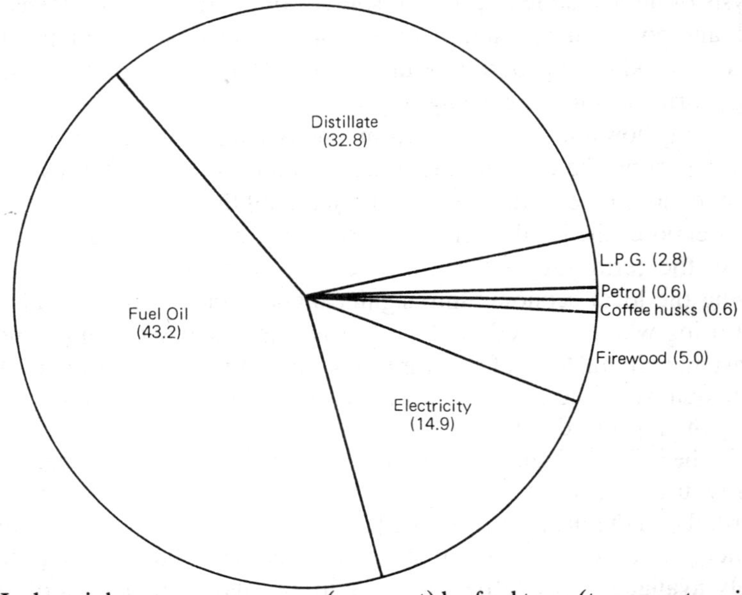


**Figure 8—4.** Industrial sector: energy use (per cent) by fuel type (transport omitted)

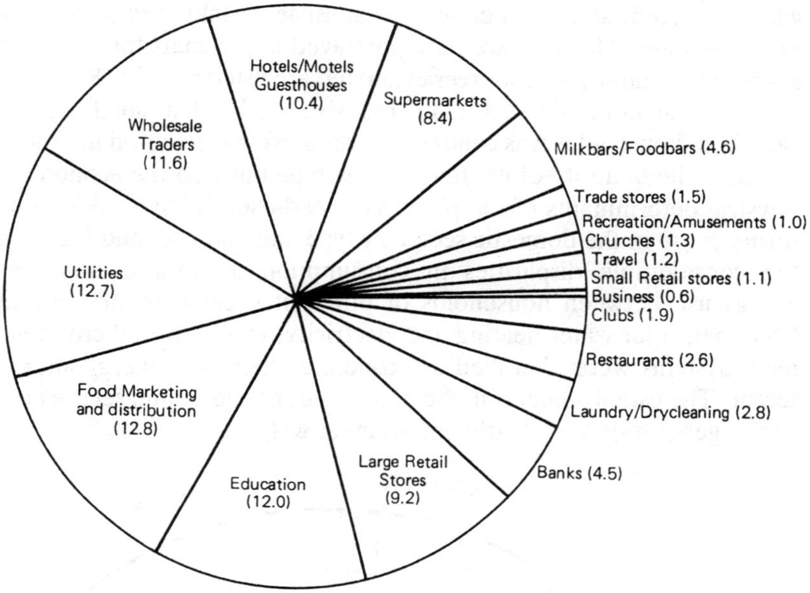


**Figure 8—5.** Commercial sector: energy use (per cent) by activity

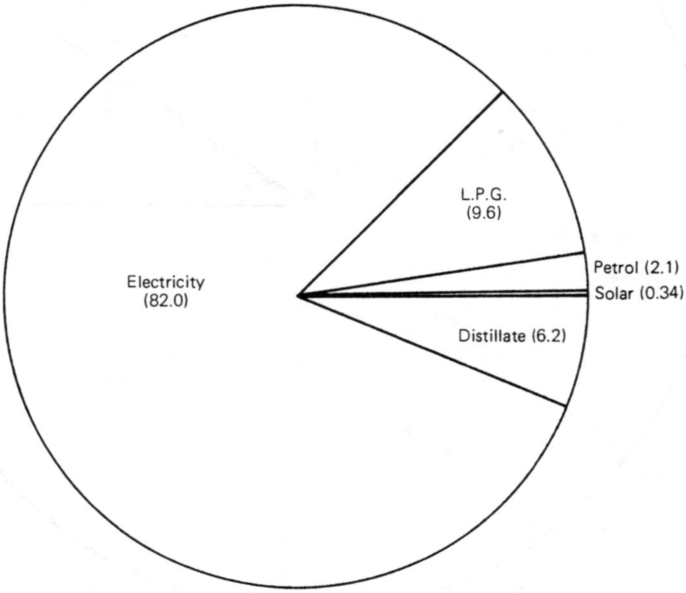


**Figure 8—6.** Commercial sector: energy use (per cent) by fuel type (transport component omitted)

In the transport sector a month long survey of road and foot traffic patterns was completed, and matched with total useage established from consumer surveys and sales data. The data are here portrayed in pie chart form in Figures 8—7 and 8—8 and the alternative sources evaluated are listed in Table 8—2.

For the domestic sector an intensive study of a small, but representative, sample of households was conducted over a two week period in order to understand not only the quantity of energy used by type but also the economic, cultural, and physical determinants of the patterns of end-use. Table 8—3 identifies the energy forms in use in the domestic sector by type and end-use, and Figure 8—9 indicates the considerable disparities in consumption in both quality and quantity of energy-use between households of different socio-economic and ethnic groups. Charcoal, solar water heating and electricity, and fuelwood cropping and management systems were identified as economic renewable energy alternatives for this sector. The establishment, in the hinterland, of a local and secure energy source for power generation was clearly important as well.

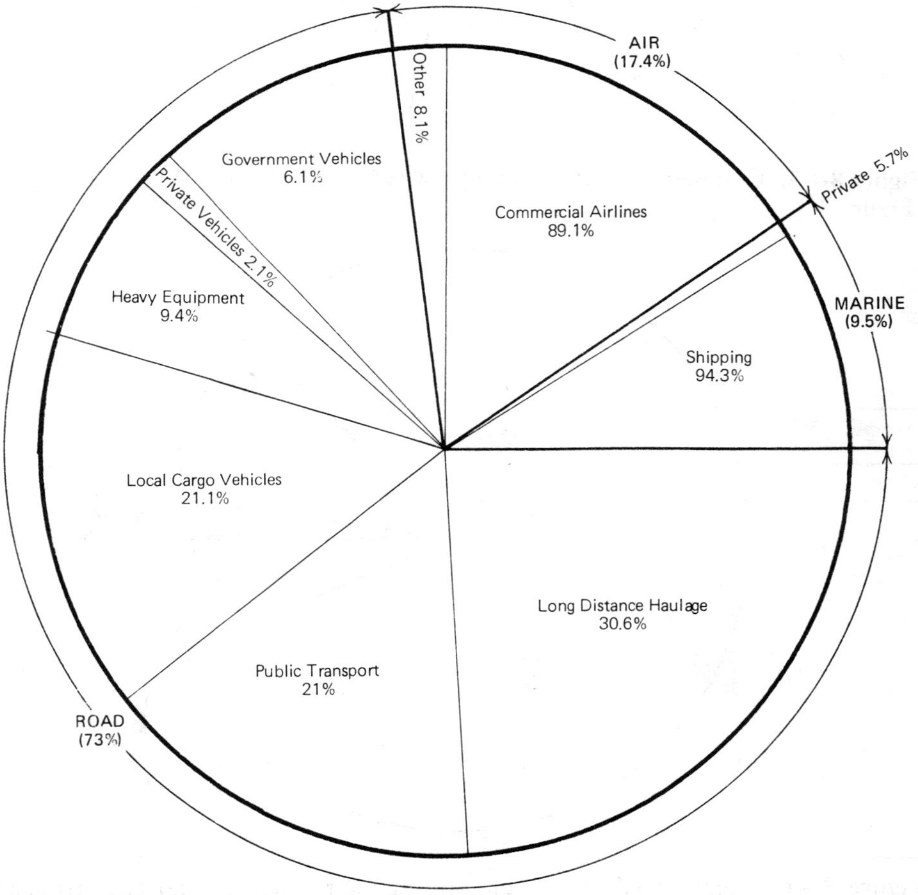


**Figure 8—7.** Energy use in the transport sector

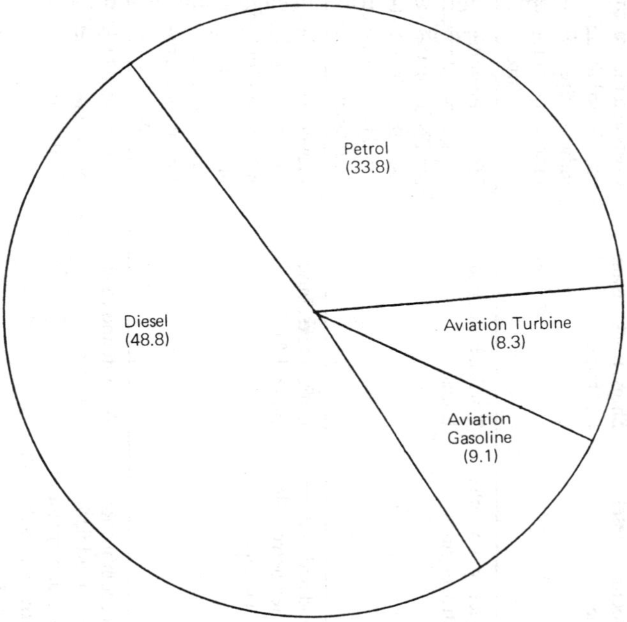


**Figure 8—8.** Transport sector: energy use by fuel type

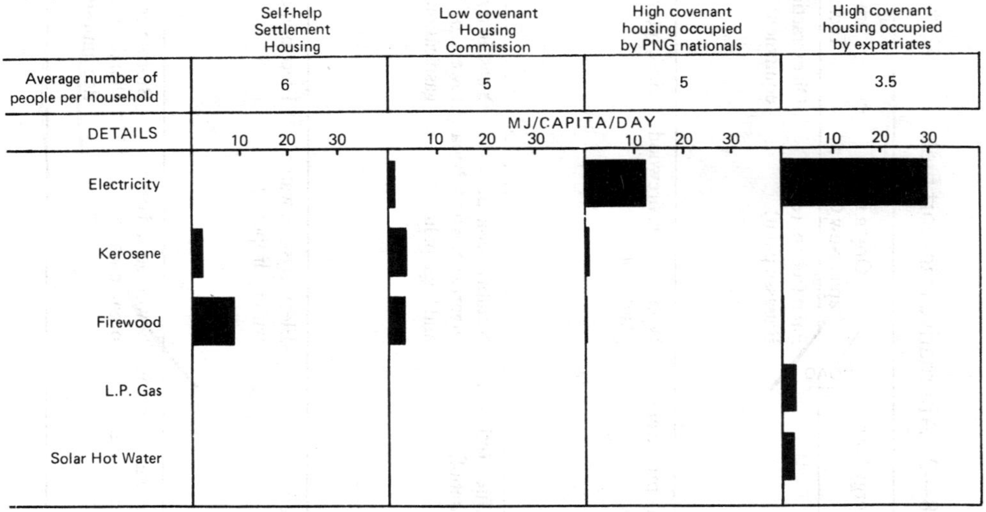


**Figure 8—9.** Energy use in the sample households

**Table 8—2.** Alternative transport fuels

| Energy Form | Origin in Papua New Guinea | Conversion Technology | Technology of end-use | Suitable transport end-use | Comments on potential |
|---|---|---|---|---|---|
| Palm oil | Palm tree sp. (e.g. Babassu palm) | Oil extraction equipment | Unmodified diesel engine | Ships, trucks, trains and buses | Costs as a fuel yet to be established though industry already exists producing oil for food at 29 toea litre. |
| Charcoal–producer gas | Wood wastes, firewood plantations | Kilns, retorts | Producer gas coupled to petrol engines | Trucks, buses and trains | Proven technology, costs to be established for PNG for this end-use, though full costs are 11t/litre of petrol equivalent. |
| Alcohol fuels (ethanol and methanol) | Notably cassava, sugarcane, wood, Nipa and Sago palm | Yeast in fermentation towers, or wood gasification to methanol | Unmodified and modified petrol, diesel and jet engines. | Cars, trucks, buses (aircraft?) | Proven technology, costs for mass production 16t/litre for alcohol from molasses CIF LAE, 25-30t/litre for cassava derived alcohol. |
| Hydrogen | Electrolysis of water with hydropower | Electrolytic cells | Direct combustion internal modified combustion engine or electric motors through fuel cells | Aircraft and road vehicles | Commercial technology not available, perhaps 5-20 years. |
| Methane Gas (Biogas is 60-70% methane) | Sewerage and other organic waste | Bioconversion chambers for anaerobic decomposition | Standard internal combustion engines with modified fuel feed systems | All road vehicles | Proven technology, costs competitive in PNG. Soon to be demonstrated. |

**Table 8—3.** Energy use in the domestic sector, Lae, 1976-77

| | Electricity | | Kerosene | | L.P.G.[a] | | Solar | Firewood | | Total |
|---|---|---|---|---|---|---|---|---|---|---|
| Details | kwh×$10^6$ | MJ×$10^5$ | litres×$10^6$ | MJ×$10^5$ | kg×$10^3$ | MJ×$10^5$ | MJ×$10^5$ | kg×$10^3$ | MJ×$10^5$ | MJ×$10^5$ |
| Total input in each form | 12.36 | 445.24 | 1.25 | 459.76 | 273.87 | 134.48 | 26.72 | 7,700 | 996.0 | 2062.2 |
| Percentage | | 22 | | 22 | | 7 | 1 | | 48 | 100 |
| Average per capita per day for population of 45,100 | 0.75 | 2.70 | 0.07 | 2.79 | 0.02 | 0.82 | 0.16 | 0.68 | 6.06 | 12.55 |
| Population served by each energy form: | | | | | | | | | | |
| 1. Households | 3,573 (act) | | 7,000 (est) | | 2,320 (est) | | 189 (act) | 5,150 (est) | | |
| 2. People (est) | 18,000 | | 36,000 | | 9,300 | | | 31,000 | | |
| Energy use for population served: | | | | | | | | | | |
| 1. Per capita per day | 1.9 | 6.8 | 0.1 | 3.5 | 0.05 | 2.7 | 6.5 | 0.7 | 8.8 | |
| 2. Per household per day | 9.4 | 34.1 | 0.5 | 18.0 | 0.3 | 15.9 | 38.7 | 4.1 | 53.0 | |

[a]Liquefied Petroleum Gas.

Finally, these data were collated and the findings integrated sector by sector with the local energy resource inventories, and with the results of related work in this MAB project, to portray the existing situation and a potentially viable energy future. Figure 8—10 is the energy flow chart for Lae at the time of the survey 1977/78, and Table 8—4 indicates the extent to which it was then believed possible (1978/79) to displace the existing imported petroleum usage with economically viable, locally available energy sources from within the city and its nearby hinterland.

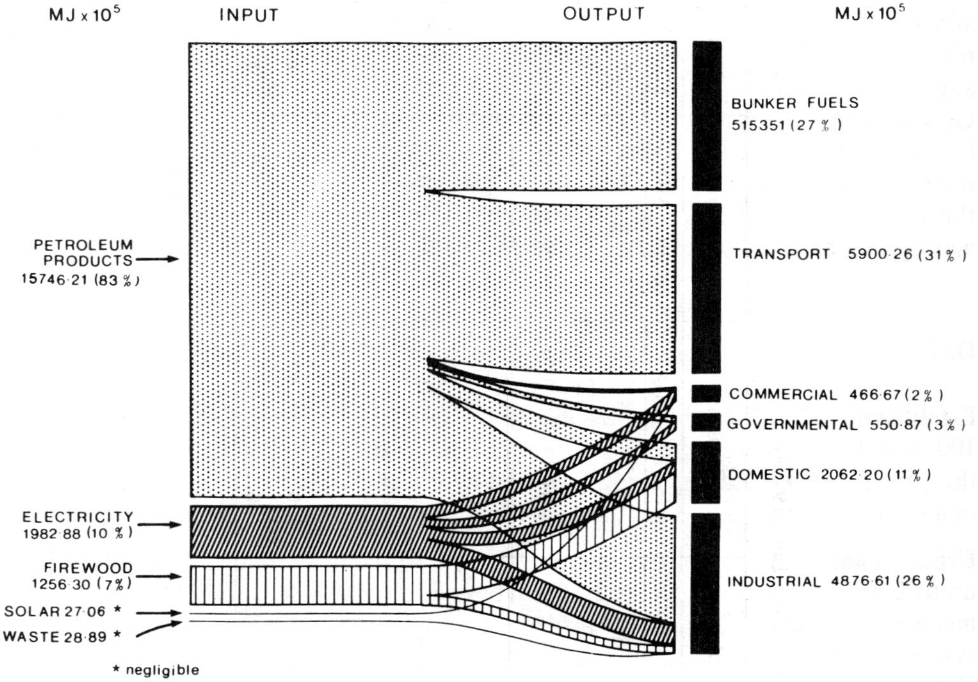


**Figure 8—10.** Energy flow chart for the city of Lae, 1977

## ENERGY POLICY, PLANS AND ACTION — 1979 TO 1981

It should be appreciated that the selection of Lae and its hinterland as the city for research and development under MAB (Project Area 11) was made at the direction of the National Government Planning Departments, and following consultation with the Local City Government Authority. Lae was thought to be fairly typical of the experience of urbanization and change in Papua New Guinea, hence any policies and programmes arising from the "Lae Project" should be applicable, within reason, to the nation as a whole. Fortuitously the staff concerned with the energy analysis of Lae, and of the populous Province of Simbu in its hinterland, were quite soon engaged in the national government service (Department of Minerals and Energy) in formulating the national energy policy and programmes of implementation, and so the findings of the Lae Project contributed heavily to the policy framework and provided the early initiatives for government funding.

The transition from theory and analysis to policy and action was ideal and allowed Unesco to fulfil its objective of meaningful contributions to change as a result of the second phase of MAB Project Area 11. The National Energy Policy and Strategy for implementation was adopted by the Cabinet of the Somare Government in February 1977 (Ministry of Minerals and Energy, 1979). This policy has been reviewed as successful and worthy of ongoing support by the former prime minister (Somare, 1981). For Lae, the MAB pilot project, and subsequent support through the national energy policy and related funding has had an impact in several action programmes, as described in the following paragraphs.

## Development of an urban-agro-ecosystem

**Establishment of allotment gardening**: Approximately 400 allotments ranging from 100 to 400 $m^2$ have been allocated from unused land to low-income families within the city. These were drained and fertilized by the city council. These were the start of a major programme to take the pressure off the surrounding forested hills.

**Urban composting**: Initially manual methods have been introduced for producing up to 2,000 t $yr^{-1}$ of compost, but ultimately the aim is to produce 11,000 t $yr^{-1}$ by mechanized means. The MAB project team and the city council designed this system and it was funded by national and local governments. The prime outlet is allotment and backyard gardens, plus council gardens and reserves.

**Fruit-tree production**: The city council funded and established an herbarium of all useful tropical fruit and nut trees, multiplied their stocks, and now plant regularly in and around the city in parks and along streets. The aim is to augment food supplies to low-income groups as well as to serve as landscaping and shade trees.

**"Atzera" project**: This initiative is named after the Atzera hill system which is the subject of ecological rehabilitation, conservation, fuelwood cropping, and subsistence level agroforestry. This was designed by the MAB team (Harris, 1978; Newcombe and Pohai, 1981) and implemented by the local government with guidance from a project committee of local and national government interests. The nationals administering the project and the city council staff successfully implemented the five-year project, supported with US$350,000 funding from the national energy programme (Dube, 1979; Seki, 1979). About 70 ha of fuelwood crops, 100

**Table 8—4.** Proportion of total energy demand which can be derived from renewable energy sources, Lae, 1977-2000

| Sectors | 1977 | | | | 1985 | | | | 2000 | | | |
|---|---|---|---|---|---|---|---|---|---|---|---|---|
| | Total Energy-use ($MJ \times 10^8$) | Sectoral end-use % | Proportion from Renewable Sources % | Renewable Energy Sources ($MJ \times 10^8$) | Total Energy-use ($MJ \times 10^8$) | Sectoral end-use % | Proportion from Renewable Sources % | Renewable Energy Sources ($MJ \times 10^8$) | Total Energy-use ($MJ \times 10^8$) | Sectoral end-use % | Proportion from Renewable Sources % | Renewable Energy Sources ($MJ \times 10^8$) |
| Domestic[a] | 2.06 | 11 | 71 | 1.46 | 3.56 | 11 | 77 | 2.74 | 9.96 | 10 | 80–100 | 7.97–9.96 |
| Industrial-Commercial[b] | 5.90 | 31 | 31 | 1.83 | 8.70 | 27 | 85 | 7.40 | 18.09 | 19 | 100 | 18.09 |
| Transport[c] | 11.05 | 58 | 0 | 0 | 20.84 | 62 | 17 | 3.46 | 67.55 | 71 | 27–83 | 18.24–56.07 |
| TOTALS | 19.01 | | 17 | 3.29 | 32.60 | | 42 | 13.60 | 95.60 | | 46–88 | 44.30–84.12 |

[a] Future energy-use is determined by calculating the product of 2% p.a. growth in per capita energy use and 5% p.a. growth in population.

[b] Includes government sector. Future energy use is determined by calculating a 5% growth in total energy use.

[c] Future energy use is the produce of 3% p.a. per capita growth in energy use and 5% p.a. growth in population. Aviation fuels are included but no alternative is expected to be found for them by 2000.

Source; Table 12, Newcombe *et al.*, 1980.

ha of conservation enrichment planting, and 5 ha of developmental *Leucaena*-short-rotation garden systems have been established. The latter is particularly interesting and successful, and has won overseas funding for parallel research into the nitrogen fixation, mineral nutrient cycling and biomass production (food/fuel) aspects under controlled conditions. The aim of this project is to stabilize the peri-urban hill system, reducing soil erosion, deforestation, and restoring the area as a permanent gardening and fuelwood production zone with buffer zones of native forest for maintaining wildlife, floristic diversity and water catchment control.

**Sewage treatment**: Industrial scale anaerobic digestion is being developed for recovery of biogas for electricity production and limited transport usage, sludge for composting, and liquid effluent for irrigation if desired. A 270 $m^3$ system is established, and, after teething problems, operating successfully. The economic component of power production and sale to the PNG Electricity Commission is currently under development with further aid funding. The object of the plant is to detoxify nightsoil and sewage entering the local river system, and to recover energy and nutrients in the process. This complex of components making up the Lae urban-agro-ecosystem is portrayed in Figure 8—2.

## Fuelwood distribution programmes

In the city some 50,000 tonnes of wood waste are generated, from sawmilling, veneer and timber dressing, only to be burnt. In 1977, it became obvious that the deforestation at one end of the city, in part through fuelwood scavenging, was more than matched in quantity by acceptable firewood burnt-off at the other end as a means of disposal. The city council agreed to establish five fuelwood distribution centres in each of the major low-income settlement areas to distribute the offcuts otherwise burnt at the sawmill. These were complete and in good operation by early 1980. Figure 8—11 portrays the situation with fuelwood distribution in 1977 and 1979. Note that pyrolysis has only been built on a pilot scale (3 oven dry tonne per day).

## Charcoal production

The production of charcoal was initiated in Lae on a trial basis early in 1980, and some 18 months later charcoal was in regular use. Initially production began under Department of Commerce supervision at the expense of the Department of Minerals and Energy, and later was expanded by the government's new Energy Development Company, and finally community groups took control of the industry under political direction from the Morobe Provincial Government in 1981. About a tonne per week is now used there in retorts and stoves made and designed in Papua New Guinea. Before 1980, Lae had never used charcoal as a domestic fuel. Interest and patronage is growing rapidly nationwide. Charcoal is mainly directed at kerosene replacement.

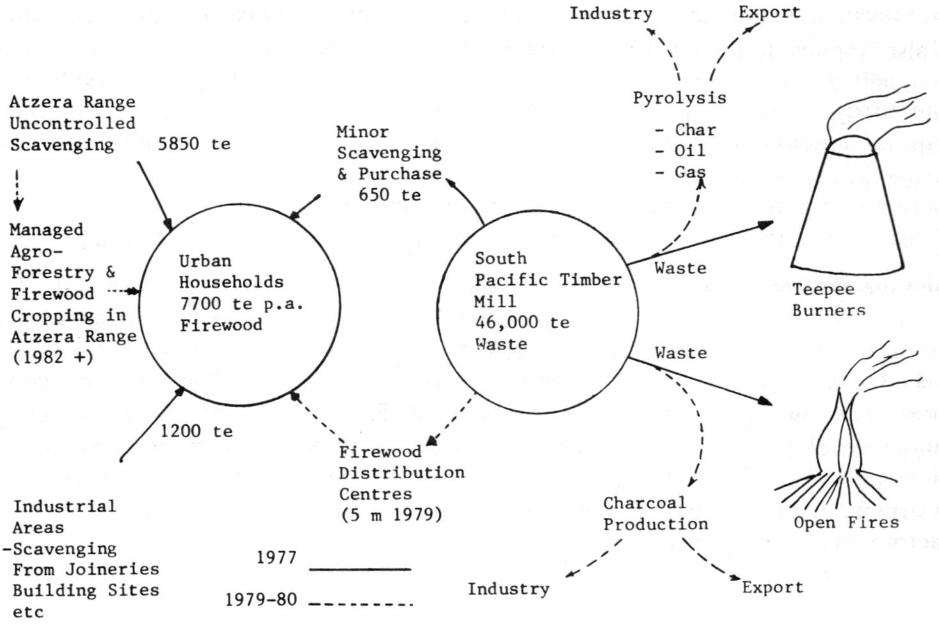


**Figure 8—11.** Firewood distribution, Lae

At the Lae Foundry in August 1981, production began of locally designed industrial charcoal water boilers and grills for commercial kitchens in boarding schools, colleges etc., and this equipment will replace more expensive electricity generated from imported diesel.

## Wood for industry

The use of wood as a replacement for diesel and fuel oil in heating is a major initiative in the national energy programme. Wood and wood-derived fuels such as charcoal, producer gas, and pyrolytic oils have been examined in great detail and test production and end-use facilities built and demonstrated in cooperation with the local University of Technology in Lae. In 1981, three major industries in Lae made commitments to shift to wood-fired boilers, representing approximately 19 per cent of the industrial use of fuel oil and distillate in the city.

## Solar energy

The use of solar energy for water heating was encouraged in the early 1960s, but not since. Solar hot-water has been heavily promoted in the national energy programme. In 1981, the cabinet banned electric hot water systems in industry and commerce, and limited electric back-up to solar hot-water to 500 watts peak per 300 litres system, simulating natural insulation. A 100 per cent tax write-off in one year is also applicable for a solar hot-water system and both electricity tariff and solar loan incentives are being considered. About 18 per cent of industrial heating by oil, and 10 per cent of electricity consumption can be replaced by solar systems. The national government has decreed in 1981 that all of its housing will have solar hot water systems inside five years.

## Alcohol fuels

These fuels represent a major transport option for Lae. The national government has advised private industry close to Lae on cassava feedstock trials for alcohol fuel production, and has funded a final design feasibility study for replacing the equivalent of 35 per cent of the city's road transport fuel (10 million $1\ yr^{-1}$) with ethanol, blended and straight. This industry is expected to be economically viable, and to proceed to construction in 1982/83. Already a 2 million $1\ yr^{-1}$ cassava ethanol factory and farm complex is under construction in the Western Highlands of Papua New Guinea as part of the national alcohol fuels programme.

## Energy from wastes

The use of wastes as an energy source offers a small but useful contribution to the balance of the local economy. A list of waste materials identified during the energy analysis of the city is provided in Table 8—5. Of these, more detail was later provided by the local University of Technology (Blowers, 1980) on the availability of waste oil and subsequently this resource has been more fully utilized: similarly coffee husks and paper wastes.

## Energy management

Programmes of energy management have been devised and publicized by the national government and examples of significant savings in the manufacturing industry in Lae can be cited as a result (e.g. 50 per cent reduction in diesel fuel used in the local match factory).

Finally a great deal of effort has been placed on securing the supply of hydropower to the city of Lae. During 1980 and 1981, resource inventories for hydropower, wood and locally produced natural gas promise to switch electricity production away again from diesel to locally available sources, further stabilizing both the price and supply of the electric power for the city. In the period 1979 through 1983, roughly US$ 1 million have been either spent or committed by national and local government to the above work, though about one third of the expenditure is not specifically to the benefit of Lae.

**Table 8—5.** Energy potential of locally available renewable energy sources other than for industrial and commercial end-uses

($MJ \times 10^5$)

| Energy source | Hydro-electricity[a] Near term—to end of 1980 | Medium term—1985 | Long term—by 2000 |
|---|---|---|---|
| Agricultural residues (coffee husks) | 29 | 43 | 89[b] |
| Cardboard | 176 | 260 | 541[b] |
| Paper | 12 | 18 | 37[b] |
| Waste oil | | 200 | 416[b] |
| Tyres | | | 104[b] |
| Wood waste— | | | |
| co-generation | 111 | | |
| pyrolysis | 1013 | 10,093[c] | 18,300[d] |
| Solar— | | | |
| 50°C | 15 (75 substitution for renewable hydro-electricity) | 90 | 180 (est) |
| 50°C—150°C | | 280 | 560 (est) |
| Air conditioning and refrigeration | | (180 substitution for renewable hydro-electricity) | |
| Electricity, from photovoltaic cells | | | (very large) |
| Totals | 1356 | 10,984 | 20,227 |

Combined Industrial/Commercial Energy Demand, 1976-77, excluding transport = 5543.37.
Combined Industrial-Commercial Energy Demand, 1976-77, non-renewable only = 4103.11
Energy Sources Available for Industrial-Commercial End-uses = 1090.

[a]Hydro-electricity is assumed to be developed at a sufficient rate in the hinterland to meet a 5% per annum growth rate in industrial demand.

[b]Wastes other than wood residues are assumed to escalate in volume at 5% per annum in line with anticipated industrial production.

[c]Includes utilization of $400 \times 10^5$MJ off-gas for in-plant electricity generation at SPT, Lae.

[d]A minimum figure for sustainable yield of pyrolytic fuels. This way not all will be used for industry for by then ethanol production from wood-waste will be economic at the 200 ODT/day production level.

## DISCUSSION AND CONCLUSION

From the original plans drawn up in 1979 for the establishment of a renewable energy base for Lae in the two final decades of the century and an urban-agro-ecosystem directed at improved subsistence for low-income urban dwellers, there is need for only minor diversion and re-adjustment. For example, there is no longer an expectation that pyrolytic oils will contribute to industrial energy use in Lae, although the pilot plant established to evaluate the technology and energy form is being deployed in the copra and cocoa industry with some considerable hope of success.

Similarly, biogas will not be used for transport but for power production: the economics of its production are more marginal than previously believed. Elsewhere the constraints have been political and economic, relating to the turbulence of changing governments and lack of capital for demonstration and development. However, despite some delays, the progress in two years of implementation has been considerable, with the introduction of two entirely new energy forms to commercial production, charcoal and biogas, and the enhanced use or production of two more, fuelwood and solar radiation. Wastes are being recycled at much higher levels than previously, including lubricating oils, coffee husks, paper, card-board, and industrial wood wastes being burnt as fuels for heat and steam-raising.

The agroforestry component of the ecological management of the peri-urban hill system (the Atzera project) is adaptable to parts of the hinterland of Lae. There are now many areas where the Department of Minerals and Energy, in cooperation with other departments and authorities, is planning to apply the same principles for subsistence level agriculture and forestry (Mt. Gilawe region, Southern Highlands, Arona Valley, Eastern Highlands, Port Moresby Region, Central Province and National Capital District). Furthermore, the concept of building an urban agro-ecosystem to improve urban management and urban subsistence for the low-income majority has become established with the funding of a parallel programme to that at Lae for the national capital of Port Moresby by the Port Moresby City Council, coordinated with the Energy Planning Unit of the national Department of Minerals and Energy.

The re-organization of any well-established *modus operandus* will bring conflict. Social and administrative systems are inherently conservative and resistant to change. It has been surprising, then, to find that the adoption of an ecological approach to urban planning and development has been so readily accepted. The only noticeable resistance has been from professions with a tradition of practice related to a western context and its standards or precepts, namely engineering and agriculture.

The major controversy that has arisen in Lae related to the implementation of policies arising from MAB research is with respect to the political control of a new energy production and distribution system, rather than opposition to its development (to wit the issue of community and smallholder control of the charcoal industry). It is worthy of note that the ecological management programme for the city of Lae is now coordinated and implemented by competent national scientists (Beka Seki and Funge Samana) with the backing of local and regional government. This is not only an essential transition in line with national goals but is consistent with a basic tenet of the Unesco MAB Programme, that local decision-makers and planners are engaged from the beginning with analysis and change.

In conclusion, ecology constitutes an approach to the planning and management of urbanization that facilitates an awareness of physical and social constraints, and their interaction, as they ultimately impinge upon human well-being. The human ecology of cities is not a discipline in its own right but a conceptual approach which enables a thorough evaluation and definition of economic and political choices, by illuminating the biological and social consequences of present and potential ecosystem behaviour. Economic wealth and political power, often synonomous, will always be the main determinants of change, barring acts of God, in human ecosystems. However the wise use of power, accepting benevolence as the norm, is predicated on a knowledge of the determinants of the existing situation and the consequences of change in any given ecosystem. It is through the evolution of this understanding that the MAB Programme has made, and continues to make, a meaningful contribution. The Lae project is a useful example of the opportunities for change, deemed desirable within the context of a national development strategy, which can arise through an ecological approach to the analysis of a city's problems and their solution.

## References


Blowers, M. (1980), *Note on the Recycling of Waste-Oil in the City of Lae, for the Alternative Technology Development Institute*, University of Technology, Lae, Papua New Guinea.

Boyden, S. (1984), Integrated Studies of Cities Considered as Ecological Systems. In F. di Castri, F. W. G. Baker and M. Hadley (Editors), *Ecology in Practice:* Vol. II, *The Social Response,* Tycooly International, Dublin, 7-29.

Boyden, S., S. Millar, K. Newcombe and B. O'Neill (1981), *The Ecology of a City and its People; the Case of Hong Kong*, Australian University Press, Canberra.

Christie, M. (1981), *Changing Consumer Behaviour in Papua New Guinea: its Social and Ecological Implications*, Report No. 3, Unesco-UNEP-CRES Papua New Guinea Human Ecology Programme, Australian National University, Canberra.

Cook, R. E. (1977), Raymond Lindeman and the Trophic Dynamic Concept in Ecology, *Science*, 189, 22-7.

Dalton, K. J. (1979), *Chimbu People under Pressure: The Social Impact of Urbanization.* Report No. 2, Unesco-UNEP-CRES Papua New Guinea Human Ecology Programme, Australian National University, Canberra.

Dube, V. N. (1979), *Firewood Cropping, Food Gardening and Conservation Planting. A Three Dimensional Strategy for Displaced Rural Communities: The Case of the Atzera Hills*, Mimeographed, Lae City Council, Lae.

Harris, J. (1978), *The Atzera Project, Ecological Management of the Atzera Range, Lae, Papua New Guinea, An Initial Proposal*, Document PNGE/T4. Unesco-UNEP-CRES

Papua New Guinea Human Ecology Programme, Australian National University, Canberra.

Jeffries, D. J. (1978), *From Kaukau to Coke: A Study of Rural and Urban Food Habits in Papua New Guinea*, Report No. 1, Unesco-UNEP-CRES, Papua New Guinea Human Ecology Programme, Australian National University, Canberra.

Meier, R. L. (1974), *Planning for an Urban World, The Design of Resource Conserving Cities*, MIT Press, Cambridge, Mass.

Meier, R..L., S. Berman and D. Dowell (1978), *Urbanization and Energy in Development Regions*, University of California, Berkeley.

Ministry of Minerals and Energy (1979), *White Paper on Energy Policy and Planning in Papua New Guinea*, Government Printer, Papua New Guinea.

Newcombe, K., K. Bowman, M. Christie and J. Pokris (1980), *Energy for Development. The Energy Policy Papers of the Lae Project*, Report No. 4, Unesco-UNEP-CRES Papua New Guinea Human Ecology Programme, Australian National University, Canberra.

Newcombe, K. and T. Pohai (1981), The Lae Project: an Ecological Approach to Third World Urbanization, *Ambio*, 10 (2-3), 73-8.

Odum, E. P. (1969), The Strategy of Ecosystem Development, *Science*, 164, 262-70.

Seki, B. F. (1979), *Food Production in the Lae City Area*, Paper presented to the workshop on Urban Food Supply, The Waigani Seminar, September 24-8, 1979, Port Moresby.

Somare, M. T. (1981), *Energy and Equity in the Third World; the Case of Papua New Guinea*, Paper presented to the ANZAAS Congress, May 1981, Brisbane, Office of the Leader of the Opposition, Port Moresby.

Stearns, F., and T. Montag (1974), *The Urban Ecosystem: A Holistic Approach*, Dowden, Hutchinson and Ross, Stroudsburg, Penn.

Unesco (1976), *Task Force on Integrated Ecological Studies on Human Settlements within the Framework of Project 11*, MAB Report Series No. 31, Unesco, Paris.

CHAPTER 9

# Urban-Hinterland Interactions: Urban Wastes and the Lake Suwa Ecosystem

*Tokio Okino*

Abstract

Lake Suwa is a shallow eutrophic lake located in a fault basin in central Japan. In recent years the great increase of population and industrial activities around it have led to much more nutrient input into the lake, resulting in considerable acceleration of eutrophication and regular appearance of heavy blooms of *Microcystis* every summer. In 1971, the construction of the sewage system was planned in order to improve the condition of the eutrophicated lake, and a part of it was opened in 1979. The results of our simulation studies show that the eutrophic state of the lake will be improved so that the level of Secchi disc will be about 70 cm and the chlorophyll amount will be reduced to half of the present level in summer by elimination of 50 per cent of nitrogen and 70 per cent of phosphorus from the present loads. The lowering of loading levels will be obtained by improved and more efficient treatment of the sewage.

*Résumé*

*Le Lac Suwa est un lac peu profond et eutrophisé situé dans un bassin tectonique dans le centre du Japon. La croissance de la population et le développement industriel autour du lac ces dernières années ont augmenté le taux d'éléments nutritifs dans le lac, provoquant ainsi une eutrophisation accrue et l'apparition régulière de populations de* Microcystis *chaque été. En 1971, la construction d'égouts a été entreprise afin d'améliorer l'état du lac eutrophisé et une partie des travaux a été terminée en 1979. Les études de simulation montrent que la condition eutrophisée du lac s'améliorera avec un niveau du disque 'Secchi' de 70 cm et un taux de chlorophylle réduit de moitié de la valeur actuelle par l'élimination respective de 50 pour cent des concentrations actuelles d'azote et de 70 pour cent de phosphates. Ce taux d'élimination sera obtenu par un traitement plus efficace et plus complet des eaux usées.*

## THE LAKE SUWA ECOSYSTEM

THE purpose of the Lake Suwa study is to clarify the mechanisms of eutrophication in lakes and to gain a better understanding of the relation between a natural ecosystem and human activities. Lake Suwa is a shallow eutrophic lake located in a fault basin in central Japan. It is at a height of 759 m above sea level and has a maximum depth of 6.5 m, a mean depth of 4.1 m, a surface area of 13.30 $km^2$, a volume of 0.06 $km^3$ and a fertile catchment area of 512 $km^2$. There is a single outflowing river, the Tenryugawa. The renewal time of the lake water is about 50 days.

Many reports have been presented on the limnological and biological characteristics of this lake, beginning with Tanaka (1914, 1918). Study of the biological productivity of this lake was carried out from 1969 to 1973 as a part of International Biological Programme (Sakamoto *et al.*, 1975). We have studied the relation between the watershed (used here in the sense of catchment area) and lake ecosystem since 1977.[1] In recent years the great increase in the population and industrial activities around this lake have brought much greater amounts of nutrients into the lake, resulting in a remarkable acceleration of eutrophication and a regular appearance of heavy blooms of *Microcystis* every summer. Transparency in summer has decreased with the years and has been fluctuating between 20 and 30 cm in the last few years (Figure 9—1). The dissolved oxygen content in summer is above 200 per cent of saturation near the surface and is reduced to almost nothing at the bottom.

The physical features of this lake have undergone artificial transformation in the last decade and the biological and chemical characteristics of the lake have changed. About $13 \times 10^5$ $m^3$ of mud and sand have been dredged from the bottom, and an embankment along the shore line was completed in 1980. As a result, the macrophytes have been reduced to 17 per cent in terms of area and to 24 per cent in terms of biomass compared with 1966, the year of their greatest development. The submerged plants have been more affected than other macrophytes (Kurasawa *et al.*, 1979).

[1]Most of the original reports of these studies appear in Japanese in Progress Reports of the Watershed Ecosystem Study of Lake Suwa, No. 1 (1978) to No. 7 (1981), published by Shinshu University.

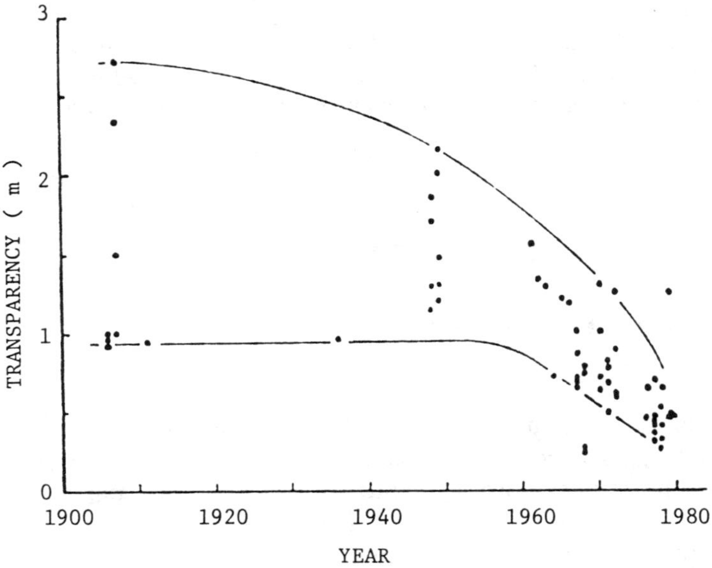


**Figure 9—1.** Changes of the transparency of Lake Suwa since 1900

## Methods

In this study, regular observations and sampling were carried out at the centre of the lake (about 6 m in depth) at 06:00 to 08:00 hours at intervals of 10 days in 1977. Water samples were taken from various depths with a non-metallic Van-Dorn sampler, being transferred to polyethylene bottles for chemical analysis and plankton counting, and to glass bottles for measuring primary production. Zooplankton samples were collected from successive vertical hauls at one-meter intervals from just above the bottom to the surface using a closing net (XX-15). The water samples for chemical analysis were filtered on a previously ignited Whatman GF/C filter exclusive of total amount of nutrients. Particulate carbon and nitrogen were determined with an automatic CHN-analyser. The calculation of the nutrient-loading of this lake based on measurements carried out in the catchment of Lake Suwa are shown in Table 9—1.

## Nutrient loading and land use

In 1977 the mean concentration of the total nitrogen and phosphorus in the water column was 1 to 3 mg $l^{-1}$ and 0.1 to 0.4 mg $l^{-1}$ respectively. The levels of nutrients have not changed much during the last 10 years, but the pattern of the seasonal changes has. Dissolved inorganic nitrogen has diminished in summer but phosphate phosphorus still remains in whole layers (Figures 9—2 and 9—3). It suggests that

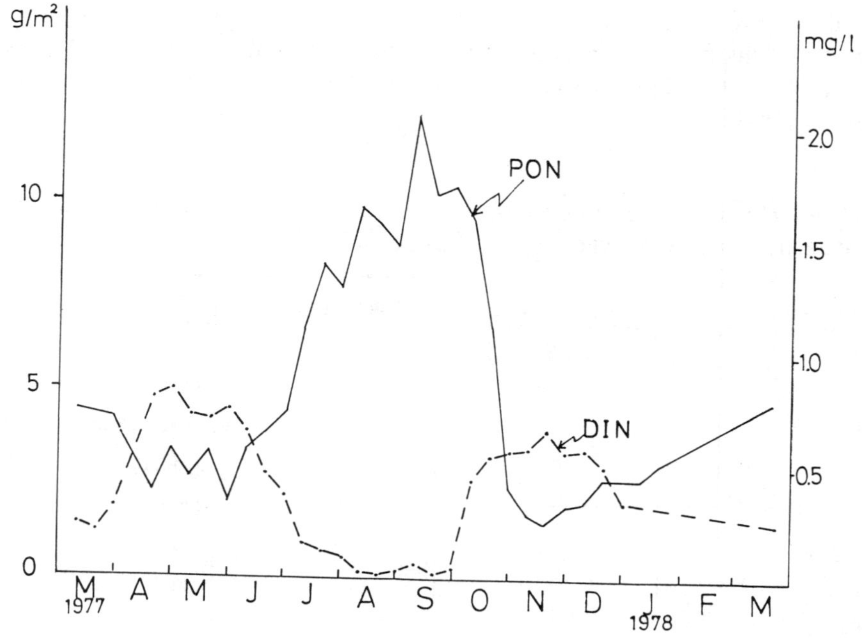


**Figure 9—2.** Seasonal changes in the amounts of PON (Particulate Organic Nitrogen) and DIN (Dissolved Inorganic Nitrogen) in Lake Suwa from 1977 to 1978

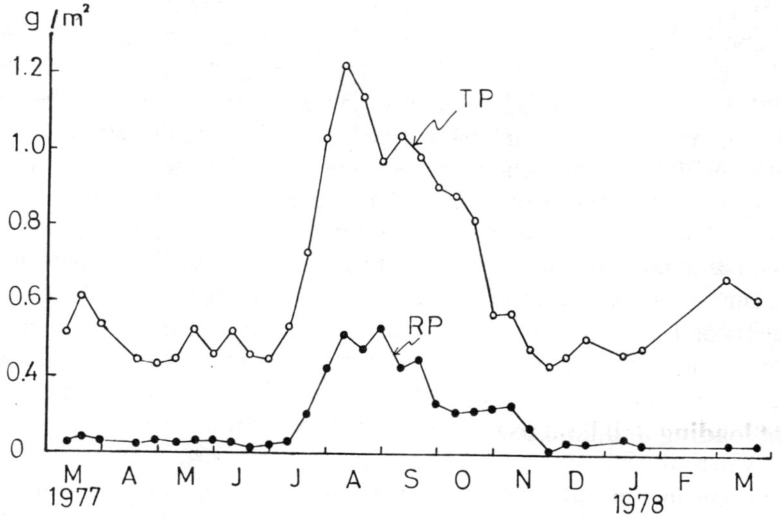


**Figure 9—3.** Seasonal changes in the amounts of TP (Total Phosphorus) and RP (Reactive Phosphorus) in Lake Suwa from 1977 to 1978

the algal growth in Lake Suwa is limited by the amounts of dissolved inorganic nitrogen. This finding was supported by the results of the modified bioassay by Nakamoto (1980) and of the chemical composition of the algal cells grown in summer (Watanabe *et al.*, 1978).

**Table 9—1.** Basic values used for the calculation of nitrogen and phosphorus loading from each source in the watershed of Lake Suwa

| Loading source | Total nitrogen | Total phosphorus | |
|---|---|---|---|
| Rainfall | 36.5 | 1.4 | g $ha^{-1}yr^{-1}$ |
| Forest | 3.6 | 0.12 | kg $ha^{-1}yr^{-1}$ |
| Field | 28.0 | 2.0 | % against fertilizer |
| Cattle | 290 | 50 | g ind. $day^{-1}$ |
| Pig | 40 | 25 | g ind. $day^{-1}$ |
| Hot spring | 2.4 | 0.3 | g $m^{-3}$ |
| Gas well | 107 | 1.7 | g $m^{-3}$ |
| Permanent resident | 3.0 | 0.71 | g man-$day^{-1}$ |
| Tourist | 3.0 | 0.43 | g man-$day^{-1}$ |
| Industry | From the water quality of waste water by each industry | | |
| Sanitary plant | 11.4 | 0.62 | g man-$day^{-1}$ |

In the summer of 1977, the mean amounts of chlorophyll-a in the water column reached about 200 mg $m^{-3}$, and the annual gross production and the community respiration were 746 g C $m^{-2}$ $yr^{-1}$ and 634 g C $m^{-2}$ $yr^{-1}$ respectively. The mean energy efficiency for gross production was 0.52 per cent. The sedimentation rate of particulate matter ranged from 0.04 to 10.01 g $m^{-2}$ $day^{-1}$ in dry weight. The higher rates were obtained in the autumn of the same year after heavy algal blooming (3 to 10 g $m^{-2}$ $day^{-1}$). Based on these data, the annual sedimentation rates were calculated as 751 g $m^{-2}$ (dry weight) and 81.5 g $m^{-2}$ (carbon base).

Land use in the Lake Suwa catchment includes forests (58 per cent), grasslands (12 per cent), paddy and vegetable fields (18 per cent), rural and urban settlements (8 per cent) and water bodies (3 per cent). The forests of the catchment area are composed of three main types, artificial larch forest, broadleaved deciduous forest and subalpine coniferous forest (Tadaki, 1980). The human population of the watershed is about 180,000 and the main industries are food processing and precision machinery near the lakeside.

The construction of the sewerage plant was planned in order to improve the condition of the eutrophicated lake in 1971. The plant is still under construction, although a part of it was opened in 1979. It is to be completed by 1985. The sources of nitrogen and phosphorus in the catchment can be grouped into ten categories — rainfall, forests, fields, livestock, hot springs, gas generated from the bottom of the lake, permanent residents, tourists, industries and sanitary plant — and the contribution from each source has been calculated on the basis of the results obtained by the project team (Okino *et al.*, 1981). The total amount of nitrogen moving from the whole watershed is 4,266 kg day$^{-1}$ and that of phosphorus is 311 kg day$^{-1}$, but the actual inlet amount into the lake is calculated as 3,226 kg day$^{-1}$ and 237 kg day$^{-1}$ owing to the loss in the river (Figure 9—4). The major sources of the phosphorus loading are the wastewater from permanent residents and the sanitary plant; that of the nitrogen loading are the rice and vegetation fields where fertilizer is used. The ratio of effluents to fertilizer inputs for this region was estimated to be 28 per cent for nitrogen and 2 per cent for phosphorus (Watanabe *et al.*, 1978).

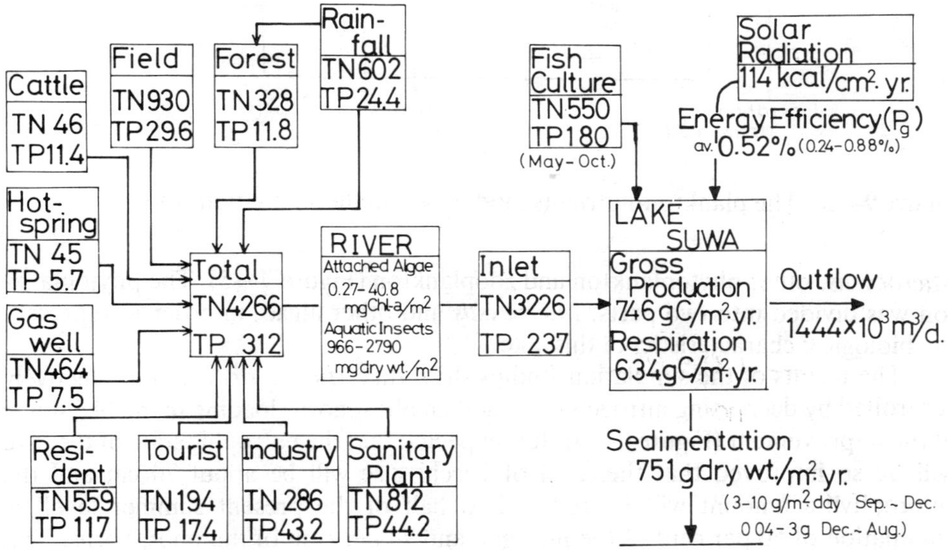


**Figure 9—4.** The amounts produced by each source in the catchment of Lake Suwa and the inlet amounts of phosphorus and nitrogen into the lake. The right part of the figure shows the matter balance of the Lake Suwa ecosystem. The fish culture box shows the loading amounts from the culture of carp in a floating net in the lake for about 150 days beginning in May

## Simulation study

When the sewerage treatment is completed, it may be possible to collect 50 per cent of the nitrogen and 75 per cent of the phosphorus from the present loads. A simulation study was carried out to estimate the effect of sewerage treatment on the lake. A box model was developed, divided into two layers with seven components such as DIN (Dissolved Inorganic Nitrogen), DON (Dissolved Organic Nitrogen), DIP (Dissolved Inorganic Phosphorus), DOP (Dissolved Organic Phosphorus),

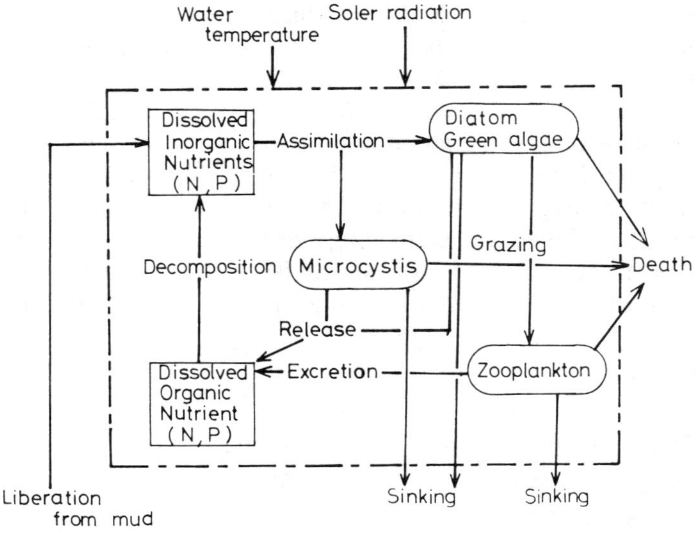


**Figure 9—5.** The plankton-nutrients model used in the simulation study

*Microcystis,* other phytoplankton and zooplankton (Figure 9—5). The phytoplankton was divided into two parts, *Microcystis* and other algae, in order to represent the biological characteristics of this lake.

The results of the simulation studies show that *Microcystis* is more effectively controlled by decreasing nitrogen loading than phosphorus loading in the beginning of the improvement (Figure 9—6). It is expected that the eutrophic state of the lake will be so improved that the level of Secchi disc will be about 70 cm and the chlorophyll-a amount will be reduced to half of the present summer level by elimination of 50 per cent of the nitrogen and 70 per cent of the phosphorus. This rate of lowering of loading levels will become equal to the elimination rate obtained by the sewerage plant, when the improved methods of nitrogen and phosphorus treatment are adopted (Figure 9—7).

## References

Kurasawa, H., T. Okino and H. Hayashi (1979), Chronological Changes of Distribution and Standing Crop of Rooted Aquatic Plants in Lake Suwa, In *Progress Report 3 of Watershed Ecosystem Study of Lake Suwa,* Shinshu University, Suwa City, 7-26.

Nakamoto, N. (1980), Evaluation of Biologically Available Nutritional Matters and the Activity of Phytoplankton in the Water of Lake Suwa by Use of MBOD Bioassay, In *Progress Report 5 of Watershed Ecosystem Study of Lake Suwa,* Shinshu University, Suwa City, 92-9.

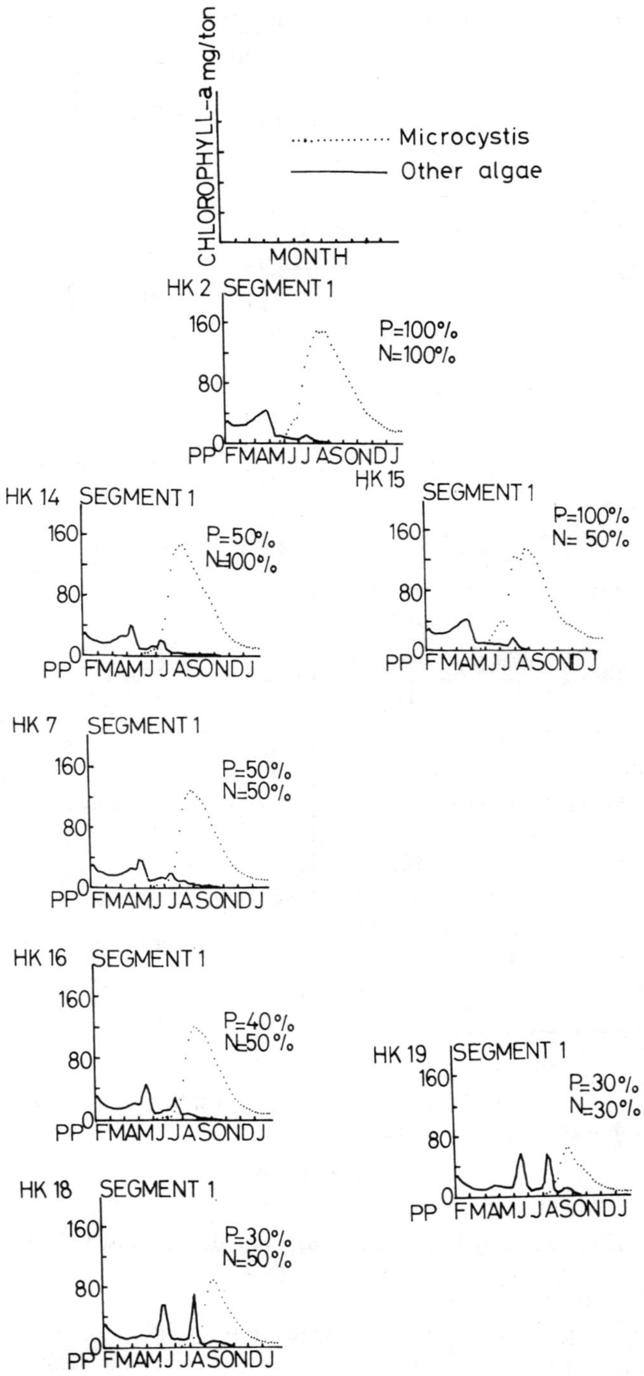


**Figure 9—6.** The results of the simulation study

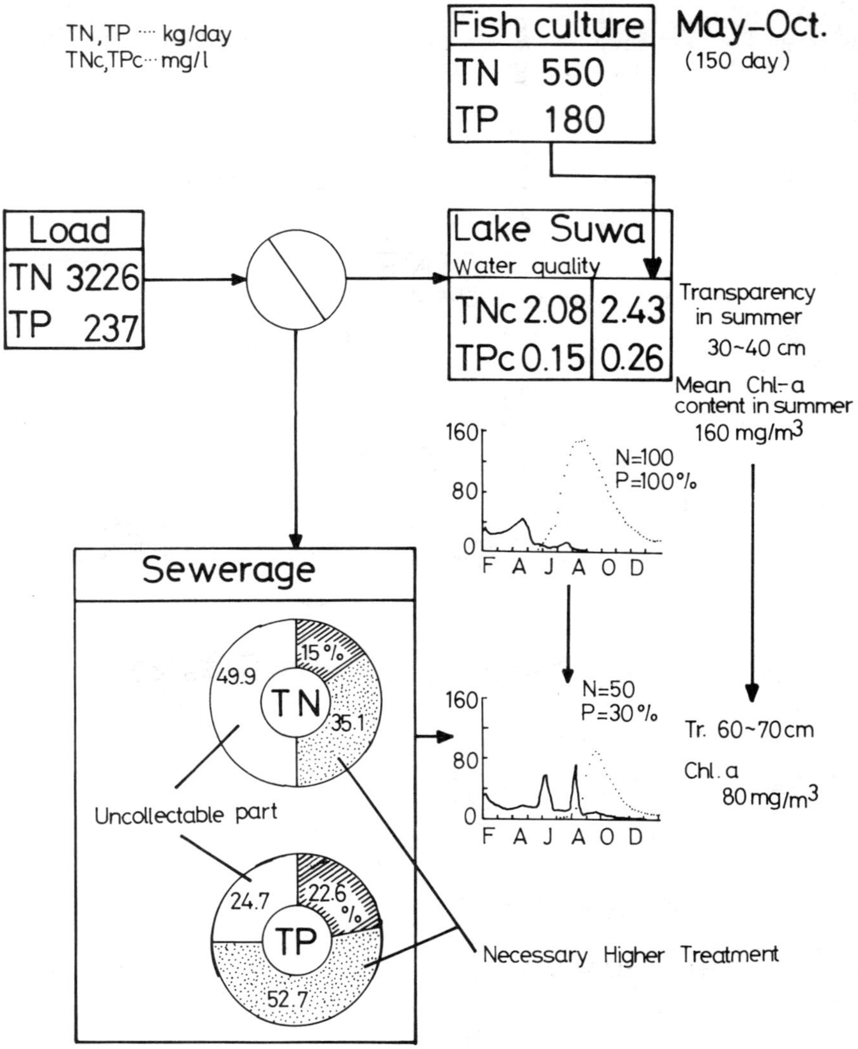


**Figure 9—7.** The results of the simulation study on the effect of sewerage treatment on the eutrophic state of Lake Suwa

Okino, T., S. Hiratsuka and T. Tanaka (1981), Simulation Model of Matter Cycle on Lake Suwa II, In *Progress Report 7 of Watershed Ecosystem Study of Lake Suwa,* Shinshu University, Suwa City, 63-85.

Sakamoto, M., H. Kurasawa and T. Okino (1975), Productivity and Nutrient Metabolism of Communities in Lake Suwa, In S. Mori and G. Yamamoto (Editors), *Productivity of Communities in Japanese Inland Waters,* JIBP Synthesis, Vol. 10, Japanese Committee for the International Biological Program, University of Tokyo Press, Tokyo, 107-47.

Tadaki, Y. (1980), Amount and Distribution of Leaves of Terrestrial Vegetation in the Watershed of Lake Suwa, In *Progress Report 5 of Watershed Ecosystem Study of Lake Suwa,* Shinshu University, Suwa City, 1-16.

Tanaka, A. (1914), *Lake Suwa,* Iwanami-shoten, Tokyo.

Tanaka, A. (1918), *Limnological Studies of Lake Suwa,* Iwanami-shoten, Tokyo.

Watanabe, Y., T. Okino and T. Tanaka (1978), The Relation Between the Eutrophication of Lake Suwa and the Fertiliser Levels in the Watershed of Lake Suwa, In *Progress Report 1 of Watershed Ecosystem Study of Lake Suwa,* Shinshu University, Suwa City, 29-46.

Watanabe, Y., M. Yamamoto, H. Hayashi, T. Okino and T. Tanaka (1980), Seasonal Changes of the Various Nitrogen Fractions in Lake Suwa, *Reports of Suwa Hydrological Station of Shinshu University,* 3, 47-52.

CHAPTER 10

# Managing Urban Space for Children: the Toronto Experience

*Howard Andrews*

Abstract

Conventional planning approaches in North America are characterized as emphasizing a pluralist resolution of conflicting needs and demands of local interest groups, through the discrete provision and regulation of urban land uses and facilities. As one such interest group, children in urban Canada are diminishing in numbers and adult priorities are elsewhere: urban planning is therefore oriented towards adults' needs and priorities, which are contrasted to those of children. Preliminary results of a large sample of young adolescents' perceptions of their residential environments are described and the outlines of an alternative planning philosophy emphasizing the ecology of human development and preventive intervention are suggested. Examples of work in progress related to this alternative framework are briefly summarized.

*Résumé*

*Les approches classiques de la planification en Amérique du Nord sont caractérisées par l'accent mis sur l'élaboration de solutions pluralistes des besoins antagonistes et des demandes des groupes d'intérêt locaux. Ces solutions proviennent de l'application de dispositions peu contraignantes et de la réglementation de l'utilisation du terrain des zones urbaines et de leurs installations. En tant que représentant un tel groupe d'intérêt, les enfants dans les zones urbaines du Canada diminuent en nombre et les demandes prioritaires des adultes se portent ailleurs: la planification urbaine s'oriente en conséquence vers les besoins des adultes et leurs priorités qui sont opposés à ceux des enfants. Les résultats préliminaires d'un echantillonnage à grande échelle de la perception par de jeunes adolescents du milieu dans lequel ils se meuvent (vivent) sont décrits et les grandes lignes d'une alternative de la philosophie de la planification, mettant l'accent sur l'écologie du développement de l'homme et l'intervention préventive, sont suggérées. Des exemples de travaux en cours entrant dans le cadre de cette alternative sont brièvement résumés.*

## INTRODUCTION

FROM a largely conventional point of view of the planning process as it is manifested in all Western industrialized societies, the notion of "managing urban space for children" turns out to be just one of a plethora of concerns we might imagine on the agenda of any social or physical planning agency embedded in the governing bureaucracies of cities throughout North America or Western Europe. Consider the following lists of words, for example, and select at random any combination of three terms, one from each of the columns, reading from left to right:

| I | II | (for) | III |
|---|---|---|---|
| Designing | housing | | elderly people |
| *Managing* | urban infrastructure | | teenage mothers |
| Providing | recreation facilities | | *children* |
| Improving | employment opportunities | | single parents |
| Organizing | *urban space* | | workers |
| Regulating | education | | the physically handicapped |
| Supervising, etc. | health care, etc. | | the unemployed, etc. |

More than 340 discrete activities or concerns are suggested by such an exercise — and these are simply a small subset of a much larger range: they share a common set of defining attributes which can be used to characterize the nature of conventional or mainstream urban planning. First, implicit in the terms listed in column I above is a view of the planning process as an ameliorative, goal-oriented, facilitative or prescriptive activity. Second (column II), there is implied a view of material resources — including space and time — as epi-phenomena, somehow abstracted from specific socio-historical contexts, whose locational and qualitative characteristics may be manipulated or altered for normative ends. Third (column III), there is an identification of classes of human populations with presumably common interests, needs or demands, on whose behalf planning is to be undertaken, and for whom, given the two preceding points, the material circumstances of life are to be enhanced.

In parallel with the characterization of contemporary urban planning in terms of content, there exists a complementary set of models or semi-formal articulations about planning as a process — that is, about how planning proceeds and how it is related as an activity to the wider domain of political processes or urban government (Scott and Roweis, 1977). A common strand running through most of the variants associated with mainstream or functional views of urban politics (see, for example, Goldsmith, 1980; Hawkins, 1971; Kaplan, 1968; Lineberry and Sharkansky, 1978) describes urban planning as some kind of reactive process, responding to the needs or demands expressed to policy-makers and decision-makers in urban governments by the interest groups, communities, higher levels of government, and so on, which comprise the external environment in which the local political system operates. To present an extremely simplified scenario of the process in action, we might imagine a citizens' group voicing opposition to proposals for an urban expressway to be located in their community because of anticipated negative environmental effects, leading eventually to a redesign of the facility or perhaps its relocation. Cities throughout North America have witnessed this kind of event in the last decade or so (e.g. Draper, 1970; Wolpert *et al.,* 1972).

Among the critical variables associated with the effective participation of citizens in this process of planning and policy-making are the level of organization, access to resources and size of the group articulating their needs or demands to planners and policy-makers (e.g. Pickvance, 1976). To consider the agenda-item "managing urban space for children" within this framework, we should, therefore, from a strategic or pragmatic point of view, examine these variables in the Canadian and Toronto contexts.

## DEMOGRAPHIC FUTURES AND PUBLIC CONCERNS IN URBAN CANADA

For several decades now the experience of childhood in Canada has been predominantly an urban experience. In opening their discussion of *Urban Canada,* Simmons and Simmons (1974) used to good effect a remark drawn from the final report of the Gordon Commission on Canada's Economic Prospects (1957):

> . . . Canadians have flocked to cities, but their institutions, their habits of mind, and especially, perhaps, their mythology, have lagged behind.
>
> The jut-jawed outdoorman, still vivid against a prairie sky, a rocky coastline or a stand of black spruce, still works long hours as a national symbol. To a degree, this is very well: such people exist, and their race will, we profoundly trust, endure, providing a flesh-and-blood link with the pioneer past. But the unromantic fact is that most Canadians today are not like this at all. They live and work in cities and towns; their environment, for most of the year at least, is an urban and largely man-made one.

By 1976, over 70 per cent of children in Canada were living in urban areas;

with over one-half of all children under 14 years of age in the 23 census metropolitan areas: one in three of all children under 14 years of age in Canada in 1976 live in urban centres of 500,000 population or more. In terms of the age structure of Canada's population as a whole, considerable changes occurred during the decade just passed. Absolute numbers of children in the 5 to 14 age group declined significantly, and in relative terms, this age group declined from approximately 54 per cent of all children under 19 years in 1971 to about 48 per cent at the end of the seventies (representing a drop from approximately 21 to 16 per cent, respectively, of the total population). Projections for the decade ahead imply growth particularly in the 0 to 4, preschool age group and in the 5 to 9 age group (Table 10—1). In relative terms, however, Canada's child population is declining, from about 30 per cent of the total population in 1971 to a projected 20 per cent at the end of the century (children 14 years and under). In Ontario, for example, the proportion of the total population represented by the under 15s is projected to decline from 32.5 per cent in 1961 to 16.9 per cent by 2001 (Ontario, Provincial Secretary for Social Development, 1979).

**Table 10—1.** Age structure of children in Canada (in '000s)

| Age Group | 1971 | | 1979 | | 1990 | | 2000 | |
|---|---|---|---|---|---|---|---|---|
| | No. | % | No. | % | No. | % | No. | % |
| 0-4 | 1,796.2 | 21.2 | 1,740.6 | 21.7 | 2,018.0 | 26.5 | 1,761.4 | 22.7 |
| 5-9 | 2,254.0 | 26.6 | 1,796.2 | 22.5 | 1,976.5 | 26.0 | 1,904.5 | 24.6 |
| 10-14 | 2,310.7 | 27.3 | 2,075.9 | 25.9 | 1,807.8 | 23.8 | 2,046.3 | 26.4 |
| 15-19 | 2,114.3 | 24.3 | 2,389.6 | 29.9 | 1,806.4 | 23.7 | 2,035.0 | 26.3 |
| Total 19 and under | 8,475.2 | 100.0 | 8,002.3 | 100.0 | 7,608.7 | 100.0 | 7,747.2 | 100.0 |
| Total population | 21,568.3 | | 23,680.1 | | 26,520.9 | | 28,521.8 | |
| % of pop. under 14 | 29.5 | | 23.7 | | 21.9 | | 20.0 | |
| % of pop. under 19 | 39.3 | | 33.8 | | 28.7 | | 27.2 | |

SOURCE: Children's Environments Advisory Service (1979, Table 1)

Throughout Canada, it seems likely that the population will be even more concentrated in urban areas, with something like 60 per cent of those aged 14 years or under living in metropolitan areas by the end of the century. However, as urban populations age significantly in the next 20 years, the child population as a proportion of the total metropolitan population will probably decline significantly, from about 30 per cent in 1966 to about 20 per cent in 2001 (Children's Environments Advisory Service, 1979).

If childhood in Canada is largely an urban experience, it is also very likely to have been predominantly a *sub*urban experience during the 1960s and 1970s, with an average growth of some 24 per cent in the child population of fringe suburban localities recorded between 1966 and 1976 across all metropolitan areas: the same period witnessed a significant decline by an average of 45 per cent in the child populations of inner city areas.

The situation in Toronto does not differ in any significant way from this national picture, triggered largely by declining birth rates, smaller household sizes and a shift by families with children from central city locations to suburbs. In absolute terms, the population of children (0 to 19 years) in the central area of Toronto declined *annually* by approximately 6,300 between 1967 and 1980, a 30 per cent decline from some 27,400 to approximately 19,200: the decline in numbers in this age group in the remainder of the City of Toronto was only slightly less in relative terms (−27 per cent), from approximately 159,100 to some 116,900 over the same period. The numbers of children in elementary schools in the City of Toronto (ages 5 to 12 years) have hence declined consistently through the 1970s and by 1980 the enrollment in these schools in the central area of the City was less than 70 per cent of their potential capacity (McLaughlin, 1981).

Assuming no dramatic reversal of these population trends, therefore, the particular population group of children whose interests are at issue will decline in both absolute numbers and in relation to other age groups in Canadian urban areas over the next 20 years. Children have never been formally franchized in western societies: they do not, as adults do, throughout their natural lives, have a constitutional right to express their wishes in the polling booth in the normal way of electoral politics. If the practice of planning in urban industrial societies is largely the sets of responses of the state to the varieties of demands placed upon it by interest groups with varying abilities to mobilize political pressure, then we are faced with the obvious realization that children as one of many different population classes with identifiable needs have never had formal access to significant power. Further, to the extent that one component of political power may be measured by the sheer number of individuals belonging to a particular class, the demographic future of urban Canada implies that the interests of children as a class are likely to be lowered in the priorities of planners even further in the future, in favour of the interests of older populations, and the aged in particular.

Without access themselves to political power in any meaningful way, children thus depend on adults as their advocates, as their spokesmen in the planning process. There is not, however, a particularly strong indication that, as a group of adult citizens, Canadians are especially concerned about those facilities and institutions in their cities which might be more clearly relevant to the activities and interests of their children. Presented with a check list of some 26 urban "problems" in late 1978, a sample of some 11,000 residents of 23 major urban centres in Canada ranked "improving recreational facilities" and "improving entertainment and cultural facilities" as 19th and 22nd priorities respectively (Table 10—2). "Improving the education system" and "protecting parks and open spaces in the city" fared somewhat better, being ranked as eleventh and eighth in priority, respectively. As noted earlier the relative proportions of young children and of older people in Canada's population is changing significantly. Perhaps an indication of this is revealed in the priorities which Canadian urban residents gave to "improving senior citizens' facilities" and "improving day care facilities" in the survey: tenth and

**Table 10—2.** Priority ranking of urban problems in Canada

| Urban Problem | Perceived priority ranked by residents in 23 Canadian Cities | in Toronto |
|---|---|---|
| Reducing inflation | 1 | 1 |
| Reducing unemployment | 2 | 2 |
| Reducing crime | 3 | 3 |
| Improving the income of the poor | 4 | 7 |
| Attracting new industry | 5 | 14 |
| Protecting agricultural land from urban development | 6 | 5 |
| Improving the use of property taxes | 7 | 4 |
| Protecting parks and open spaces in the city | 8 | 11 |
| Involving people in Government decisions | 9 | 10 |
| Improving senior citizens' facilities | 10 | 9 |
| Improving the education system | 11 | 13 |
| Reducing energy consumption | 12 | 6 |
| Reducing traffic congestion | 13 | 12 |
| Reducing air pollution | 14 | 8 |
| Improving medical and health care | 15 | 15 |
| Improving public transportation | 16 | 17 |
| Improving streets and sidewalks | 17 | 23 |
| Reducing income differentials across regions of Canada | 18 | 22 |
| Improving recreational facilities | 19 | 24 |
| Improving downtown | 20 | 16 |
| Improving existing buildings and homes | 21 | 20 |
| Improving entertainment and cultural facilities | 22 | 25 |
| Improving daycare facilities | 23 | 19 |
| Reducing noise from traffic, etc. | 24 | 21 |
| Stopping the growth of the city | 25 | 18 |
| Building more freeways | 26 | 26 |

SOURCE: CMHC (1979; Table 1). The complete data for Toronto made available from the Institute for Behavioural Research, York University, Downsview, Ontario.

twenty-third respectively, of a total of 26 items (Central Mortgage and Housing Corporation, 1979). In assessing attributes of their neighbourhoods, on the other hand, the provision or quality of children's playgroups received the lowest evaluation of 15 possible items: as much as 45 per cent rated neighbourhoods as "poor" or "fair" in this regard and less than one-quarter as "very good" or "excellent". The second and third lowest rated neighbourhood attributes were recreation facilities and traffic safety, respectively (Central Mortgage and Housing Corporation, 1979). In addition, while a greater frequency of positive comments regarding parks, open space and recreation facilities were recorded by residents in new suburban and exurban neighbourhoods than by those residing in inner city and older suburbs closer to the city centre, the distribution of *negative* evaluations also followed this pattern, with *fewer* negative assessments recorded by residents in inner city neighbourhoods than by those living in the new suburbs and the urban fringe (Central Mortgage and Housing Corporation, 1979).

Faced with these demographic prospects and expressions of public priorities, advocates of Canadian children's interests cannot afford to be sanguine about the likelihood of significantly-increased attention being paid to children's needs and concerns by policy-makers and planners in the years ahead. Indeed it is not until relatively recently that questions of environmental design and urban planning in relation to research on child development and children's activities have been confronted in any systematic fashion, and with a few notable exceptions it is very likely that the level of communication between the community of researchers working on these questions and the community of planners and officials with major responsibilities for impacting the built-environment of urban children, remains either non-existent or abysmally low. As Lynch found in the comparative study of adolescents in the urban settings in Argentina, Australia, Mexico and Poland, "the officials are, for the most part, well ignorant of what the children do" (Lynch, 1977). Less harsh, though similar conclusions have been recorded after interviewing Toronto's planners on the implications of specific neighbourhood and site-plans for children's activities (Muscovitch, 1980).

## ACTIVITIES OF CHILDREN AND URBAN PLANNING

The facilities, services and environmental setting of the neighbourhood in which children and adolescents live undoubtedly have an impact on the quality of their lives. Children living in different parts of the city, for example, are not exposed to the same structure of opportunities in their neighbourhood, the same services, the same schools, parks, landscapes, shopping facilities, densities, and types of neighbours. While evidence gathered from several disciplines is beginning to show the effects of some of these factors, a great deal remains to be learned about how the environmental settings of children and teenagers influence how they spend their time, their achievement, their mental and physical health, alcohol and drug abuse, delinquency, their range of experiences, their life chances, and so on. In attempting to understand some of these relationships, researchers often find that the information necessary to describe the environment and the opportunities experienced by children living in different neighbourhoods is simply not available in a suitable and consistent form.

Further, governments at all levels are directly and indirectly involved in providing many of the facilities, services and environmental settings which children and teenagers use. Governments therefore have a responsibility to ensure that these are provided equitably, efficiently, and in accordance with the needs of people living in different areas. Through a great many policies having to do with zoning, planning, standards, income redistribution, and the direct provision of services, for example, governments have an important influence on the sorts of environments in which children and teenagers find themselves. If governments examine the environments which are in part consequences of their actions and policies, their decisions in this area may hopefully be better informed.

It was these types of considerations that led the Child in the City programme (University of Toronto) to attempt to undertake as one of its research projects a systematic compilation of services and settings relevant to children in Metropolitan Toronto (Michelson *et al.*, 1979). With the cooperation of the various school boards in Metropolitan Toronto, some 3,000 senior students completed a two week curriculum-unit in which they collected material for the project. Working within areas of a few city blocks, students prepared an inventory of facilities, services, environmental settings and opportunities thought to be relevant to children and teenagers. They recorded not only the existence of these various places and so on, but their thoughts on how well they met the needs of children: they also evaluated each neighbourhood as a place for children to live and made suggestions for improving each area to meet the needs of children more appropriately.

While an enormous amount of information was collected in this fashion with evident positive effects because of the students' participation (Hill and Friendly, 1980), there is far too much variation in the thoroughness and the quality of the data collected for these to form the basis either for a fully comprehensive catalogue as originally hoped, or for systematic analyses of the distribution of resources for children throughout the metropolitan region. The information for certain parts of the urban area has been supplemented by further research and results of these analyses and a detailed discussion of the methodological problems involved in the exercise have been completed (Hill, in press).

Far more revealing, however, has been the results obtained from a lengthy questionnaire completed by some 7,000 younger students (ages 13 to 14), collected as part of the original design for the catalogue of children's services and facilities. Certainly for the first time in Metropolitan Toronto, it is now possible to describe in some detail the variations in how, and where young adolescents spend their time out of school, how often and with whom they participate in a variety of activities, their likes and dislikes about their neighbourhoods and what facilities or other attributes of living environments they feel are missing. Apart from preliminary descriptive analysis of the data (Hill, 1980), a number of more specific questions are now being examined, such as, for example, the extent to which neighbourhood perceptions, activities and perceived opportunities of young people living in state-subsidized (public) housing differ systematically from those of comparable adolescents living in private sector housing (Phillips, 1981). Research is also underway to examine the extent to which children living in low-income (public) housing projects located in neighbourhoods of differing income and social mix perceive and utilize locally-available facilities in different ways (Hill, 1981; Saldov, 1980). Some general comments on certain aspects of the results follow (summarized from Hill, 1980), with the caveat that finer-grained analyses of specific topics and of the data for specific smaller areal units are still very much "in progress".

Among the most interesting questions included in the survey were three open-ended ones asking students to describe what they liked best and disliked most about their neighbourhood. Responses to these questions were sometimes quite long, and allowed us to code up to seven items for each of the three questions. The development of the coding system itself was a challenge, given the possible variants of each of the several dozen areas of concern which the students could and did mention. An effort was made to capture as many of the subtleties of each answer as possible. The detailed codes were grouped into four broad categories: (1) the accessibility or quality of facilities or "things to do"; (2) aspects of the physical environment, both natural and man-made, including traffic; (3) characteristics of the people, and (4) other aspects which, although they may incorporate elements of the three previous categories, were felt to deserve a special category. Safety, law-and-order, the police, privacy, and pets are the most typical of this residual category.

Table 10—3 shows the numbers of students in a 3 per cent, random sample of the total data set, who mentioned various items in response to each of the three questions, grouped into the four broad categories. For both boys and girls, the "accessibility of facilities" group ranked first in terms of what students liked best and what they would like to see added — the wording of the latter question being especially conducive to responses in this group. As a dislike, however, the "people"

**Table 10—3.** What students like, dislike, and feel is missing in the neighbourhood[a]

| Category | Items liked | | Items disliked | | Items missing | | Total | |
|---|---|---|---|---|---|---|---|---|
| | Boys | Girls | Boys | Girls | Boys | Girls | Boys | Girls |
| Accessibility & facilities | 335 | 343 | 95 | 87 | 256 | 288 | 686 | 718 |
| Physical environment | 271 | 276 | 106 | 120 | 47 | 56 | 424 | 452 |
| People | 239 | 318 | 171 | 208 | 74 | 102 | 484 | 628 |
| Other | 19 | 30 | 68 | 81 | 9 | 7 | 96 | 118 |
| Total | 864 | 967 | 440 | 496 | 386 | 453 | 1,690 - | 1,916 |

[a]Numbers in the table are the number of times which the various items in each broad category were mentioned in response to each of the three questions. Up to 7 items were coded in response to each of the questions by each student (457 boys; 443 girls).

SOURCE: Hill (1980, Table 18).

group was most often mentioned by both sexes although girls were more concerned about characteristics of the people than were boys.

Consider first the specific aspects of their neighbourhood students liked best. The top 20 answers for each sex are shown in Table 10—4. The most popular answer by both sexes was that their neighbourhood was quiet, calm, or peaceful, an attribute mentioned by almost a quarter of both boys and girls. In second place for boys was the availability of public open spaces (usually parks), but girls mentioned friendliness and simply "the people" or the "nice people", as well as shopping facilities more often than open spaces. These were important to boys, too, but boys stressed access to friends somewhat more than friendliness of the people. Sports

**Table 10—4.** What students like best about their neighbourhood

| Boys: Aspect | Boys: No. of times mentioned | Girls: Aspect | Girls: No. of times mentioned |
|---|---|---|---|
| Quiet, calm, peaceful | 109 | Quiet, calm, peaceful | 96 |
| Public open space (e.g. parks, school yards) | 65 | Friendliness | 90 |
| Access to friends | 57 | The people (general) | 67 |
| Shopping facilities | 49 | Shopping facilities | 68 |
| The people (general) | 43 | Public open space (e.g. parks, school yards) | 54 |
| Friendliness | 43 | Access to friends | 54 |
| Things to do, activities, excitement | 39 | Things to do, activities, excitement | 45 |
| Clean | 35 | Clean | 39 |
| General access to facilities | 28 | General attractiveness, "nice" | 38 |
| No traffic | 23 | Lots of teenagers | 34 |
| Schools | 22 | Transit | 32 |
| Access to opposite sex | 22 | General access to facilities | 30 |
| Trees | 19 | Schools | 25 |
| Transit | 18 | Kindness, politeness | 15 |
| General attractiveness, "nice" | 16 | Trees | 13 |
| Community centre, recreation centre, recreation facilities (general) | 15 | No traffic | 13 |
| Spaciousness | 14 | Community centre, recreation centre, recreation facilities (general) | 11 |
| Kindness, politeness | 14 | Spaciousness | 9 |
| Lots of teenagers | 13 | Scenery, view | 9 |
| Sports facilities | 13 | Ravines, river | 9 |
| | | Attractiveness of opposite sex | 9 |
| | | Individual neighbours | 9 |

SOURCE: Hill (1980, Table 19)

facilities, both in general and for specific types of sports, account for a fairly large number of responses to the "like best" question, although no specific types of sports facilities rank in the top 20 responses. Schools and public transit were the most popular specific non-sport facilities mentioned. Girls seemed to value good transit service more than boys. (The data also show that boys rely more on bicycles than do girls.) Over a third of the responses in the "physical environment" grouping were accounted for by the "quiet" category already noted. Cleanliness, general attractiveness, lack of traffic, and trees followed. Girls were much more likely to mention attractiveness (e.g. "it's nice"), while boys mentioned the absence of traffic more often than did girls. As noted before, it is in the "people" grouping of responses that girls' greater concern for the social environment of the neighbourhood is evident, especially in terms of friendliness. It was also much more probable for girls than boys to mention liking the presence of many other teenagers.

Turning now to what students disliked most about their neighbourhood, note first that the number of things mentioned is only about half as large as the number of likes. This difference is in keeping with the generally favourable attitude of students towards their neighbourhood which was evident from responses to other questions not reported here. Although the "accessibility and facilities" grouping accounted for fewer responses than either the physical environment or the people, a general lack of things to do, activities, and excitement, typified by comments such as "it's boring" or "it's dull", was the most common of all the responses to this question (Table 10—5). No other response in the accessibility and facilities grouping approached it. Complaints about the schools by girls and the lack of quality of the open spaces by boys ran distant seconds in this grouping. Regarding the physical environment, noise and traffic — sometimes undoubtedly the noise of the traffic — were the major dislikes. The apartments or townhouses, dirt, and construction were the next most frequent responses in this grouping. Girls were also concerned about crowding and the boys with the conditions of the streets. The most common response in the "people" group was a dislike of individual neighbours. Second to individual neighbours came a lack of teenagers, a concern which was especially important to girls. Too *few* teenagers of the opposite sex was also a fairly common response, but less so than too *many* younger children: the age structure of the neighbourhood population is important to teenagers. Unfriendliness was also a common complaint. When taken together, responses indicating dissatisfaction with the neighbourhood's ethnic composition amounted to substantial numbers. Six per cent of both boys and girls mentioned some aspect of ethnic or racial composition when asked to describe what they disliked *most* about their neighbourhood. It was in the response to this dislike question that most of the answers in the "other" category were recorded. A lack of personal privacy (e.g. "nosey neighbours"). pets, gangs, vandalism and a general concern about safety and law-and-order were the most common responses here. A lack of privacy was especially disliked.

When asked what is *missing* in their neighbourhood which they would like to see added, students most often responded in terms of facilities although even here people-related answers such as "friendly people" or "more teenagers" were not at all uncommon, especially among girls. About 60 per cent of the answers, however,

**Table 10—5.** What students dislike most about their neighbourhood

| Boys | | Girls | |
|---|---|---|---|
| Aspect | No. of times mentioned | Aspect | No. of times mentioned |
| Lack of things to do, boring dull | 32 | Lack of things to do, boring, dull | 40 |
| Individual neighbours | 28 | Noise | 32 |
| Noise | 22 | Not enough teenagers | 32 |
| Too much traffic | 18 | Individual neighbours | 31 |
| Gangs, bullies, toughs | 15 | Lack of personal privacy, "nosey" neighbours | 24 |
| Pets, pet facilities | 14 | Too much traffic | 21 |
| The people (general) | 12 | Unfriendliness | 19 |
| Not enough teenagers | 12 | Too many younger kids | 12 |
| Too many younger kids | 12 | Gangs, bullies, toughs | 12 |
| Unfriendliness | 11 | Pets, pet facilities | 12 |
| Other individuals | 11 | Dirt, litter | 11 |
| Vandals | 11 | Other individuals | 11 |
| Public open space (e.g. parks, school yard) | 9 | Schools | 10 |
| Apartments, townhouses | 9 | Apartments, townhouses | 9 |
| Ethnic diversity, "foreigners | 9 | The people (general) | 9 |
| Unsafe | 8 | Lack of access to friends | 9 |
| Construction, redevelopment | 7 | Not enough boys | 9 |
| General condition of the streets | 7 | Crowded | 7 |
| Not enough girls | 7 | The kids (general) | 7 |
| Too many police, behaviour of police | 7 | Not enough younger kids | 7 |
| | | Italians | 7 |
| | | Vandals | 7 |

SOURCE: Hill (1980, Table 20).

were in the "accessibility and facilities" grouping. Rather vague answers such as "recreation facilities" or a "community or recreation centre" were the most popular, although swimming pools ran a close second (Table 10—6). Parks, "things to do", shopping facilities, tennis courts, movie theatres, playing fields, and pinball machines followed in that order. Once again the girls' greater concern for socializing is evident in the relative numbers of boys and girls wanting roller-rinks, discos, and restaurants. The fact that many more girls also mentioned youth centres, youth programmes and drop-in centres as well as community or recreation centres could also reflect a greater unmet need for girls' recreation facilities and programmes. Responses to other parts of the questionnaire suggest that fewer girls participate in community or recreation centre programmes, and those who do participate seem to enjoy them less than boys. In this context, the larger number of girls reporting that they would like their neighbourhood to have more of these opportunities suggests that boys' needs are being met better than those of girls. The physical environment

**Table 10—6.** What students feel is missing from their neighbourhood

| Boys | | Girls | |
|---|---|---|---|
| Aspect | No. of times mentioned | Aspect | No. of times mentioned |
| More girls | 29 | Community centres, recreation centres, recreation facilities | 42 |
| Swimming pools | 25 | More teenagers | 34 |
| Other specific sport & recreation facilities | 24 | Swimming pools | 32 |
| Public open space (e.g. park school yards) | 23 | Public open space (e.g. parks, school yards) | 29 |
| Things to do, activities, excitement | 22 | Things to do, activities, excitement | 22 |
| Community centres, recreation centre, recreation facility (general) | 21 | Community centres, recreation centre, recreation facility (general) | 22 |
| Shopping facilities | 20 | Friendliness, friendly people | 18 |
| Tennis courts | 14 | Shopping facilities | 16 |
| More teenagers | 14 | Theatre, movies | 15 |
| Pinball | 12 | Trees | 15 |
| Skating, hockey, arenas | 12 | Roller-skating rink | 13 |
| Trees | 12 | Restaurants, fast food | 12 |
| Playing fields (baseball, football, soccer) | 11 | Tennis courts | 12 |
| Theatres, movies | 10 | Youth centre, drop-in youth programme | 11 |
| Sports organizations & activities | 9 | More boys | 11 |
| Friendliness, friendly people | 8 | Other specific sport & recreation facilities | 9 |
| Sports facilities (general) | 7 | Friends | 9 |
| Basketball courts | 6 | Playing fields (baseball, football, soccer) | 8 |
| Transit | 5 | Sports facilities (general) | 7 |
| Pool-halls | 5 | Sports organizations & activities | 7 |
| Quiet, calm, peaceful | 5 | The people (general) "nice people" | 7 |
| Improvements to the general condition of the streets | 5 | | |
| Friends | 5 | | |
| Attractive girls | 5 | | |

SOURCE: Hill (1980, Table 21).

grouping accounted for a fairly small percentage of the responses to the question about what was missing. The most popular answers in this category were trees and, when added together, various aspects of the streets: sidewalks, lights, general repair, traffic controls, and modifications to the street pattern. As noted earlier, friendliness and simply more teenagers were the most popular answers within the "people" grouping, especially among girls. Boys, on the other hand, more often mentioned that they wanted more girls in their neighbourhood. In fact, among boys this was by far the most common answer to this question in the "people" category and was their top response overall.

Differences between boys and girls in the proportion of their friends who live in their neighbourhood provide further evidence that neighbourhoods are not perceived or utilized in quite the same way by boys and girls. More boys than girls maintained that half, most, or all of their friends lived in their neighbourhood, while 20 per cent more girls reported that only a few of their friends live there (Table 10—7).

These differences could have several interpretations. Girls may have a narrower geographical definition of their neighbourhood, they may have different ideas about who qualify as friends, or they may indeed have more friends outside and/or fewer inside their neighbourhood as defined objectively. From the data at hand there is no way of knowing the contribution of each of these possible factors which could explain the quite pronounced differences between boys and girls in the proportion of their friends living in their neighbourhood as subjectively defined by the respondents. But the fact that for most girls, most of their friends do not live in their neighbourhood makes the greater emphasis which girls place on the characteristics of the people in their neighbourhood all the more significant. The relative importance of "access to friends" and "friendliness" as responses to the "like best" question for boys and girls is also consistent with the sex differences in relation to where their friends live.

To sum up, students were found to have generally favourable opinions about their neighbourhood along most dimensions which they consider important. The most notable exception was that as many considered their neighbourhood boring as interesting. When given a free rein to say what they liked and disliked about their neighbourhood, the responses demonstrated the importance of a great *variety* of aspects, of the people, the physical environment, and the opportunities available for many types of recreational activity. A common trend running through the responses to most of the neighbourhood-related questions was the greater impor-

**Table 10—7.** Proportion of students' friends living in their neighbourhood

| Proportion of friends living in their neighbourhood | Percentage of students | |
|---|---|---|
| | Boys | Girls |
| All | 9 | 5 |
| Most | 37 | 25 |
| About half | 22 | 18 |
| A few | 23 | 43 |
| None | 5 | 6 |
| I don't have any friends | 1 | 3 |
| Not stated | 4 | 3 |
| Total | 100 | 100 |

SOURCE: Hill (1980, Table 22).

tance of social concerns among girls, and in particular, a greater concern with the characteristics and numbers of people in the neighbourhood. This difference, however, is one of *relative* importance, and the major concerns of boys and girls about their neighbourhood are basically the same, with only the order varying somewhat. Both boys and girls appreciate a quiet, clean, attractive neighbourhood with "nice", friendly people, friends nearby, and "things to do", but the desired combination of ingredients in such a neighbourhood varies among individuals, between the sexes, and, to anticipate results of future analyses to be reported, between ethnic groups and social classes as well.

This kind of research may be related in fairly straightforward fashion to the conventional concerns of the urban planner, as described in the introduction to this contribution. The location of public and private facilities, the discrete uses of spaces in urban areas regulated by land-use zoning policies, the design of residential environments, and so on, are all fairly traditional issues and problems confronted at the political decision-making level via the conventional wisdom of planning practice. The results of these practices reflect the concern and activities of adults in an adult-oriented world: they do not necessarily provide a good "fit" for the interests and activities of children. As a good example of this point, recall Abernathy's description of provision of children's play space in new housing developments as SLOAP — "space left over *after* planning". As adults our conduct and behaviour have become readily fragmented, as we live out the daily round, in ways which guide our self-presentation (or guard our self-preservation) and provide rules for our interactions with others. The structuring of our time and the space and environments in which we operate are all direct manifestations of this fragmentation of existence. There are daily, weekly, seasonal and longer cycles of patterns of different activities; there are spatial patterns in the arrangement and distribution of places which provide the settings for our fragmented activities (e.g. Carlstein, 1978). Similarly, the complexity of providing or supporting the provision of services to individuals and families in industrial societies requires a functional, or sectoral division of responsibilities, a logical outcome of the paradox of the increasing personal autonomy of the individual, simultaneously with the closer dependence of the individual on society. Children, too, are faced with this same paradox but the ongoing development of personal identity through childhood and adolescence is accompanied by a growing *in*dependence of the child from his/her immediate social world, of the family, and the immediate physical world of the home and neighbourhood. In the earlier stages of childhood, the functional and fragmented views of activities, of other people (as roles), of places and of time, which are more or less well-learnt and practised by adulthood, are by no means the manner in which the child naturally structures the world around him/her and derives meaning from it. "Managing urban space for children", therefore, as an agenda-item split-off from *all* of those other typical planning activities suggested by the exercise in combinatorics provided in the introduction, only serves to reinforce the functionalist or sectoral view of urban planning — a view that generally works in the opposite direction to the needs and interests of young children, which require a more sensitive, and above all, an ecological approach to planning and design.

## HUMAN DEVELOPMENT AND URBAN PLANNING: ECOLOGICAL PERSPECTIVES

The "needs and interests of young children" in terms of the environments in which they live and learn may be expressed in a number of ways. With respect to the urban physical environment for example, some of these needs have been expressed as follows:

1. A safe and healthy setting;
2. Accessibility to the range of facilities and social groups within the urban area;
3. Suitable stimulation and challenge, including the opportunity to exercise relevant control over the physical and social environment;
4. Appropriate sizes of outdoor spaces, relative to the physical size and the physical, cognitive and emotional development of children: variety is important in this connection;
5. Sufficient environmental flexibility and diversity.

There are a number of considerations involved in such a statement of needs. For example, some are more specific than others. All depend in one way or another on facilitating responses or the provision of services from agencies or governments outside of the immediate control of families and individuals. Finally, all may be shown to rest on the conclusions derived from past research and demonstration projects examining the activities of children and families, the ingredients necessary for the healthy development of the whole child and the characteristics of the environments in which this development occurs (Michelson and Roberts, 1979). The focus for this particular statement of children's needs was an international seminar in June 1979, organized by the Child in the City programme and called "Managing Urban Space in the Interests of Children". The conference was sponsored by the Unesco-MAB Programme and attended by some 40 delegates and observers, mostly selected by their home MAB programmes (Michelson and Michelson, 1980).

There are, then, a number of principles on which programmes or proposals to address the needs of children and families through the ecological design and management of urban spaces should be based, with a primary focus on the development and well-being of the *whole* child for his *total* environment. Such a focus prepares the way for a countervailing approach to the partitioning and fragmentation of time and space which dominates the adult's world. Children are regular and natural users of the total environment, *not* just the spaces which adults — planners, or otherwise — provide especially for them. Children also *use* the environment in ways that are qualitatively different from the fragmented or functional activities of adults. Similar conclusions may be stated about the temporal dimension of children's activities and their use of time: their activities do not coincide "naturally" with the cycles and temporal world of adults, nor does their use of perception of time display much concordance with the values placed by adults on time (Martensson, 1977).

One of the workshops at the MAB seminar in June 1979 expressed the consequences of this perspective as follows:

1. The extent to which the immediate area around a child's home is safe for play and exploration influences his or her freedom and development: if the child must be confined to the dwelling, gross motor activity will be restricted;
2. The ability to manipulate and control things by themselves, to be able to do things on their own, enhances the cognitive development of children and their ability to grasp abstract concepts;
3. The ability to take risks and learn that one can manage within the physical environment is likely to strengthen a child's self-confidence and help to develop competence in dealing with the social environment;
4. A child who has opportunities for privacy can more easily develop independence;
5. High density may limit space available for children, and may result in overcrowding: low density, on the other hand, may result in few companions and little environmental opportunity;
6. To avoid defining as vandalism children's normal use of areas not specifically set aside for them, areas must be planned with unintended uses in mind, and adults must develop more tolerance for children;
7. Control over the physical environment may lead to a feeling of responsibility and concern for its maintenance.

Each of these conclusions may be related to the two basic concepts of control and identity, and the dialectical relationships for which the emergence and the development of a sense of control and a sense of personal identity represent the continuous resolution. They are referred to as "basic" concepts because a surprising number of additional ideas about how children live and learn and how individuals make sense of the world about them may be traced through to the circumstances surrounding the individual's sense of control and sense of identity. If these are well-formed and not ambiguous, enduring and not fragile, we have, it is believed, the essential product of healthy and healthful child development in the experience of childhood. Among other prerequisites, this requires special attention to the character of reciprocity among family members and the nurturing of an environment producing and supporting the development of competence (control) and of positive self-regard (identity) (Garbarino, 1980). Finding ways of enhancing and reinforcing the young child's sense of control over the spaces and time that frame his activities thus becomes a crucial consideration for environmental design (see also Fookes, 1980).

It also implies a vastly-expanded agenda for the physical and social planning of urban environments, beyond a functional view of human and material resources to embrace a deeper understanding of the ecological dynamics of human development and the contexts and settings in which activities occur. "What we seek," wrote Lynch (1977), "is the comprehensive description of indoors and outdoors as a used landscape — an ecology of children" (1977). Such a description might well serve as a starting point to view the *settings* of activities: the *contexts* in which they occur requires a more fundamental appreciation of the social ecology of child development (Little and Ryan, 1979).

There is a rapidly growing literature on ecological dimensions of children's activities and development (e.g., Catalano, 1979; Cobb, 1977; Eggleston, 1977). Seminal amongst this literature is the work by Bronfenbrenner (1979), which provides a systematic review of previous research organized by the principal ecological dimensions of human development. Space precludes a detailed discussion of this work and the intricacies of Bronfenbrenner's conceptual framework, but the author gives major significance to the patterns of activities, roles and interpersonal relations experienced by the developing individual in particular environmental settings. Further, these settings are organized by scale and increasing complexity and are interrelated both horizontally and vertically through second-order effects, direct and indirect linkages. Bronfenbrenner is less concerned with the physical or design attributes of environmental settings than he is with the quality, nature and effects of activities, roles and interpersonal structures occupying such settings. Adopting such a perspective for our present concern, an appropriate question derived from this work would be to ask: how do characteristics of urban environments and urban social organization enhance or diminish, stimulate or inhibit the effects (motive power) of interpersonal structures, roles and activities conducive to healthy human development? This order of question clearly supercedes questions concerning the "management of urban space for children" — complex as these latter questions may appear — since the cultural values, social structures and institutions delineating the characteristics of urban space and social organization are fundamentally more significant in influencing children's lives than the locational attributes and characteristics of urban spaces and facilities, *per se*.

It is in such a context that the research projects making up the Child in the City programme should be properly viewed, with a vital additional focus linking basic research to the policy-making process in the realm of preventive intervention and programme evaluation. A philosophy of prevention related to child development amounts in practical terms to a strategy which aims "to improve the environment in which the intellectual, emotional and physical growth of children occurs" (Rae-Grant, 1979). The visible artifacts and organization of the urban built-environment are thus just one of a variety of components making up the "environment" in which children grow. As a strategy, research with a preventive focus implies a five-fold set of tasks in bringing knowledge to bear on practice with fruitful results:

1. The description and measurement of the child's "intellectual, emotional and physical" development;
2. The description and measurement of the environmental settings in which this development takes place;

3. The description and measurement of the characteristics or attributes of these environments thought to be positively and negatively related to the child's intellectual, emotional and physical development;
4. The identification of ways to isolate and measure the effects of environmental settings on child development, and, moreover;
5. The development, implementation and evaluation of strategies to enhance those attributes of environmental settings with positive effects on child development, and to diminish — ultimately to eradicate — the influence of those attributes of environmental settings which have negative consequences for child development and well-being.

The fifth task is critical and is essentially the mechanism which translates the other four, largely "basic" research activities, into "applied" research or practice, with significant implications for a truly ecological planning thrust. As Weinberger (1980) has noted:

> . . . preventive intervention forces attention on the wide band of identified variables that significantly bear on the individual's mental health, including physical, social, educational, cultural and economic aspects. The preventive approach, in effect, presents one of the strongest arguments for a truly "ecologic" or "wholistic" perspective on man and his mental health.

As selected examples of research questions being addressed by the Child in the City programme, with relevance for such an expanded ecological planning framework, the following may be noted.

There is first, a cluster of concerns relating to the social and institutional constraints on the ways in which families function and the implications these have for the health and well-being of children. The effects of family structure and employment status and accessibility to family and child-oriented services have significant implications for the quality of life in childhood (Michelson, 1980, 1981; Michelson *et al.,* 1978). The availability and quality of child care facilities have become critical concerns in urban areas where mothers of young children are entering or rejoining the labour force in increasing numbers (Friendly and Johnson, 1981). The effects of social isolation of mothers both on their health and that of their children as well as their activities are being examined, together with the development of locally-based support mechanisms aimed at reducing some of the negative consequences (Hodgson, 1981a; Hodgson and Birnberg, 1980) and alternatives to existing ways of providing social services at the community level (Hodgson, 1980, 1981b).

Another set of concerns may be loosely aggregated around the effects of belief systems and attitudes as they relate to practices and behaviours impacting children's health and development. Toronto's population is made up of an unusually broad mixture of ethnic groups and many families are relatively recent arrivals from diverse cultural backgrounds: approximately half of the "heads of households" were born outside Canada and about one in four children do not learn English as their first language. There are numerous implications of this multicultural setting for the activities of children and the provision of various services and facilities. There are, for example, clear variations across ethnic groups in practices and beliefs about nutrition and diets (Ziegler, 1979a), and in mothers' preventive beliefs and practices concerning health care and environmental threats for their children (Ziegler and Kelner, 1981), and in the level and extent of institutional response to the needs of an ethnically diverse population, for example, from hospitals and other social agencies (Ziegler, 1980a). The extent of within and between ethnic-group friendship patterns of young children and adolescents have also been examined, and the development of strategies designed to promote inter-group cooperation and tolerance and to reduce nascent prejudice amongst children of various ethnic backgrounds (Ziegler, 1979b, 1980b, 1980c, 1981a, 1981b). Many recently-arrived children from other countries have particularly difficult tasks in adjusting to a new, urban and heterogeneous cultural milieu, and a school- and community-based intervention programme is about to be designed to diminish a variety of the problems that such children face (Ziegler, 1981c).

Inter-ethnic variations in attitudes and beliefs are by no means the only dimension-impacting practices and behaviours affecting children and families. Access to health care services, particularly for adolescents and minors, is a good example where considerable confusion exists on the parts of both professionals and clients (Catton *et al.*, 1979, 1980); many families with children are denied access to housing by restrictive practices of landlords maintaining "adult-only" accommodation (Barnhorst and Hill, 1981). Questions relating to the attitudes and beliefs of children concerning the law, as well as practices affecting children in the "eyes of the law" are also being addressed (Barnhorst, 1981, in press; Bolger, 1979; Catton and Leon, 1977; Leon, 1978).

A final cluster of research examines two dimensions of the transition between childhood and adulthood. This concerns, first, the transitional experiences of young adolescents from the environment of school to that of work (Koulack and Young, 1981; Larter *et al.*, 1978; Levine, 1980); and second, the transitional experiences of young adolescent girls into parenthood (Finkelman, in press; Hodgson, 1981c; Weizmann *et al.*, 1980).

## CONCLUSION

The approaches and content of this book — and indeed the MAB Programme itself — forces a confrontation between the domain of scientific knowledge and that of human practice. With certain limited, though by no means trivial exceptions, the domain of scientific knowledge in the field of urban systems and urban planning is itself that of human practice. Thus there is less concern here with knowledge of the natural physical or biological environments in which human activity occurs, as with understanding the social, political and institutional environments of human development. The variables, therefore, have less to do with those of biomass, soil nutrition, and food-chains, and so on, than with social structures, political and institutional settings and economic organization.

The practice of contemporary urban planning characterized in the introduction to this contribution holds little potential for the future development of urban environments in which children's interests are safeguarded. As one of many "interest groups" in a pluralist political setting, children are diminishing in urban Canada, and adults, generally, place higher priorities on tackling other concerns, even though confronting and solving these other problems may eventually and in indirect fashion enhance the lives of their children. In place of a planning philosophy dominated by a functional view of discrete activities, programmes and facilities, the outlines of a planning thrust with an ecological and human development focus have been suggested. A genuine concern for the interests of children in urban settings implies that a far greater sensitivity is needed by planners and policy-makers to the changing circumstances in which families function and "manage" their lives on a day-to-day basis within urban-industrial societies.

## References

Barnhorst, S. (1981), *The Role of the Lawyer representing Minors in Family Court: Professional Conduct Implications,* Child in the City programme, Toronto.

Barnhorst, S. (in press), *Children's Understanding of and Attitudes toward Law,* Child in the City programme, Toronto.

Barnhorst, S. and F. Hill (1981), *The Rights of Families with Children to Equal Access to Housing,* Child in the City programme, Toronto.

Bolger, J. (1979), *A Study of the Legal Drinking Age Policy in Ontario,* Child in the City Report No. 2, Child in the City programme, Toronto.

Bronfenbrenner, U. (1979), *The Ecology of Human Development,* Harvard University Press, Cambridge, Mass.

Carlstein, T. (1978), A Time-Geographic Approach to Time Allocation and Socio-Ecological Systems, In W. Michelson (Editor), *Public Policy in Temporal Perspective,* Mouton, The Hague, 78-119.

Catalano, R. (1979), *Health, Behavior, and the Community: an Ecological Perspective,* Pergamon Press, New York.

Catton, K. and J. Leon (1977), Legal Representation and the Proposed Young Persons in Conflict with the Law Act, *Osgoode Hall Law Journal,* 15 (1), 42-67.

Catton, K., W. Graham and E. Koulack (1979), *Doctor's Understanding of and Practices Regarding the Law of Minors' Medical Consent,* Child in the City Report No. 3, Child in the City programme, Toronto.

Catton, K., V. Farrer and W. Graham (1980), *Adolescent Beliefs and Practices Regarding the Law of Minors' Medical Consent,* Child in the City Report No. 7, Child in the City programme, Toronto.

Central Mortgage and Housing Corporation (1979), *Public Priorities in Urban Canada: a Survey of Community Concerns,* CMHC, Ottawa.

Children's Environments Advisory Service (1979), *Housing Canada's Children: a Data Base,* Central Mortgage and Housing Corporation, Ottawa.

Cobb, E. (1977), *The Ecology of Imagination in Childhood,* Columbia University Press, New York.

Draper, J. A. (1970), *Citizen Participation: Canada,* New Press, Toronto.

Eggleston, J. (1977), *The Ecology of the School,* Methuen, London.

Finkelman, S. (in press), *Consequences of Teenage Motherhood,* Child in the City programme, Toronto.

Fookes, T. (1980), Children's Needs and Child-Space Relationships, In W. Michelson and E. Michelson (Editors), *Managing Urban Space in the Interests of Children,* Canada/MAB Report 14, Ottawa, 244-53.

Friendly, M. and L. Johnson (1981), *Perspectives on Work-related Day Care,* Child in the City Report No. 11, Child in the City programme, Toronto.

Garbarino, J. (1980), *Who are the Puzzlemasters?* Paper presented at the 6th annual meeting, Canadian Association for Young Children, Calgary.

Goldsmith, M. (1980), *Politics, Planning and the City,* Hutchinson, London.

Hawkins, B. W. (1971), *Politics and Urban Policies,* Bobbs-Merrill Co., Indianapolis.

Hill, F. (1980), *The Lives and Times of Urban Adolescents: Activity Patterns and Neighbourhood Perceptions,* Child in the City Report No. 8, Child in the City programme, Toronto.

Hill, F. (1981), *Public Services in Low and Mixed Income Housing Environments: the Research Design,* Child in the City programme, Toronto.

Hill, F. (in press), *The Whole City Catalogue: Methodology and Selected Results,* Child in the City programme, Toronto.

Hill, F. and M. Friendly (1980), The "Whole City Catalogue" project, *Ekistics,* 47 (281), 138-9.

Hodgson, S. (1980), *The Provision of Children's Services in Metropolitan Toronto,* Child in the City programme, Toronto.

Hodgson, S. (1981a), *Social Isolation: a Perspective on Primary Prevention,* Child in the City programme, Toronto.

Hodgson, S. (1981b), *Citizen Participation in Human Service Planning,* Child in the City programme, Toronto.

Hodgson, S. (1981c), *A Teenage Mother's Decision of Whether or Not to Keep Her Infant,* Child in the City programme, Toronto.

Hodgson, S. and P. Birnberg (1980), *Developing Informal Social Supports for High Risk Families with Preschoolers: a Preventive Strategy,* Child in the City programme, Toronto.

Kaplan, H. (1968), *Urban Political Systems,* Columbia University Press, New York.

Koulack, E. and V. Young (1981), *The Student Employment Experience Centre: an Evaluation of a Work Experience Project for out of School Young Adolescents,* Child in the City Report No. 10, Child in the City programme, Toronto.

Larter, S., J. Fitzgerald and M. Friendly (1978), *Students' Attitudes to Work and Unemployment,* Report No. 151, Toronto Board of Education Research Department, Toronto.

Leon, J. (1978), *Legal Representation of Children in Selected Court Proceedings: the Capacity of Children to Retain Legal Council,* Child in the City Report No. 1, Child in the City programme, Toronto.

Levine, S. (1980), *The Psychological and Social Effects of Youth Unemployment,* Child in the City Report No. 9, Child in the City programme, Toronto.

Lineberry, R. L. and I. Sharkansky (1978), *Urban Politics and Public Policy* (3rd edition), Harper and Row, New York.

Little, B. R. and T. J. Ryan (1979), A Social Ecological Model of Development, In K. Ishwaran (Editor), *Childhood and Adolescence in Canada,* McGraw-Hill, Toronto, 273-301.

Lynch, K. (1977), *Growing up in Cities: Studies of the Spatial Environment of Adolescence in Cracow, Melbourne, Mexico City, Salta, Toluca and Warszawa,* Unesco-MIT Press, Cambridge, Mass.

Martensson, S. (1977), Childhood Interaction and Temporal Organization, *Economic Geography,* 53, 99-115.

McLaughlin, S. (1981), *Loss of Family Housing in the Central Area,* Planning and Development Department, City of Toronto.

Michelson, W. (1980), Spatial and Temporal Dimensions of Childcare, *Signs: Journal of Women and Culture in Society,* 5 (3), 242-7.

Michelson, W. (1981), *Some Changing Factors in Urban Context with Implications for the Behavior and Well-being of Parents and Children,* Paper presented to the symposium, "The Child in the City: Families and Children in Urban Communities", Tokyo.

Michelson, W., S. Hodgson and S. Ziegler (1978), *Childcare under Constraint: Conditions of Daily Life as influenced by Family Structure, Employment and Children's Services,* Child in the City programme, Toronto.

Michelson, W., S. V. Levine and A-R. Spina (1979), *The Child in the City: Changes and Challenges,* University of Toronto Press, Toronto.

Michelson, W. and E. Michelson (Editors) (1980), *Managing Urban Space in the Interests of Children,* Canada/MAB Report 14, Ottawa.

Michelson, W. and E. Roberts (1979), Children and the Urban Physical Environment, In W. Michelson, S. V. Levine and A-R. Spina, *The Child in the City: Changes and Challenges,* University of Toronto Press, Toronto, 410-77.

Muscovitch, A. (1980), *Child's Perception of the Neighbourhood,* Children's Environments Advisory Service, CMHC, Ottawa.

Ontario, Provincial Secretary for Social Development (1979), *The Family as a Focus for Social Development,* Toronto.

Phillips, E. (1981), *Neighbourhood Perceptions of Young Adolescents in Public Housing,* Child in the City programme, Toronto.

Pickvance, C. G. (1976), On the Study of Urban Social Movements, In C. G. Pickvance (Editor), *Urban Sociology: Critical Essays,* Tavistock Publications, London.

Rae-Grant, N. (1979), *The State of the Art: a Background Paper on Prevention,* Children's Services Division, Ministry of Community and Social Services, Ontario.

Saldov, M. (1980), *Public Services and Facilities in Low and Mixed Income Housing Environments — their Perceptions and Utilization by Children from Low Income Families in Metro Toronto,* Child in the City programme, Toronto.

Scott, A. J. and S. T. Roweis (1977), Urban Planning in Theory and Practice: a Reappraisal, *Environment and Planning A,* 9, 1097-119.

Simmons, J. and R. Simmons (1974), *Urban Canada* (2nd edition), Copp Clark, Toronto.

Weinberger, A. (1980), Contours of Primary Prevention, *Canada's Mental Health,* 28 (4), 59-63.

Weizmann, F., M. Friendly and G. Gonda (1980), *Evaluation of the Montrose Infant Daycare Centre,* Child in the City programme, Toronto.

Wolpert, J., A. Mumphrey and J. Seley (1972), *Metropolitan Neighbourhoods: Participation and Conflict over Change,* Resource Paper 16, Association of American Geographers, Washington, D.C.

Ziegler, S. (1979a), *Food and Ethnicity Resources Kit,* Child in the City programme, Toronto.

Ziegler, S. (1979b), *Demographic Influences on Adolescents' Cross-ethnic Friendship Patterns,* Child in the City Report No. 4, Child in the City programme, Toronto.

Ziegler, S. (1980a), *Report on the Survey of Ontario Hospitals and Social Agencies regarding Services to Non-English Speaking Patients,* Child in the City programme, Toronto.

Ziegler, S. (1980b), Adolescents' inter-ethnic friendships, *Children Today,* 9 (2), 22-4.

Ziegler, S. (1980c), Measuring Inter-Ethnic Attitudes in a Multi-Ethnic Context, *Canadian Ethnic Studies,* 13 (3), 61-7.

Ziegler, S. (1981a), The Effectiveness of Cooperative Learning Teams, *Human Organization,* in press.

Ziegler, S. (1981b), *Diagnosing and Treating a Community Illness: Primary Prevention of Racism in Ethnically Heterogeneous Communities,* Child in the City programme, Toronto.

Ziegler, S. (1981c), *A Proposal for a School- and Community-based Prevention Programme for West Indian Children,* Child in the City programme, Toronto.

Ziegler, S. and M. Kelner (1981), *Mothers' Preventive Beliefs and Practices: some Ethnic Variations,* Child in the City programme, Toronto.

SECTION 2

# Use of Scientific Information for Environmental Education Purposes

CHAPTER 11

# Introduction

*Francesco di Castri, F. W. G. Baker and Malcolm Hadley*

THERE is little disagreement today as to *why* environmental education is essential, but there is less agreement on *what* is to be taught or communicated by such education, and *how* this is to be achieved.

Existing scientific information is often not used, or is poorly used, in environmental education. And when such information is put to good use, there is often a long delay between information being obtained and applied in education programmes. In spite of environmental education's claim to be interdisciplinary and non-sectoral, in practice, it has remained fragmented and divided with little communication or contact between the different disciplines or sectors involved.

Environmental education programmes do not always make use of local examples, and they often ignore local populations' tested, intuitive appreciation of the environment. Many environmental education programmes fail to clearly define their target audiences, and often overlook the needs of large segments of the population who are unable to read. Furthermore, they often do not deal with such basic environmental problems as poverty and health.

Information materials which aim to transfer technical knowledge on the environment to specific audiences are often ineffective. Typical faults include: a too limited perception of the audience by the designer; an excessively static conception; an impression of certainty about scientific issues which may, in fact, be poorly understood and constantly evolving; information which dates rapidly and gives the materials too short a life span; excessively culture-bound or complex concepts and descriptions used for populations with low literacy rates; insufficient use of locally-relevant information with which the target audience can easily identify. Also, professional communicators, education specialists and scientists may fail to form a sufficiently integrated group during the preparation of materials for environmental education, so that technical information may become distorted in the transfer process. Jargon may make the final product too difficult to understand by the target audience; for example, words such as "ecology" or "biosphere" are often inadequately explained in environmental education programmes.

But in spite of these perceived difficulties and shortcomings, considerable progress has been made in recent years in bridging the gap between scientific research and environmental education. Scientists in many countries have become convinced of the need to render information "accessible", that is to make it available in a form that can be adapted and used by those concerned with environmental education.

Such changes in attitudes of the scientific community are a reflection of more general public concern for the environment, as described in the opening contribution (Chapter 12) to this section of the book. Antonio Moroni and Oscar Ravera argue that the so-called crisis in man's relations with the environment is essentially a cultural one, and consider that environmental education holds one key for man's awareness of his responsibility for such problems and their solution. The concepts, methods and results of research initiatives such as MAB can play an important role in shaping programmes of environmental education, both within and outside the school context. Moroni and Ravera consider that the effective transfer of experience and results from research to education should be one of the priorities for the MAB Programme in the 1980s, and suggest a number of ways in which such a transfer might be approached.

The eight contributions that follow are based on specific efforts to promote environmental education at local, national, regional and international levels. Two contributions take up the key question of communication of scientific information on the environment. The poster-exhibit "Ecology in Action" provides the focus for a case study described by Bruno Lefèvre. This exhibit attempts to highlight some of the major problems of land use and resource management in the world today, and to suggest ways through which research can help solve them. The exhibit is an experiment in form and content, and results from a joint effort of scientists, communications specialists and graphic designers. Lefèvre describes some of the features of the system of communication used in preparing the exhibit, and outlines some of challenges and difficulties that were encountered. Initial reactions to the exhibit as a tool in environmental education have been largely favourable, and it is hoped that evaluation of the effectiveness of such exhibits will provide a basis for their testing and development.

The central thesis of Chapter 14 by Justin Murphy and Maurice Woodward is that communication of the results of research involves transmission, reception and understanding not just transmission. Drawing on their experience within the CSIRO of Australia, and within a particular MAB field project in northern Kenya, they argue that the techniques and approaches of professional communicators should be considered as an integral part of the research process. A number of the key steps in communication are described, including identification of audience, interpretation of the research findings, timing of the communication exercise, selection of the right forms for transfer and importance of assessing the effectiveness of the communications exercise.

In Mexico, as in most developing countries, a priority problem is to find inexpensive ways of increasing food output using systems that take advantage of the characteristics of the local environment and that can be sustained without the need for large-scale inputs in terms of fertilizers, equipment, capital, etc. Luis Morales describes one promising approach to this problem developed by scientists from the Institute for Research on Biological Resources in Jalapa, who recognized the special nature and success of a traditional agricultural system, the *chinampa*, and decided to experiment with it in order to create a new method for meeting food needs. As a result of the scientific assessment of the system, trials have been undertaken for developing mixed agro-piscicultural systems that can be combined with small-scale forest industries to produce multiple products from farm and forest. The initial promising results owe much to the activities of training, demonstration and environmental education that have accompanied the research effort. Field training seminars have been organized for and by the farmers, based on their own knowledge and experience and reinforced by their active participation in field work. Information is thus exchanged informally, in everyday language, reflecting local perceptions and values. The seminars have enabled the farmers to visit other distant areas and compare similarities and differences in production systems and community experience. The outreach programme developed by the Mexican scientists also includes extension bulletins, posters and slide-sound materials as part of the overall strategy for harnessing scientific knowledge to local development need.

Historical dimensions of land use, and differing perceptions of environmental quality, are two important elements in environmental education brought out in the case study by F. G. Bernaldez, F. El Cadi and R. Perelman (Chapter 16). A cooperative programme for studying and interpreting the Mediterranean environment has been set up in contrasting areas in Tunisia (coastal zones on the Isle of Djerba), France (upland area in Western Pyrenees National Park) and Spain (wooded areas in the Collado-Mediano northwest of Madrid). Characteristics of the programme include the organization of multidisciplinary workshops for conceiving and preparing educational materials and identification of leitmotivs linking environment, land use and employment as a means of arousing the interest and active participation of local people in educational programmes. The results of environmental perception studies have shed light on attitudes and behaviour towards the environment, and have also provided clues on how to tailor educational materials in a form suited to different cultures and interests. One conclusion is that perception studies are practically indispensable if environment education is to facilitate two-way communication between different groups of the community.

Chapters 17 and 18 deal with two mountain situations where the traditional land use practices have been disrupted, and where the traditional environmental know-how of the local people has been eroded and lost to a greater or lesser extent. In both cases, scientists have sought to learn from local knowledge about the mountain environment and to understand what has happened to the mountain people's forms of land use with a view to helping guide programmes of environmental education and appropriate technology. The contribution by Milka Castro, Alberto Veloso and Eduardo Bustos deals with the Chilean Altiplano, settled by pre-Inca Aymara people. The Aymara have been subjected to many outside influences over the centuries, but have managed to maintain some of their ancient cultural practices and systems of land management. The effect of modern communications and education systems is problematic, and a key question is whether the acute perception by the Aymara of their local environment can be harnessed in the development both of new productive systems and of programmes of environmental education in the region.

The case study described by Jiro Kawakita is also concerned with a mountain people (in the Himalayas), whose culture, social solidarity and land use practices have eroded in recent years. Serious environmental and social problems have arisen, particularly in respect to pollution of water sources and the need to transport firewood from long distances. Identification of the nature of the key problems has led to an overall scheme for tackling these problems, based on use of "appropriate technologies" such as use of a ropeline for providing access to remoter forests and forest resources. A method for integrating qualitative information in a scheme for correcting ecological inbalance is described, based on the principle that data must be allowed to speak for themselves without being distorted to fit any preconceived theory or hypothesis.

In the German Democratic Republic, a sustained effort has been made to promote public environmental awareness and environmental protection. Egon Seidel outlines some of the approaches and activities contributing to the systematic diffusion of information on the environment, both within and outside the formal system of education (Chapter 19). Activities include the preparation of public information materials, training of environmental engineers and other environmental specialists, the incorporation of environmental concerns in agricultural exhibitions, community projects for improving living and working environments. Cooperation between the scientific community, professional associations, social organizations and the information media helps ensure continuity in promoting public interest, awareness and involvement concerning environmental issues. One example is the field of eutrophication and restoration of surface waters, which has been the focus of a sustained information campaign in magazines, press, radio and television.

In Indonesia, a group of scientists have become increasingly involved in measures for environmental education, as one avenue for putting to work scientific information on the Indonesian environment. Pilot schemes have been developed at both formal and non-formal levels. At the formal level, attention has focussed on introducing environmental dimensions within the existing school subjects rather than developing separate curricula on environmental issues, and the rationale for such an approach seems to have been borne out by experience in a number of other countries. The Indonesian case study also illustrates the possibilities for introducing environmental education within programmes for teachers training as well as for the in-service training of established teachers. The case study also highlights the results that can be achieved through imaginative efforts at the non-formal level, such as writing and drawing contests on environmental issues within children's magazines, contests which attract 700 to 800 entries each.

CHAPTER 12

# Trends and Perspectives in the Contribution of Science to Environmental Education, and MAB's Role Therein

*Antonio Moroni and Oscar Ravera*

Abstract

Cultural reductionism and man's desire to dominate his environment, regarded as extraneous, have been invoked as underlying causes of the present crisis in the relationship between human communities and the environment. This crisis is, in other words, essentially a cultural one. The aim of environmental education is therefore to make man see the environment as a system of living relations with its own special operating rules and history, and to convince him that he has a personal involvement in ensuring its correct use and management. To this end, Unesco has promoted research, information and education concerning the environment. The MAB Programme is one of the most significant examples. Ten years after the launching of this programme, it would seem appropriate to examine what it represents and the contributions it has made, with specific reference to environmental education. The increasing demand, in many countries, for projects and experiments for the promotion of environmental education, calls for an organic transfer of the results of studies and the working methodologies developed by MAB to the context of school education, as well as that of lifelong education.

*Résumé*

*On a dit que la crise actuelle des rapports entre la communauté des hommes et l'environnement avait pour cause profonde le "réductionnisme" culturel de l'homme et son comportement vis-à-vis de l'environnement, à l'égard duquel il se montre étranger et dominateur. Il s'agit par conséquence d'une crise essentiellement culturelle. L'objectif de l'éducation relative à l'environnement est donc aussi bien de faire percevoir l'environnement comme un système vivant de relations, ayant son fonctionnement et son histoire propres, que d'inculquer à l'homme les motivations voulues pour qu'il se sente personnellement impliqué dans les processus de fonctionnement de l'environnement, afin que sa gestion soit correcte et témoigne d'un sens des responsabilités. A cet égard, l'Unesco a mené une action de promotion de la recherche, de l'information et de l'éducation relative à l'environnement. Le Programme MAB est l'un des exemples les plus significatifs de cette action. Dix ans après sa mise en route, il paraît opportun d'en analyser la nature et les résultats, en s'attachant tout particulièrement à l'éducation relative à l'environnement. Face à la demande croissante, dans de nombreux pays, de projets et de données d'expérience propres à promouvoir l'éducation relative à l'environnement, il semble que l'on devrait proposer un transfert bien structuré des résultats des études et des méthodes de travail mises au point par le MAB, surtout en faveur du monde de l'enseignement et de celui de l'éducation permanente.*

## CULTURE AND ENVIRONMENT

MAN'S attitude towards his environment is measured by the awareness he shows in the management of this relationship. This is the origin of culture, that is the system of concepts, motivations, instruments and methods whereby a human community expresses its relationship with its own context of life, and gives it form and identity.

Different cultures have come into being in time and space, each one distinguished by its own particular social, philosophical, economic, political, ethical, aesthetic and religious expressions. Various cultures have adopted different attitudes, for example, towards the relations between man and his environment. Western culture traditionally, more especially since industrial development, has been profoundly marked by two characteristics: first, an existentialist attitude, regarding man not as part of nature and the environment, but outside and above them, occupying a position of absolute mastery and domination; second, an extremely sectoral perception and management of the environment, taking little or no account of the systemic character of the environment itself. Modern technology has meant that attitudes deriving from these cultural premises have now spread to a large part of mankind, and have in turn dramatically accentuated the crisis of man's relations with his environment, which is a feature of our times.

In contrast, there are other cultures that have developed a harmonious man and environment relationship. The system of communal ownership of the land, found in African and some other cultures, had the meaning and merit of maintaining the balance between man and nature, and so ensuring the survival of both. Some of these cultures, despite having access to the Western way of life, have nevertheless adhered in some measure — varying from country to country — to their own particular concept of man's relations with his environment. The models elaborated by Western civilization, with its advanced technology, have however influenced all other cultural regions and are continuing increasingly to do so. This being so, we feel justified in basing our proposed analysis on the natural and human environment in the Western world, where the environmental crisis is most acute.

## BASES FOR A NEW RELATIONSHIP BETWEEN MAN AND THE ENVIRONMENT

### Increased awareness of the environment

Public awareness of the nature of the environment and of the growing environmental crisis is recent, and its extent varies from country to country. Generally speaking, recognition of the ecological crisis emerged in the late 1960s and early 1970s as a result of certain ecological disasters and the widespread deterioration of natural resources and of the general quality of life. Analysis of these environmental problems, many of them worldwide, was not at that stage based on scientific ecological research. Alerted by various writers and journalists about the problem, protest groups and environmental protection movements were formed, contributing in turn to awakening interest of the public in environmental isues. There was an increase in research on the environment, which first found effective expression in ICSU's International Biological Programme (1964-74). Ecology continued to develop its own identity as a scientific discipline, studying the dynamics of natural and human environmental systems in space and time, under normal and pathological conditions (Odum, 1977). The need for closer links between the natural and social sciences became more apparent.

As a result, the often purely emotional approach — which nevertheless had and still has a contribution to make towards the perception of the environment — has gradually given way to a more mature and scientific view of the environment. In this connection, a significant set of concepts on the environment have been developing. They include the following:

1. Every environmental system, whether natural or human, is unique, being the outcome of interrelations between the physical, chemical, biological and in some cases also the cultural factors of which the system is composed. This system is perceived by a living organism, a population or a community as a unitary set of conditions which directly influence its development and

activities, and to which the individual, population or community adapts itself, successfully or otherwise;

2. Each environment, like all other levels of biological organization, is a living system with basic vital processes — birth, growth, maturity and ageing — and can fall ill, recover or cease to exist as a functioning environment (Feibleman, 1955);
3. All environments, from the forest to the village, from the metropolis to the biosphere, are alike in possessing a structure, a function and a time-sequence;
4. Promotion of the quality of the environment involves estimation of its carrying capacity, moves towards economy and balance in man's use of natural resources, checking of environmental degradation, cautious managment of change, conservation and promotion of biological and cultural diversity;
5. Man is not above or outside nature, but part of it;
6. The dynamics of the human environment are governed by fundamental processes similar to those that characterize the natural environment;
7. Every human intervention, on whatever scale, exercises an influence, positive or negative, on environmental systems. The physical, chemical and biological mechanisms which nature has available for restoring the quality of the environment, and re-establishing balances that are in jeopardy, are limited both in quality and in quantity.

Full recognition of these basic aspects of the dynamics of the environment — in practical as well as in intellectual terms — would greatly contribute to the establishment of a more correct relationship between human communities and their related living conditions.

## Interdisciplinary methodology and systemic interpretation of the environment

At the root of the crisis in man-environment relations is chiefly a lack of real understanding either of the profound interactions between water, air, soil, plants, animals and the human community, or of the fact that the birth and growth of the human environment, like the natural environment, is based on "relations", and that its quality depends on the "quality of the relations" which man, in the course of time, establishes with the natural environment and with his fellows (Lindeman, 1941; Schultz, 1967). This lack of understanding has led to a sectoral, restrictive idea of the environment, the consequences of which are well known.

Compared to developments in specialized disciplines, there is a lack of means for identifying the links between the various factors and processes which characterize the dynamic nature of the environment. In other words, the analysis of an overall situation has been reduced to the study of its respective parts, on the assumption that the whole has the same characteristics as the sum of its parts. But since the separate data have not been related to the appropriate system of interactions, the true nature of environmental reality has been rendered incomprehensible. The main responsibility for this disastrous cultural approach lies with schools of all types and at all levels, from the kindergarten to the university.

An interdisciplinary approach, by definition dynamic, would appear to provide the most appropriate means for: firstly, fitting specialized data on separate aspects of the environment into a system of relations constituting the environment as a whole (di Castri, 1978; Moroni, 1978; Piaget, 1972; Rowe, 1951); and secondly, discarding the use of faulty sectoral methodology for the investigation of environmental systems as a method on which to base action in respect to environmental issues and natural resources management. A restrictive approach to the environment forms the basis of many of the political, socio-economic and ethical options of the present day, and it is clear that the adoption of a systemic approach and of interdisciplinary methods of work constitutes a challenge to deeply rooted cultural systems as well as to important interests of various kinds.

This is why it is important that environmental education, both in schools and within the context of lifelong education, should promote scientific knowledge of the environment, designed to facilitate the transition from a narrow sectoral approach to a systemic interpretation. In a similar way, environmental education should encourage people to show greater respect for natural balances, and to take active part in measures for improving man's relations with the environment and in developing human environments (towns, districts, places of work, educational institutions) of which individuals and communities are both promoters and users.

## The human environment

The human environment is born of the continuing interaction between the natural environment, the biological characteristics of human populations, and the social, economic, political, aesthetic, philosophical, ethical and religious expressions of human culture, which are in constant evolution. Whereas plant and animal organisms contribute deterministically to the equilibrium of the environment, man — because his relations with the environment are established consciously — bears the responsibility for improving the quality of the environment or determining its gradual deterioration.

The objectives of environmental education in respect to the human environment are twofold: first, to promote valid scientific knowledge of the structure, functioning and history of the human environmental system; and second, to make a critical reassessment of man's relations with nature, with a view to identifying a more coherent ethical approach. Structurally, the human ecosystem is made up of physical, chemical, biological and cultural components and factors. Dynamically, the situations and processes occurring in the human environment, are similar to those in the natural environment, for example with regard to ecological niche, limiting factors, carrying capacity, nutrient cycling, energy flow, succession and climax, relationship between diversity and stability, mechanisms of environmental homeostasis. The only difference is that man today intervenes deliberately to some extent in the functioning of his own environment, transforms these factors into so many problems which politicians, sociologists, planners, scientists and even whole communities are called upon to discuss and solve.

Studying the functioning of the human environment over a period of time constitutes a third contribution to a better understanding of the true identity of a village, a city or a country. This is particularly important at the present stage of cultural transition (the phase now reached by a large part of mankind), when scholars of various disciplines are accusing the world of having neither historical memory nor a sense of responsibility for the future. Environmental education, in this connection, stresses the need for reclaiming the positive, valid factors of the past and integrating them in initiatives for cultural development for the future.

In regard to the past, it is certainly not proposed either to restore a "historical framework" — characterized, in many cases, by shortages, a short life-span, epidemics, a high rate of infant mortality, etc. — or to concentrate solely on folklore, works of art, the humble implements of everyday life, songs, dialects or poetry. On the contrary, the aim should be to examine and reconstruct all those elements which contributed, in the past, to making those communities human, and intensifying their pleasure in living together, thanks to active involvement in the main events of human existence and the organization of community life.

It is especially important to give thought to the future. Indeed, planning for the future is a relatively recent feature in the management of the human environment, and one to which projects for land use planning still pay little heed. Reference might be made to Descartes, who in his *Discours de la méthode* wrote that our attention should extend beyond the present. It is well to set aside things designed to bring profit to the living in order to lay the foundations of other things designed to benefit posterity.

## Relations between man and the environment

There is sometimes a tendency to take an oversimplified view of man's presence in the environment, and to represent nature as a harmonious, uncontaminated scene wherein man, directly he appears, proceeds to destroy animals and plants, to upset the natural balances, to set off processes of environmental degradation, and so on. But in reality the earth, even before the appearance of man, had been the scene of immeasurable disasters, upsetting the existing order. And moreover, not all the results of man's presence are negative. Art, poetry and music, the organization of the community, the sharing of the joys and sorrows of others, friendship and love, the fight against scourges, such as famine, plagues and penury, which have for thousands of years decimated the world's population, are all part of a process of progress which has resulted in enriching and adding new aspects to the environment in which man rightfully lives.

At the same time, however, man has become increasingly estranged from the realities of his environmental system: the air, the soil, plants and animals, natural elements and, increasingly, also his fellow-men, themselves transformed into a means for extending his own power and dominion (Black, 1970). Instead of being an integral part of nature, man has come to exercise absolute dominion over it, causing its material degradation; he has become the key factor of the ecosystems in which he lives, developing his capacity to make increasingly serious and rapid changes in the major processes governing the life of the planet. This has led to conflict not only between human communities and nature, but also between economics and ecology, science and technology, quantity and quality.

Opposed to this, environmental education aims to awaken man's awareness of his relations with the environment, and of his personal involvement in the web of relations that constitutes his surroundings and has its own independent values, not necessarily dictated by nature alone (Koslovsky, 1974; Leopold, 1949). This concept of man's relations with nature and of the management of the human environment represents a further stage in the study of ecology, covering problems not only of conservation but also of social and economic development; or, as Dasmann has put it, development with due regard to nature and conservation with due regard to man.

The study and implementation of this concept is probably one of the most important tasks of environmental education in the next twenty years. It is a case of education for authenticity, in which intention and freedom are not necessarily synonymous with domination over nature or exploitation of natural resources, but rather with the realization that man forms part of nature and has a responsibility for the correct management of the human and natural environment. By its very nature, this task concerns formal education first and foremost, but also the vast field of lifelong education. Human communities are the laboratories for testing these new cultural trends — human communities which have a critical, purposeful experience of their own daily problems, but which are imbued with social, economic and political ideas devised elsewhere and in many cases alien to their own cultural dynamics. A further reason why particular importance attaches to the community is because it is now realized that the individual approach to the environment must give way to a community-oriented approach.

## PROMOTION OF ENVIRONMENTAL RESEARCH AND TRAINING BY UNESCO AND THE MAB PROGRAMME

### Priority goals

It is clear from analysis of the present state of the natural and human environment, and of likely developments over the next twenty years, that both individual countries and the international community as a whole will have to give urgent consideration to the crisis in the relations between man and his environment. It is now generally agreed that this is essentially a cultural crisis. Thus although, in the short term, technological intervention will be necessary, the problem can only really be solved by ethical, social and political measures designed to foster a new type of relationship between human beings and to inculcate respect for the environment. The institutions involved in this process are many and varied. The question increasingly asked is what part scientific research and the human sciences can play in this matter. As a response, the three following priority goals may be proposed for the 1980s:

1. To organize high-quality scientific research, both fundamental and applied, on the natural and human environment, using systems approaches and interdisciplinary methodologies and adopting a problem-based and community-oriented focus;

2. To identify more effective methods for making the research results of natural and social sciences accessible to planners, environmental specialists, decision-makers in general, schools and communities, with the twofold aim of ensuring continuity of policy and advancement in methodologies in the management of resources, and promoting effective dissemination of reliable scientific information;

3. To promote, among the people at large, attitudes and behaviour so as to ensure continuation of the interest in the environment now displayed by a growing number of citizens.

In this connection, the United Nations has taken effective measures for the promotion of environmental research, dissemination of ecological information and environmental education. Through the United Nations Conference on the Human Environment (Stockholm, 1972) and other subsequent conferences, it has disseminated information and developed guidelines on the main problems relating to the natural and human environment. Unesco, through the Intergovernmental Conference on Environmental Education (Tbilisi, USSR, 1977) claimed its right to play a leading role in environmental education. The results of this conference, in fact, define the objectives, contents and methodology of environmental education for the next twenty years. And through its MAB Programme, Unesco has sought to bridge the gap between research, training and information activities in developing "the basis within the natural and social sciences for the rational use and conservation of the resources of the biosphere and for the improvement of the global relationship between man and the environment; to predict the consequences of today's actions on tomorrow's world and thereby to increase man's ability to manage efficiently the natural resources of the biosphere" (Unesco, 1972).

After ten years of MAB's existence, it would seem timely to examine to what extent and with what degree of success the MAB Programme has contributed to the attainment of the three goals set forth above. It is thought that this also might help in formulating a strategy for the future development of this programme, particularly in regard to environmental education.

## Scientific research on the environment in the MAB Programme

It might be useful to recapitulate the context within which the MAB Programme has evolved and the stages through which it has passed between 1971 and the present day.

In the 1970s, in the face of the accentuation of environmental problems, it was realized that the results of much scientific research on ecological problems were of little use for the analysis and management of environmental systems. Di Castri and Hadley (1979) proposed three major reasons why only a small proportion of available data is applied in solving practical problems of land use in areas where these problems are most acute — namely in countries of the tropical and subtropical zones.

First, existing knowledge about different geographical areas is unevenly spread (concentrated mainly in the industrialized countries) and information is incomplete in many technical fields. Second, many sound data are not applied, either because there is a lack of political will or of adequate administrative machinery in countries, or because data are not widely available or easily accessible, or because data are not presented at a time and in a form that can be readily used by planners and decision-makers. Third, the very nature of many data is such that they are just not applicable or that, if they are applied, they will lead to unpredictable or adverse effects.

The MAB Programme was initiated in this cultural context and has attempted to promote and develop research which responds to such shortcomings in existing information. The approach adopted within the programme has included analysis of the structure and functioning of ecological systems and their reaction to human intervention, and was designed to yield reliable scientific data for rational management of the resources of the biosphere. Within MAB, emphasis has been given to research encouraging studies in regions where data were scarce or unavailable, and to making research an integral part of programmes for training local researchers and of development projects. By paying due attention to the solution of local environmental problems, MAB avoided making the mistake — all too dramatically evident in some other environmental research programmes — of transferring environmental management projects and techniques designed for one area mechanically to another area with different climatic, biological and cultural conditions.

While the ecosystem (in the sense of holistic) approach has been a constant feature of MAB, the use of ecosystems as the principle units of research has given way to one based on human use systems, defined as systems through which and by means of which resources are managed (e.g. a delta with its interfaces between the dry land and fresh, brackish and salt water; an island, containing both land and aquatic environments, etc.). This shift in emphasis reflected the increasing recognition that human uses of the environment rarely coincide with ecosystems, and that human use systems provide a suitable methodological and spatial unit for integration of the work of the natural and social scientists and decision-makers. It is certainly an appropriate choice from the standpoint of the management of resources, and it represents an innovation as regards the scientific analysis of the environment, to which due importance should be accorded.

The organization of scientific research by problems — with the problems identified jointly by researchers, land use planners and local populations — has led to fruitful ties between the disciplines of the natural and human sciences, between fundamental and applied sciences, between ecological principles and economic development. Such links and interfaces have been important in the consequent development of MAB pilot projects of research and training, focussing on problems of national importance but also having regional and international significance. In this way, the MAB Programme has tried to respond to the demand of various countries, with different socio-economic and cultural conditions, for adequate support to research designed to underpin the use of their natural resources. At the same time, the programme has sought to develop the logistic bases, infrastructures and scientific capabilities required for long-term research on present or future problems.

The international character of research, training and information exchange within MAB helps ensure genuine cooperation, comparison of findings and realistic assessment of the transferability of results from one country to another. The systems approach and interdisciplinary methodology adopted in MAB as basic criteria for the planning and execution of research provide a coherent framework for scientific research on the environment.

It is quite true that this approach may present difficulties both for researchers involved in MAB field projects and for outsiders, and that these difficulties may well have delayed the execution of certain projects or held up progress towards interdisciplinarity. But this will not disconcert those who have really grasped the fact that the objectives and methodology of the MAB Programme bring new weight and a sense of innovation to natural and social science studies on the environment.

In conclusion, critical analysis of the achievements of the MAB Programme to date indicates that this is a noteworthy attempt to use coherently advanced scientific approaches and methods for the analysis of the environment — approaches and methods that even the academic world, for a variety of reasons, still finds difficulty in applying. In addition, the authors are convinced that the sorts of approaches to environmental research being tried out within MAB could have a positive impact on the development of ecological research as a whole.

## Transfer of the scientific findings of the MAB Programme

The MAB Programme has set itself the specific aim of developing research projects and devising research and operational methods both applicable to the problems of specific countries and transferable to disinterested users. These results are naturally designed for researchers, administrators, planners and technicians responsible for the management of natural resources as well as for the local communities concerned.

The transfer of results provides in itself an invaluable occasion for adult environmental education as well as an opportunity for systematic research in the environmental education field. It should also be possible to collect and publish case studies based on experience within MAB, on the ways whereby ordinary people, research workers and the users of research results have succeeded, through participation in the various phases of the research process, in scientific projects aimed at improving the quality of their local environment. The use of the findings and facilities of the MAB Programme in the training of environmental researchers and local technicians is intended to ensure the satisfactory execution of the planned research projects but also to provide continuity to scientific research as such. In the 1980s, it will also be necessary to ensure that, despite the increasing diversity of positions and cultures in the different countries, efforts geared to the transfer of research results and the training of specialists should be such as to guarantee the continuation of contents and methods which have already proved their worth, and which give the MAB Programme its special character (di Castri, 1979).

## Environmental education in the MAB Programme

Environmental education is a factor of the culture of human communities arising from reflection on the structure, functioning and history of the environmental system and from the close interconnection between the natural and human environment in the context of the stage of cultural development of a particular country and, more generally, of the international community as a whole. Its general aim is to integrate information on the environment into the educational process, whether at school or in life long education.

More specifically, the Intergovernmental Conference on Environmental Education organized by Unesco in cooperation with UNEP in Tbilisi, USSR, in 1977, defined the objectives of environmental education in the following terms (Unesco, 1978a). Environmental education is the result of the reorientation and dovetailing of different disciplines and educational experiences which facilitate an integrated perception of the problems of the environment, enabling more rational action, capable of meeting social needs, to be taken. A basic aim of environmental education is to succeed in making individuals and communities understand the complex nature of the natural and the built environments resulting from the interaction of their biological, physical, social, economic and cultural aspects, and acquire the knowledge, values, attitudes, and practical skills to participate in a responsible and effective way in anticipating and solving environmental problems, and the management of the quality of the environment. A further basic aim of environmental education is clearly to show the economic, political and ecological interdependence of the modern world, in which decisions and actions by the different countries can have international repercussions. Environmental education should, in this regard, help to develop a sense of responsibility and solidarity among countries and regions as the foundation for a new international order which will guarantee the conservation and improvement of the environment.

Since the Tbilisi Conference, the major thrust of Unesco's activities in the environmental education field has been the responsibility of a special unit within the Education Sector of the organization. The contribution of MAB in Unesco's total effort in this field has been a relatively minor one, even though environmental education has been a part of MAB since its inception. There has been an interesting evolution in the role of environmental education within MAB, as the programme has developed. The International Co-ordinating Council of the Programme on Man and the Biosphere, at its first session in Paris in 1971 (Unesco, 1972) recognized the need for studying this question, and indicated several possible areas for action. Subsequently, an expert panel was convened in 1972 to examine the form and content of educational and training activities that might be promoted under the MAB Programme (Unesco, 1973). The panel dealt mainly with the subject of training specialists to take part in MAB research projects. The panel felt that MAB national committees should be prepared to collaborate, as regards environmental education, with the relevant national bodies responsible for restructuring teaching curricula. But the panel also considered that the programme could participate only indirectly in measures for popularizing and disseminating information on environmental education and promoting new teaching approaches at the primary and secondary teaching levels. These aspects were considered to be the responsibility at the national level of national education authorities, in actively introducing and strengthening environmental ideas and disciplines, and at the international level of groups or organizations — including Unesco — dealing with education.

Some years later, in 1977, the Tbilisi Conference on Environmental Education recommended that the research results of the MAB Programme be turned to account for projects and activities designed to promote environmental education. And again in 1977, the MAB Council at its fifth session in Vienna (Unesco, 1978b) assigned to MAB the responsibility for making use of the research structures of MAB projects in the various countries for environmental education purposes, and for utilizing the research done under the programme in developing appropriate educational materials.

Even more recently, at its twenty-first session held in Belgrade in 1980, the General Conference of Unesco emphasized the need to consider in the further development of MAB, the idea of a more explicit contribution to deepening and strengthening the environmental education of all people. The General Conference also considered the need to help countries take advantage of the results of the MAB Programme for their development activities and for the preparation of teaching materials suitable for the different levels of education.

As suggested by this historical review, the whole MAB operation has potential as a meaningful example, at the international scale, of environmental education as part of lifelong education for adults. The MAB Programme has, however, given only marginal attention to the world of formal education — from nursery school to university. And yet, on reflection, certain difficulties besetting man in his cultural approach to the environment, and some of the same difficulties that the MAB

Programme has experienced in implementing various projects, owe their origin to a mistakenly academic approach to environmental education. This fact, and the growing demand for guidelines and examples for the development of environmental education at school are prompting those responsible for the MAB Programme to give careful reconsideration to the issues of the transfer of tested methodologies and the results thereby obtained.

The MAB Programme has three features which correspond quite closely to the conditions that any long-term operation to promote environmental education must fulfil. First, it has a store of experience at an advanced level concerning the objectives, scientific content and methodology needed for the study of the environment and the management of environmental resources. Second, it is firmly international, and is thus a meeting place and focal point for contributions from various kinds of culture and experience in both developed and developing countries. Third, through the structure of national committees, the general principles of MAB actions are effectively linked with the identity and the socio-political and economic circumstances of the various countries.

These observations give rise to a twofold set of requirements: that in the 1980s the MAB Programme puts transfer activities among its primary objectives; that similar interest be accorded to school education and to specific, important but hitherto neglected sectors of lifelong education. With these requirements in mind, the following objectives are suggested: reinforcement of the natural links between MAB and the general environmental education programme of Unesco; increased contributions from MAB to improving environmental education in schools and in the context of lifelong education; strengthened role of MAB national committees in promoting environmental education.

## Development of relations between MAB projects and environmental education

The contributions that MAB research projects can make to environmental education are of various kinds. The wealth of scientific, methodological and research policy experience that international and national projects produce extends beyond the straightforward use of that experience in natural resources management. If that experience were made available in suitable forms, it could prove extremely useful in both school and lifelong education.

This is a new area of interest within the purview of MAB, where it might be worthwhile to consider launching a MAB initiative to study the methods and technologies used in practice to transfer scientific findings in general, and MAB research methodologies in particular, to the formal and non-formal education sector. A second suggested initiative might focus on the theme of relations between man and the environment in various cultures, and would concern the acquisition of less piecemeal and more critical documentation on the cultural causes of the ecological crisis. This is a theme whose importance has taken it beyond the confines of academic research; it is treated in the press, in school textbooks and reference documents, and needs reliable cultural support.

It is further recommended that increased emphasis be given to MAB Project Area 13 ("Perception of environmental quality"), with a view to determining appropriate indicators of environmental perception. This is not an easy objective, as may be inferred, for example, from the different types of environmental perception in towns compared with rural areas, and from the ways environmental deterioration is perceived in countries which differ in culture and degrees of development.

## Experiments in environmental education in the context of formal education

In a culture in transition such as our own, the school, and above all the primary school, is of fundamental importance. In the throes of a fairly widespread crisis concerning the ability and willingness of the family and of other educational institutions to provide some kind of training, increasingly complex responsibilities have devolved upon the school. The school (to varying degrees, according to age-group) is today required to meet two main functions. First, it must make provision for specialization according to discipline, in order to meet among others the demand for specialized technicians and researchers. Second, it must encourage the pupil to identify and relate the interconnections between the physical, biological and cultural factors that make up the environmental system, as a basis for the pupil's own fruitful coexistence with nature and society. Otherwise, the school will lose touch with reality, will fail to develop the innovative culture that a changing society like ours requires, and will not foster the ability to participate constructively in the daily life of the community.

Three objectives can be suggested for all institutions concerned with the environment, with education and with cultural development:

1. Identification of a number of basic types of subject matter and sound methods that pupils can use as a key to understanding their environment, and acquisition by pupils of suitable skills and abilities for the purposeful management of their surroundings;
2. In-service training for teachers in how to plan a syllabus based on a problem-oriented approach and geared to the life of the community in which the school operates, but without neglecting relations with the country in question or with the international community;
3. Establishment of links between the school and other information media operating outside it.

It is easy to detect a marked resemblance between the objectives and methods of the MAB Programme and the proposed objectives of environmental education at the school level. In order to determine what the responsibility of the MAB Programme might be in this sector, it is important to recognize the increasing gap that exists between the level and quality of knowledge generated by scientific research and the scarcity of environmental education at present. The gap is obvious if one analyses the objectives, content and methods of school syllabuses and textbooks. The findings of scientific research sometimes reach educational circles very late, filtered through a whole range of intermediaries, so that the end-product lacks scientific rigour and is not expressed in terms suitable for teaching purposes.

From the methodological point of view, environmental education is usually confined to the science side of the curriculum. One consequence is that the pupil is not taught to interpret the environment in its entirety, both natural and human at the same time, and there in fact seems to be a renewed split between the two cultures, the humanities and science, which began as an intellectual debate and is now affecting a wide variety of attitudes and behaviour, in social, educational, political and other spheres. Education is still arranged by disciplines that are more or less strictly compartmentalized, and pupils are not taught at the same time how the various concepts are organized in relation to processes, and so fail to learn correctly how to use interdisciplinarity approaches and methodologies. School textbooks, therefore, tend to be either based on outdated ideas (which no longer have anything to do with the content or the methods of ecological research) or are merely translations of textbooks compiled in other countries and accordingly bear the stamp of the cultures and politics peculiar to those countries.

An effort must be made to make up for lost time by diligently seeking solutions to a situation that is not easy and is certainly not conducive to effective environmental education. From a strategic viewpoint, it is above all urgent to develop educational research with scientific research on the environment, in order to incorporate in curricula the content and findings of advanced research with scientific research on the environment, in order to incorporate in curricula the content and findings of advanced research in both the natural and the human sciences. Educational research, without a doubt, finds its natural level in interfaculty centres, university departments of education and various national research councils, for which MAB national committees could serve as a sort of clearing house. Teachers with practical experience of teaching in all types of school, and scientists specializing in a particular sector of research, should work together in special centres such as these.

A second strategic choice concerns teachers in all kinds of schools at all levels. Educational research findings concerning content and methods should be incorporated in textbooks and in-service training courses for teachers.

A third objective is to embark, with the most advanced teachers, on wide-ranging experiments in keeping with the cultural dynamics of the individual countries.

In considering the feasibility of these objectives, it becomes clear that a lead needs to be taken by the university and research sector. In this respect, MAB could promote experimentation with different forms of environmental education for various types of school and specific educational structures, could assess the results of such experiments and make the experience gained available to different countries.

Of the strategic choices to be made, the most important concerns the pre-service or in-service training of teachers. In this context, centres for specialized post-graduate training are of particular importance in providing training for teacher-educators in the environmental education sector. A further possiblity, in the light of experience within MAB of research and transfer of environmental data, would be to examine the objectives and structures of experimental centres for field work (in both natural and urban environment) and interfaculty or interdepartmental university centres for environmental education. These centres originated in the universities, but they are also specifically designed for promoting effective interdisciplinary cooperation. They are especially well placed to demonstrate that environmental education is the province not only of universities or science departments, but also of faculties of law, medicine, economics and commerce, veterinary science, engineering, town planning and human sciences.

Parallel with such activities, the MAB Secretariat and national committees should develop an effective system of dissemination, through educational networks of the results of MAB research projects presented in various forms such as audio-visual programmes (e.g. the slide-tape programme on "Man and the Humid Tropics" (Unesco, 1979)) and posters (e.g. the 36 poster exhibit "Ecology in Action" (di Castri *et al.*, 1982; Unesco, 1981)). MAB national committees could also carry out methodological experiments involving teachers and students from both science and arts faculties, and from technical and vocational areas of study, in their research and transfer activities, by encouraging them to participate in the identification and solution of a number of environmental problems typical of their surroundings. The MAB Programme should not itself be directly responsible for teacher training or environmental education for schoolchildren, but could nevertheless conduct experiments that are valid for both, and test and diffuse the results of such experiments for the benefit of the various countries. In short, the MAB Programme should have a multiplier effect.

In conclusion, we can say that in the ten years of existence the MAB Programme has established a firm base in countries with the setting up of national committees; it has conducted valid scientific research on the environment, linked with training on environmental problems for researchers, technicians and decision-makers. But the programme has not included a specific interest in the school world as such. It is unnecessary to inquire into the reasons for this. It need only be emphasized once more that in the 1980s it is both desirable and necessary that the MAB Programme should — alone or in collaboration with other bodies — seek links with the school on issues concerning management of the natural and human environment.

## Guidelines for lifelong environmental education

Increasing populations and growing demands for improved quality of life are often a heavy burden for the natural and human environment, despite the availability of ever more sophisticated and widely available technology. Meanwhile, in many countries, legislation is granting increased power over the environment to local administrators. Hence the desirability of developing lifelong education for administrators, planners, industrialists and trade unionists, magistrates and others, who have a major part to play in problems of environmental management. Though the difficulties involved are perfectly clear, approaches such as that of environmental impact assessment (Carter, 1977) can provide the basis for a practical environmental training for decision-makers.

Environmental education at the community level is another area deserving special attention. An increasingly profound and critical awareness on the part of the individual of his own natural and human environment, or of the positive meaning of conservation and the development of his own cultural identity, is certain to contribute greatly to responsible community action to enhance environmental quality. This applies both to developing countries and to many social groupings in industrialized countries, reduced almost to the status of robots, with no voice in decision-making, in built-environments where everything is planned and predetermined. The objectives of environmental education in the sector of lifelong education accordingly encompass issues that are exceptionally broad and urgent from a social point of view.

One need only think of the consequences of passive acceptance of technologies by a given country, given that technologies are inevitably the carriers of the culture and lifestyle of the exporting country. The issues of cultural diversity, of salvaging what is valuable from the past and of forecasting the future, must receive immediate attention from any community that wishes to protect and develop its identity in these days of rapid cultural change. And indeed, this is the right way for a community to improve the quality of its life and to contribute, by expressing its own cultural diversity, to the stability of a country and of the international community in general.

Progress towards this broad set of objectives will depend primarily on meticulous work in environmental education. The MAB Programme, by basing its work on national scientific programmes and using appropriate transfer methodologies, has acted in the right direction and has broken away from models for technology transfer that are more or less deterministic. In addition to obtaining results from scientific research on the environment, the MAB Programme encourages the training of specialists to gather and manage those results and enlists the participation of the community in research and management. But a further effort to educate all age groups to adopt a responsible attitude towards their own environment remains a matter of some urgency. The strategy of this kind of mass communication concerns, mainly, the schools for young people and the mass media for both young people and adults.

A third focus for environmental education concerns non-governmental associations and various kinds of environmentalist and other interest groups. In many countries, such groups have had an important influence in arousing the community's interest in environmental problems. Unesco has concerned itself directly with these groups, *inter alia,* at the Intergovernmental Conference on Environmental Education held in Tbilisi in 1977 and at such events as the national symposium on "The Environment and the Role of Non-Governmental Organizations" held in Dacca (Bangladesh) in 1981. The MAB Programme has acquired some experience of collaborating with such groups, especially in certain countries.

In many cases, such groups would welcome greater access to scientific information on the natural and human environment. Links could also be strengthened between these associations and the administrative authorities of particular areas, with a view to enlisting the involvement of qualified voluntary workers in certain types of community service (e.g. as ecological monitors, curators of museums of local culture, etc.). There would also appear merit in collecting and studying the experience of these associations in heightening environmental awareness in various communities.

### MAB national committees and environmental education

Analysis of the MAB Programme's objectives and working methods with regard to environmental education clearly reveals the strategic importance of the activities carried out by the MAB national committees. These national committees are responsible for studying and adapting Unesco guidelines for the programme to socio-political and economic circumstances at the national level. Effective transfer of the results of MAB's international and national activities to local planners, administrators and communities must take place through a mediator, to wit, an efficient national committee. One of the most important objectives of the MAB Programme for the 1980s is to strengthen the MAB national committees, so that they become active centres which serve their respective national communities by transferring to them the results and methodologies of the MAB Programme, devising whole series of proposals to suit the cultural context of individual countries, and providing an expert environmental service.

## CONCLUSIONS

"The greatest spiritual revolutionary in Western history, Saint Francis, proposed what he thought was an alternative Christian view of nature and man's relation to it: he tried to substitute the idea of the equality of all creatures, including man, for the idea of man's limitless rule of creation. . . . With him the ant is no longer simply a homily for the lazy, flames a sign of the thrust of the soul towards union with God; now they are Brother Ant and Sister Fire, praising the Creator in their own way as Brother Man does in his" (White, 1967).

If man is to achieve an authentically dynamic relationship with the environment, he must advance beyond such concepts as "absolute mastery" over the environment and "estrangement from the workings of nature". This transition can be made, in the context of a culture that is broadly innovative compared with the present, if man chooses to develop a partnership with all the physical, chemical, biological and cultural aspects that make up the real world environment and if he manages these realities in an enlightened way. Another ingredient to achieving such a relationship is for man to recognize his personal role and involvement in how the environment works. Man is not above or outside nature, but part of the network of relationships that make up his environment.

The MAB Programme's importance is that it offers the international community objectives and methods for the interpretation and management of the environment in a scientifically advanced way, while respecting the identity of individual cultures and patterns of social behaviour in various countries. At the time, it encourages a frame of mind that seeks to improve the quality of life. But there is another, less obvious but certainly no less important, contribution that the Unesco's MAB Programme is making to a better relationship between mankind and the environment. Di Castri (1981) writes as follows:

> The Programme's point of departure was the study of man from the "outside", i.e. the impact of human activity on the various ecosystems (tropical and temperate forests, savannas, prairies, tundras, lakes and rivers, mountains and islands). Subsequently, in a growing number of research projects, man came to be considered as an integral part of the ecosystem and of the biosphere, becoming in fact the central element of study.
>
> The changeover from "man the outsider" to "man the insider" is not merely a play of words. It marks a real revolution in concepts and particularly in methods, for ecology has begun to take into consideration the intangible and non-quantifiable elements of human activity and thought — the different perceptions which populations and individuals have of development and of the quality of life, their aspirations and their feelings of belonging and of accomplishment.

These words convey the significance and the promise of a programme in which ambition gives place to realistic day-to-day verification, step by step, of the logical consistency of objectives and methods; a venture that calls upon every individual, to some extent, at one time or other, to view his relationship with the environment as the bond between parent and child.

## References


Bertalanffy, L. (1968), *General System Theory,* Baziller, New York.

Black, J. (1970), *The Dominion of Man,* Edinburgh University Press, Edinburgh.

Burck, W. R. Jr. and F. H. Forman (Editors) (1975), *Beyond Growth: Essays on Alternative Futures,* Yale University, New Haven.

Canter, L. (1977), *Environmental Impact Assessment,* McGraw Hill, New York.

Castri F. di (1976), International Interdisciplinary Research in Ecology: Some Problems of Organization and Execution. The Case of the MAB Programme, *Human Ecology,* 4 (3), 235-46.

Castri, F. di (1978), Planning International Interdisciplinary Research, *Science and Public Policy*, 5 (4), 254-66.

Castri, F. di (1979), Ricerca e Formazione in Ecologia per lo Sviluppo del Terzo Mondo, *Ann. Accad, Ital. Sc. Forest.*, 23, 19-31.

Castri, F. di (1981), Ecology: the Genesis of a Science of Man and Nature, *The Unesco Courier*, 34 (4), 6-11.

Castri, F. di and M. Hadley (1978), Ecological Approaches to Land Resources Management in the Tropics: Some Case Studies from the Man and the Biosphere (MAB) Programme, In J. S. Singh and B. Gopal (Editors), *Glimpses of Ecology* (Professor R. Misra Commemoration Volume), International Scientific Publishers, Jaipur, 569-86.

Castri, F. di and M. Hadley (1979), A Typology of Scientific Bottlenecks to Natural Resources Development, *GeoJournal*, 3.6, 513-22.

Castri, F. di and M. Hadley (1980), Ecological Training for Developing Countries: Strategies and Actions of the MAB Programme, In T. S. Bakshi and Z. Naveh (Editors), *Environmental Education: Principles, Methods and Applications*, Plenum Press, New York, 57-75.

Castri, F. di, M. Hadley and J. Damlamian (1980), The Ecology of an International Scientific Project: Insights from the Man and the Biosphere Programme, *Impact of Science on Society*, 30 (4), 247-60.

Castri, F. di, M. Hadley and J. Damlamian (1982), Ecology in Action — an Exhibit: an Experiment in Communicating Scientific Information, *Nature and Resources*, 18 (2), 10-17.

Feibleman, J. K. (1955), Theory of Integrative Levels. *Brit. J. Phil. Sc.*, 5, 59-66.

Johnson, P. L. (Editor) (1977), *An Ecosystem Paradigm for Ecology*, Oak Ridge Associated Universities, Oak Ridge.

Koestler, A. and J. R. Smythies, J. R. (1969), *Beyond Reductionism*, Hutchinson, London.

Koslovsky, D. H. (1974), *On Ecological and Evolutionary Ethics*, Prentice Hall, London.

Leopold, A. (1949), *The Land Ethics, A Sand Country Almanac*, Oxford University Press, Oxford, 402-15.

Lindeman, E. C. (1941), Ecology. An Instrument for the Integration of Science and Philosophy, *Ecol. Monogr.*, 10, 367-72.

Moroni, A. (1978), Interdisciplinarity and Environmental Education, *Prospects*, 7 (4), 480-94.

Odum, E. P. (1977), The Emergence of Ecology as a New Integrative Discipline, *Science*, 195, 1289-93.

Piaget, J. (1972), The Epistemology of Interdisciplinary Relationship, In *Interdisciplinarity, Problems of Teaching and Research in Universities*, CERI (OECD), Paris, 127-39.

Rowe, J. S. (1951), The Level of Integration Concept and Ecology, *Ecology*, 42, 420-7.

Schultz, A. M. (1967), The Ecosystem as a Conceptual Tool in the Management of Natural Resources, In S. W. Ciriacy Wantrup and J. J. Parsons (Editors), *Natural Resources: Quality and Quantity*, University of California Press, Berkeley, 139-81.

Tansley, A. G. (1935), The Use and Abuse of Vegetational Concept and Terms, *Ecology*, 16, 284-307.

Unesco (1972), *International Coordinating Council of the Programme on Man and the Biosphere (MAB)*, First Session, Final Report, MAB Report Series No. 1, Unesco, Paris.

Unesco (1973), *Expert Panel on Educational Activities under the Man and the Biosphere Programme (MAB),* Final Report, MAB Report Series No. 7, Unesco, Paris.

Unesco (1976), *Moving towards Change: Some Thoughts on the New International Economic Order,* Unesco, Paris.

Unesco (1978a), *Intergovernmental Conference on Environmental Education,* Tbilisi (USSR), October 1977, Final Report, Unesco, Paris.

Unesco (1978b), *International Coordinating Council of the Programme on Man and the Biosphere (MAB),* Fifth Session, Final Report, MAB Report Series No. 46, Unesco, Paris.

Unesco (1979), *Man and the Humid Tropics,* MAB Audio-Visual Series 1, Unesco, Paris.

Unesco (1981), *Ecology in Action: An Exhibit. Thumbnail Sketches of 36 Posters,* Unesco, Paris.

White, L. Jr. (1967), The Historical Roots of our Ecological Crisis, *Science,* 155, 1203-7.

CHAPTER 13

# The Ecology of Scientific Information: Telling what we Know, Asking what we Need

*Bruno Lefèvre*

Abstract

On the occasion of the tenth anniversary of the beginning of the MAB Programme, a special exhibit was mounted to provide information on the programme. In 36 posters, the exhibit provided insights to the approaches and results of MAB, using an integrated mixture of text, photographs and drawings in both black and white and colour. A group of specialists in communications and exhibition display interacted with the MAB Secretariat to prepare a unique display which has been produced in multiple copies for easy transport and use in all parts of the world. An account is given of the evolution of thinking about the exhibit during its preparation and production. Some of the principles underlying the design of the exhibit are described, as well as some of the initial results of attempts to evaluate the effectiveness of such exhibits as a communications tool and to assess how they could be further developed.

*Résumé*

*Une exposition a été montée spécialement pour marquer le dixième anniversaire du début du Programme sur l'Homme et la Biosphère dans le but d'informer le public sur celui-ci. Composée de 36 affiches, l'exposition fournit des aperçus sur les approches et les résultats obtenus dans le cadre du MAB par l'utilisation d'un ensemble de textes, photographies et dessins tant en noir et blanc qu'en couleurs. Un groupe de spécialistes de la communication et de la préparation d'expositions s'est joint au Secrétariat du MAB pour monter une exposition qui ensuite a été reproduite en multiples exemplaires afin d'en faciliter le transport et l'utilisation dans le monde entier. L'article décrit l'évolution de la réflexion sur l'exposition au cours de sa préparation et de sa production. Quelques uns des principes qui ont déterminé la forme de l'exposition sont exposés ainsi que quelques premiers résultats des essais d'évaluation de l'efficacité de telles expositions comme moyens de communication et la façon dont elles pourraient être complétées.*

## INTRODUCTION

THE Man and Biosphere (MAB) Programme of Unesco celebrated its tenth anniversary in September 1981. To mark the occasion, a conference and a specially mounted exhibit were organized in order to present some of the main scientific results of the programme. The exhibit itself was in several respects an exceptional event. The approach adopted made it possible both to present the scientific information in an original way, and at the same time to disseminate it to hundreds of interested parties at a very low unit cost.

The exhibit was the result of a joint effort by the MAB Programme Secretariat of Unesco and Études et Planification des Communications (EPC), a small independent research group based in Paris. The following account discusses the preparation of the exhibit; the communication system used; and draws some preliminary conclusions about both the exhibit and the communication system.

## THE MAB EXHIBIT

### Objectives

The MAB Programme lays special emphasis on communicating the results of its various research projects to the scientific community, decision-makers, students in disciplines potentially involved in ecology, and interested members of the public. It does this in a number of ways, including publication of reports, technical notes, brochures, and newsletters; audio-visual presentations; and sponsorship of international conferences.

The overall aim of the exhibit produced in 1981 was to communicate and evaluate some of the findings and experience of the MAB Programme for the period 1971-81. Specific objectives were therefore:

1. To present some of the most important results of MAB, based on field experience in different ecological and climatic zones;
2. To help synthesize these results, during the actual preparation of the exhibit;
3. To illustrate essential aspects of MAB methodology and approach;
4. To emphasize the importance given by MAB to communication activities by experimenting with a new system of information transfer.

## The exhibit team

The team assembled for the project consisted of members of the MAB Secretariat responsible for specific subject areas, together with communication research specialists, graphic designers, and illustrators from EPC. The team also included photo-engravers and printers during the later stages of the project. Ten months were available to prepare and produce the exhibit from initial data gathering through design, preparation of illustrations, colour separation to printing. The time-scale in itself represented a challenge, since it normally takes museum staff two or three years to produce a scientific exhibit of similar scope.

Pressure on the team was thus considerable. Although the individual roles and responsibilities of each team member had to be clarified as work proceeded and deadlines were rescheduled, cooperation remained good throughout. The origin of any differences of opinion that arose appeared to be more institutional than educational or cultural. In retrospect, several observations can be made. First, in any future project of a similar nature, the team should be expanded to include decision-makers and users of the information, in addition to the communication specialists and scientists. This would ensure that the form and content of the exhibit would be as appropriate as possible for the various target audiences. Second, experience suggests that all the tasks to be performed by each team member should be set out clearly and completely in advance. Third, in any such project, the editorial role needs to be clearly assigned within the team. This may be difficult to achieve, but it is necessary for a successful outcome.

## The basic concept

Since the format for the MAB exhibit developed from a concept developed by EPC, it may be helpful to outline this basic concept. The problem was to find an effective and relatively inexpensive way to disseminate cultural, scientific, and technical information to a heterogeneous, multilingual and geographically dispersed audience. And to contribute to the public debate on such pressing issues as development and the environment, and to encourage participation at the local level in the decision-making process.

For various reasons the exhibition medium was chosen as an area where an innovative contribution could be made. However, traditional, "one-off" exhibitions were not the answer, since they do not provide for wide distribution nor for ease of public access that was desired and cost per person attending is prohibitive. Travelling exhibitions increase access, and improve cost effectiveness, but are still expensive. In addition, they have their own drawbacks, including handling, packing, unpacking, and shipping costs, not to mention wear and tear. They are cumbersome to handle and still fail to provide the ease of access and use that was sought.

The concept eventually adopted centred around the large-scale production of exhibition panels that were as compact and light-weight as possible. It was assumed that expertise would be available locally to display the panels at a low cost. It was further assumed that these exhibition panels would not be returned after use, but retained within local institutions for future use. In time, new and additional panels could be produced and dispatched to complement the initial stock. Thus a collection of material would become available at the local level, that could be increased and updated at minimum cost.

The decision to seek a wide distribution had further consequences: the panels themselves should be produced by mass production techniques; and creation and production of the panels should also take into account their future distribution, use and evaluation. What was envisaged, therefore, was the creation of a complete, widely available system of communication, which would start off a dynamic process of communications that could influence the very subjects being presented.

The basic system would be a starting point that could progressively encompass other communications materials and mediums, such as books, manuals, games, slides, films, and video-tapes. It could also be used in conjunction with other broadcasting media of a different nature, such as radio and television, and could be incorporated within the programmes of suitable organizational networks, such as national education systems, non-governmental groups, specialized seminars and workshops, cultural and information centres, science museums, libraries and so on.

The MAB exhibit was designed along these lines, and was eventually produced (from an original target of 100 copies) in 1,000 copies in three languages — English, French and Spanish.

## Selection of topics

The communicaiton system is intended to provide an overall perspective on a given subject. To be successful, it should briefly outline the problem, examine its various significant aspects, and discuss the main issues at stake. In doing so, the system should help guide the user towards the areas most relevant to him or her, and to the general literature on the most important points.

With these aims in mind, the selection and organization of the topics to be presented was extremely important. Each exhibit panel had to be independent and able to stand alone, both in form and content. At the same time, the panels had to be carefully planned so that, when combined in series, they presented a visual and intellectual continuum of interrelated issues.

Within the overall exhibit theme of "Ecology in Action" the 36 panels were organized in five main series:

1. Research for land use planning — ingredients, strategies, obstacles;
2. The tropical forest — a rich but fragile resource;
3. Marginal lands — coping with and taking advantage of constraints;
4. Cities as ecological systems;
5. Conservation — how, where, why;

## Design principles

Three important general principles influenced the design of panels:

1. The content of each panel must be viewed as a whole. Text and graphics can not be dissociated. Each reinforces the other to give the message maximum impact;
2. Graphics (which include illustrations as well as other graphic devices) must be visually attractive and where necessary, dramatic, first to catch the attention and then to sustain the interest of the user. As much information as possible should be conveyed graphically, to avoid the need for large amounts of text;
3. The text should be straightforward, provocative, and above all concise, or it would not be read.

## The panels

Each panel of the exhibit measures 80 by 120 cm. They were printed as posters in full colour, on heavy paper, and are plastic-coated to increase their durability. They integrate text with various kinds of visual material, including original paintings, drawings, photographs, and diagrams. All the visual materials were specially produced or chosen for the purpose, and were fitted into a carefully planned layout based on a predetermined colour code.

The text matter was divided into titles, leads (which briefly introduce the topic and its importance), subtitles (which provide a structure for the presentation), and a limited number of text paragraphs. The latter should ideally be very brief, and give a synopsis of the subject under discussion; introduce key concepts; and provide basic factual information and illustrative examples to enable the user to understand the ideas being presented.

## ASSESSMENT AND FOLLOW-UP

### Initial evaluation

An initial evaluation of the exhibit (EPC, 1981) was carried out during the Unesco-ICSU Conference-Exhibit held at Unesco in late September 1981. Given the comparatively small number of participants that answered the questionnaire (50 people from 19 countries), and the fact that they were mostly university-based scientists linked to international organizations, the results can only give a general idea of the exhibit's effectiveness. Additional comments were obtained from members of the public attending the exhibit when shown together with additional French material at the Hôtel de Ville (City Hall) in Paris from late September to early November 1981.

On the whole, the respondents felt that the exhibit provided a good synthesis of the subjects covered. Many suggestions were made on other topics that might be covered in the future, or on other ways of presenting the material. In particular, it was felt that the human dimension of the problems did not come through clearly enough, despite the importance attached to this aspect by the exhibit team.

The balance between themes was judged satisfactory, and the topics chosen within each theme were considered to be interesting. Most of those questioned felt that the number of different ideas and themes presented was not excessive. But on the other hand, most people felt that the panels were too crowded and the message too dense. The exhibit was considered, in general, to be aesthetically pleasing, with the colours well chosen and appropriate for the communication of scientific information. Illustrations and graphics corresponded well with the scientific message, and complemented the text.

The text was generally considered to provide a good synthesis and to have been well adapted to the panel format. Sentences were considered sufficiently short and the vocabulary clear (apart from a few expressions, such as somatic/extra-somatic, trans- and interdisciplinarity, which were in any case defined in the text). Some people criticized the size of the type necessary to fit in all the text, but were reluctant to suggest cutting any text. Instead, the use of subtitles and key words was recommended to improve the presentation of the panels.

The information transmitted was generally considered accurate, although a number of people questioned the precision or thoroughness of some of the panels. A few felt it was too simplified for a scientific audience. The general impact of the exhibit was considered good, and many people felt they had learned a good deal on one or more of the subjects.

Respondents considered that the primary target audience for the exhibit would be university and secondary school students. This judgement was confirmed by the large proportion of students visiting the French version of the exhibit at the Hôtel de Ville in Paris: many of them came during school hours with their teachers. Other uses for such an exhibit suggested by respondents included city-wide exhibits; permanent exhibits in science museums, universities, and large cultural centres; and as seminar material for professional training and non-governmental activities.

## An expansion of the basic system: from panels to modules

The experimental use of this basic communication system to present the results of the MAB Programme revealed the difficulty of providing, through the panels alone, a presentation that was at the same time sufficiently detailed to satisfy scientists, and also clear enough to communicate effectively to non-specialist audiences. Initially, it had been planned to combine the "Ecology in Action" panels with other information materials, such as simulation games and one-page technical handouts giving further information on specific subjects mentioned in the panels. But the time available for producing the exhibit (10 months) was too short for these supplementary materials to be prepared. Consequently, there was a tendency to overload each panel with text. This meant that although the information in the panels was more complete, it also became harder to read and to absorb. There was thus a risk that some of the text would not be read at all.

In the light of the "Ecology in Action" experience, and in order to satisfy the conflicting requirements of satisfying scientists and communicating effectively to non-specialists, EPC is now planning to develop the basic component of the system, the panel, into a more diverse set of communication tools, to be called a module. In this new form of presentation, the text on the panels will constitute the first level of information; it will be kept to the bare minimum, and will refer, whenever necessary, to briefing cards.

Briefing cards will be the second level of information, and will consist of A4 format cards printed on both sides containing further details on key issues. Each briefing card will include a title, a short discussion of the topic, an introduction to related subjects through key words, a summary of the state of relevant knowledge, an identification of existing areas of uncertainty, and a bibliography. The bibliography will refer users to the third level of information, the most important books and papers on the subject in the general literature.

Within the system, various visual aids will be developed on each topic, which will be made available on request to back up oral presentations and contribute to various kinds of seminars, training courses, conferences, etc. Each panel/module forms part of a series. Each series will be characterized by a common subject and by having a distinct visual identity, involving colours, graphic layout, etc. Initially, it is envisaged that modules will be developed in the context of specific exhibits, or of an intensive effort devoted to the creation of series of modules on a given topic.

As more modules become available, it will be possible to select them individually or combine them into series according to the needs of particular occasions or communications needs.

Experience with the MAB exhibit suggests that a series should comprise at least two kinds of modules. One would present basic facts and background information that are not in dispute. The other kind would discuss issues, presenting those aspects of a given problem that are uncertain by nature, or are perhaps the subject of controversy or disagreement amongst scientists and decision-makers — but which deserve attention.

## CONCLUSION

It is still too early to draw up a balance sheet on the "Ecology in Action" exhibit and on the underlying system of communication on which it has been based. Nevertheless, it can already be said that, even if substantial modifications of format need to be considered, the communication system holds great promise as a means of disseminating scientific and technical information to large sectors of the general public.

Economies of scale will allow the production of scientifically accurate information material in large quantities. This permits a dramatic improvement in the efficiency of the institutions acting as the source of the research effort and the scientific information, in this case the MAB Programme and its Secretariat. Furthermore, because of the approach adopted, the information source can concentrate the maximum proportion of its resources on preparation of the content of the panels, rather than on the hardware normally associated with information dissemination. Further research during the later stages of this experimental exercise should indicate the extent to which local users are able to mobilize the resources necessary for effective display.

If the "communication logic" of the source — its attitude to communication — has been dramatically improved by the use of the new system, the next priority is to focus attention on the communication logic of the users — the organizations that will receive the material and use it within their own local areas. For these organizations it can be assumed that the system's material will be most useful on a long-term basis if:

1. Modules are indeed part of a system, and not limited to the production of a single exhibition package;
2. The basic modules can be adapted readily to the needs of the various audiences that various organizations wish to address;

3. An increasing choice of modules is made available on a given subject, so as to assure comprehensive coverage;

4. A large variety of different subjects is covered;

5. The content of the modules is prepared by a variety of sources, thus providing different perspectives and presentations;

6. The system can provide the framework for a coordinated production of communication modules in a form that users can handle easily (including erection, transport and storage), and at a price compatible with their financial resources.

A great deal of further effort is therefore needed in order to determine how best the system could be developed to reach its full potential. It is clear that a continuous process of research and development is required, based on evaluation of the existing material and of specific user situations, and the creation of new, more efficient techniques of communication. In this way the basic system of communication for the dissemination of cultural, scientific and technical information will be able to satisfy better the needs of both institutional users and the general public. In that context, the use of the system by the MAB Programme constitutes an important source of experimental data and feedback. It is hoped that the communication system described in this contribution will prove to be well suited to the needs of programmes such as MAB, and that expansion of the MAB collection of modules will benefit from the continuing improvement of the basic communications system.

## Reference

EPC (1981), *Résultats de l'Evaluation de l'Exposition MAB "L'Ecologie en Action"*, Evaluation auprès des Scientifiques Présents à la Conférence-Exposition MAB-Unesco, Septembre 1981, Études et Planification des Communications, Paris.

CHAPTER 14

# Communicating Science — The Australian Experience

*Justin Murphy and Maurice Woodward*

Abstract

Various means of broadly-based appropriate communication are needed to reach the potential users of the results of research. Communication involves transmission, reception and understanding, and not just transmission. Experience in Australia has shown the importance of involving the techniques and approaches of professional communicators in this task, and some of the key steps in the professional approach can be identified. The role of a slide-tape programme as a communications tool within one MAB field project in northern Kenya is described.

*Résumé*

*Une variété de moyens de communication à large support sont nécessaires pour toucher les utilisateurs potentiels des résultats de la recherche. La communication suppose la transmission, la réception et la compréhension de l'information, non la transmission seule. L'expérience menée en Australie a montré l'importance d'impliquer les techniques et les approches des professionnels de la communication dans cette tâche et quelques unes des démarches clés de l'approche professionnelle ont pu être identifiées. On décrit ici le rôle d'un programme audio-visuel basé sur un projet du MAB sur le terrain au nord Kenya, comme moyen de communication.*

## THE ROLE OF COMMUNICATION

THE communication needs of the MAB Programme (indeed, of Unesco itself) are remarkably similar to those of CSIRO, the Commonwealth Scientific and Industrial Research Organization of Australia. In Western Australia, CSIRO has developed a communications philosophy which suits our needs well — we also believe that it is a philosophy that would suit MAB.

CSIRO is the Australian government's major scientific research body. CSIRO's purpose is to do scientific research on problems faced by Australian society — be those problems agricultural, environmental, to do with human nutrition, industrial technology, building technology and methods, the mining industry, alternative energy options, and so on. It is a big organization with a good reputation, enjoying high credibility with the Australian people. The bulk of CSIRO's problem-solving work is on contemporary issues. There is of course much long-term basic research in progress, but the majority is, and must be, relevant to today's problems.

Clearly, if it is to be effective, this research and its results must be communicated to Australia's "decision-makers" (its politicians, bureaucrats and industrialists), its educationists and its general public. But traditional methods of scientific exchange, such as the learned scientific journals, aren't appropriate for this job — they are only read by the scientific community itself. Therefore, other forms of broadly-based, appropriate communication are essential.

What is really *meant* by communication? Communication is the transfer of information from one person or group to another. It involves transmission, reception and understanding — it does not mean transmission only. It involves identifying and satisfying information needs.

CSIRO's Division of Land Resources Management (headquartered in Western Australia) works on environmental problems. In highly-urbanized and rapidly developing Australia, the environment is under some considerable pressures — from various forms of development and from increasing population concentration. Therefore, our research which deals with environmental protection, monitoring, fore-warning and rehabilitation must be communicated promptly to decision-makers and the general public. Such communication is an urgent necessity. It is an inseparable part of our research process — our science is devalued if our results are not put forward in a form that decision-makers can use.

Australia is a federation, with a central government (which funds and operates CSIRO) and seven state governments, with whom CSIRO fully collaborates. Sometimes CSIRO's environmental research results and recommendations (particularly when they sound cautionary), don't agree with the policies of the state governments (who manage and administer Australia's natural resources) and with sections of industry. At these times our communication needs a considerable degree of courage, and sensitivity. (It is worth pointing out that CSIRO is funded almost entirely from tax-payers' money. The organization therefore has a responsibility to keep the community informed about its research findings. In addition, the Act of Parliament which governs CSIRO stipulates that our research findings must be widely disseminated — even if they are at variance with opinions held by other sections of society.)

## THE INVOLVEMENT OF PROFESSIONAL COMMUNICATORS

The need for courage, the need for accuracy and the need for an appropriate and understandable form for our communication all adds up to one thing — the use of professional communicators. When communicating ecological recommendations that may conflict with development, the task is made easier and more convincing if the communication is deliberate, planned and professional in attitude and appearance. Haphazard and uncoordinated communication is almost always unproductive — at worst even counter-productive.

And yet there is not enough involvement of professional science communicators in scientific research — certainly not enough in Australia and, we suspect, not yet enough in international scientific programmes such as MAB. We feel that there is insufficient recognition that a wide range of professional help is available — using and blending what is best from a number of disciplines and professions — for example, lobbying (which is big business in Australia these days — involving common interest groups hiring experts to get their point of view across to government — and to get concessions from government); public relations; marketing and advertising; education; psychology and so on — in addition of course to the more familiar writers, journalists, broadcasters, etc. All of these have proven techniques for measuring community attitudes and aspirations, targeting audiences, and putting the chosen message into the most appropriate form.

## STEPS IN COMMUNICATION

The professional approach involves several deliberate steps[1] (at least):

1. *Identify* the audience;
2. *Translate* or *interpret* the research (paying particular attention to the removal of scientific jargon);
3. Consider the *timing* (i.e. when audiences can be expected to be most receptive);

[1]Readers may feel this list is too constrained, and that communication should be constant: that of course is true — we merely want here to spell out the *deliberate* steps necessary.

4. Select the right *form* or *forms* of transfer;
5. *Repeat* the message until it is received and understood, and then
6. See that it *has been understood*.

In all CSIRO's communication exercises great importance is placed on audience identification and assessment of the final job — to measure the job's effectiveness. The need for audience identification is obvious — different audiences (e.g. government decision-makers as opposed to, say, a local nomadic population) need different styles of communication.

Assessment is crucial because, naturally enough, it allows one to see whether the job is being done effectively. If not, there is the chance to modify what is being done until feedback tells that the message is being understood. Assessment is not easy though, and CSIRO has only started learning how to do it. In Australia, the work is being done with social and behavioural psychologists, who design assessment programmes (using interviews, questionnaires, passive observation and so on) to go hand in hand with the communication exercises.

## PUTTING THE APPROACH INTO PRACTICE

These are techniques we have used and will continue to use in our work on behalf of MAB's Integrated Project in Arid Lands (IPAL) in northern Kenya. So far one 35 mm slide/tape audio-visual documentary for IPAL has been made. Its purpose is to describe the north Kenyan desertification problem, describe IPAL's purpose and approach, communicate some of IPAL's results so far, and raise further questions about the nature of the human problem in northern Kenya. Its main audiences include delegates at scientific conferences, IPAL's funding source, MAB national committees, similarly affected countries, scientists, education authorities and students in Kenya, and, of course, the Kenyan government.

The assessment of it as an effective means of communication has now started. This style of slide/tape documentary is easy and cheap to modify — merely substitute outdated or inappropriate 35 mm slides with new ones, and record a new narration on a standard cassette. This style is extremely flexible, unlike 16 mm film, which is expensive to make initially, and expensive to update.

Slide/tape is also ideal for duplication in any one of a number of forms. A point to be stressed here is the penetration that such documentaries can achieve — the ones already made to communicate CSIRO's research in Australia have far exceeded our first expectations. They are in constant demand from many sections of the community.

But this is just the first step, for the first identified audience group. There are other audiences of course — notably the northern Kenyan administrators, and the nomads themselves, for whom the IPAL project exists. Work is now proceeding on the most appropriate forms of communication for them — they may be in documentary form, or booklet or poster form, or in a combination. But in working out the precise audience targets and the right form, our efforts will be greatly aided by IPAL's research style and approach — it is working directly and constantly with the people in the study area (which will provide constant feedback), it recognizes that communication must be sustained, and it recognizes that communication must be constantly linked with the research work. This is the most likely recipe for success.

CHAPTER 15

# Chinampas and Integrated Farms: Learning from the Rural Traditional Experience

*Hector Luis Morales*

Abstract

The "chinampa" form of agriculture used in prehispanic Mexico and integrated agriculture systems in Asia has recently served as the basis for pilot projects developed in the Mexican humid tropics. Interesting lessons have been learned through this experience concerning the limitations of conventional agricultural research and new ways of incorporating the subsistence farmer into this process. It is important that ecologists express their point of view by establishing projects which are technically and socially feasible in order to obtain the confidence of politicians as well as the subsistence farmer. These projects should be alternatives to existing tropical land use projects, which are often costly or of doubtful success. This contribution proposes some methodologies for the implementation of projects of this kind.

*Résumé*

*La pratique agricole dite "chinampa" employée au Mexique préhispanique et dans les systèmes d'agriculture intégrée en Asie a récemment servi de base aux projets pilotes réalisés dans les zones tropicales humides du Mexique. Des enseignements intéressants ont été tirés de cette expérience, notamment en ce qui concerne les limites de la recherche agricole traditionnelle et les nouvelles méthodes pour faire adopter cette pratique par les agriculteurs qui ne travaillent que pour leur propre subsistance. Il est important que les écologues expriment leur point de vue par des projets concrets et viables du point de vue technique et social afin de gagner la confiance aussi bien des politiciens que des agriculteurs. Ces projets devront constituer des alternatives aux projets déjà existants d'aménagement des terres tropicales, lesquels sont souvent coûteux et d'une réussite peu évidente. Cette comunication propose des méthodologies pour mettre en oeuvre des projets du type "chinampa".*

## INTRODUCTION

MANAGEMENT of the humid tropics in Mexico presents a serious challenge for politicians and ecologists. This challenge has been met equally by both sectors, and in practice, both the small autonomous villages and the colonists have suffered the consequences through pauperism and marginalism. To this, may be added the problem of exploitation by large oil companies, which introduces unexpected social and economic tensions, not easily assimilated by the economic social systems in the small villages.

Some alternative projects for this zone are based on the search for an application of the traditional prehispanic management of tropical swamp areas, and the practice of Asiatic systems: the integrated use of land and water together with the integration of vegetable production and animal husbandry through the management and recycling of organic matter.

The "chinampa" or "raised field" experiment of agriculture in swamp zones resulted in a series of small development projects in rural comunities, which aimed at satisfying basic needs of food, energy, employment and education. In the course of the experiment there has been a transition from "chinampas" to the integrated farm, from which certain conclusions have been drawn. It is hoped that these will assist the subsistence farmer to achieve an integrated use of his available resources within the framework of community development.

## CHINAMPAS AND INTEGRATED FARMS

The "chinampas" of the Valley of Mexico are an example of prehispanic agricultural systems in which a swamp area is used, and the natural productivity of the recycling of earth and water is recovered (Gómez-Pompa, 1978; Venegas and Meyer, 1981). This system is characterized by the following:

1. The use of bottom sludge and aquatic vegetation for the construction of "floating islands" or "raised fields" called "camellones";
2. The recovery of nutrients from the water in the form of sludge, for use as fertilizer for seedlings (or as a mud base for these seedlings);
3. Subirrigation through natural earth capillaries;
4. The intensive cultivation of plants destined for local auto-consumption (corn, beans) and varieties of vegetables having a high commercial value (Swiss chard, cabbage, cilantro or "chinese parsley", chilies, flowers):
5. Originally in Mexico water was used as a medium for fish culture (Ciprinidae) and its aquatic plant resources (*Eichornia* and *Lemna* sp.) were used as feed for ducks and chickens;
6. Vegetable wastes and cuttings are used as cattle and horse feed and all animal wastes are reintegrated as fertilizer for the fields;
7. The farms are small and vary from 500 to 4,000 $m^2$. They are usually managed by family units plus occasional salaried workers (Figure 15—1).

On the basis of these systems, experiments were conducted for their transference to the tropics and for their use in the recuperation of vast swamp areas in the states of Tabasco and Veracruz (Gómez-Pompa and Venegas, 1976; Gómez-Pompa *et al.,* 1982; Marten and Sancholuz, 1982; Morales, 1980, 1981). The government of the State of Tabasco and the National Indigenist Institute (INI) have recently constructed a 100 ha area of camellones in the town of Nacajuca, Tabasco. This area is called "Camellones Chontales" and is situated in a zone inhabited by approximately 60,000 Indian subsistence farmers. In the State of Veracruz, there are two small experimental farms presently in operation: El Castillo (Xalapa) and Buena Vista (Actopan) (Morales, 1981).

Their system of integrating agriculture, animal husbandry and aquaculture, with management of wastes through the use of digestors or composts, was inspired by the study of certain Asiatic systems. These systems (Figure 15 — 2) are designed to satisfy basic food and energy needs through appropriate technologies and the optimal use of available resources. They are attractive not only for their economy or simplicity, but also for the way they help the subsistence farmer to become self-sufficient.

## APPLIED RESEARCH AND RURAL DEVELOPMENT

The results of applied research relating to agriculture, cattle raising and aquaculture, have been widely disseminated. The search for better seeds, chemical fertilization techniques and other spin-offs of the "green revolution" has tended to emphasize

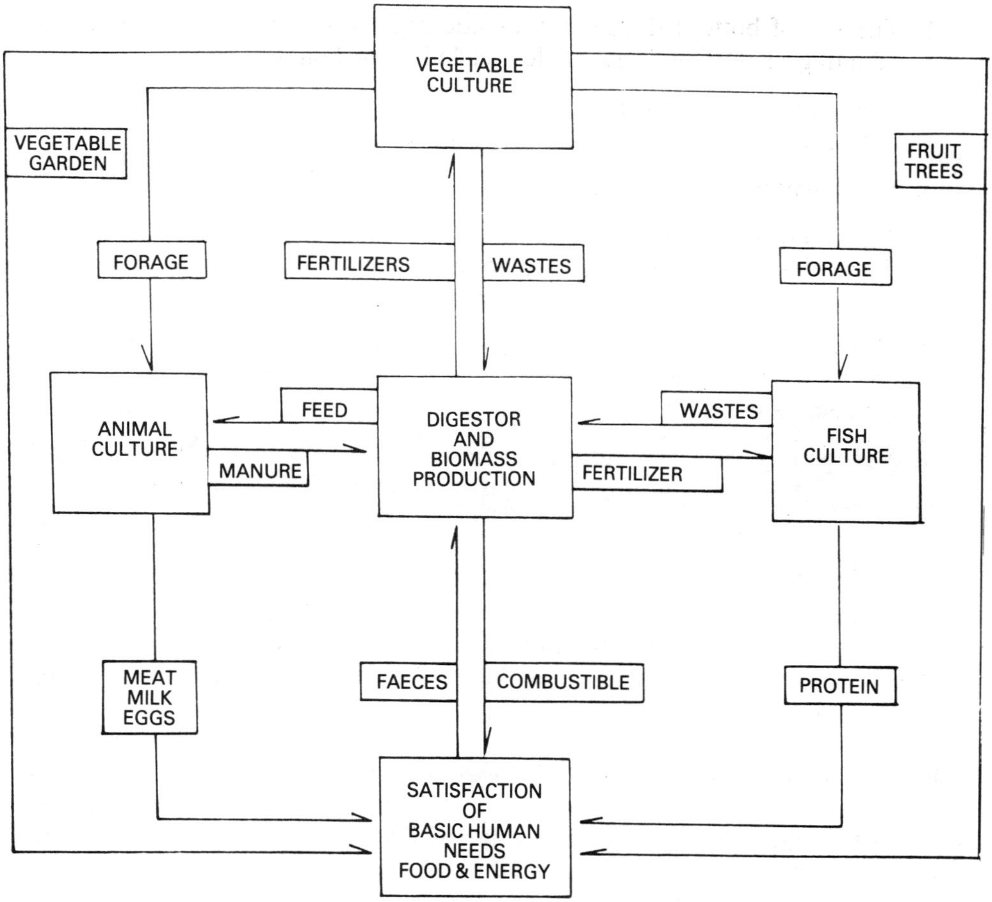


**Figure 15—1.** Integration of elements in an integrated farm

a high level of productivity through enterprises which are rationally managed and vertical in their organization. This research, focussed especially on grains (wheat, corn, rice, barley) or on export species (trout, shrimp) has had significant impact on the food sectors of many countries of the world. These results however seem more difficult to apply on a large scale among the masses of subsistence farmers.

In the developing countries, results of this type of research have usually been viewed as something external, which has been imposed on the social and productive systems of the country. But the subsistence farmer has his own system of values and production, in which the methods and techniques of agricultural practices are a result of local experience, or of the "concrete science" by which he systemizes his observations about the environment. These farmers find it difficult to accept any technological innovations because they have no confidence in outside agents and because they don't fully understand the basis and rationality of the new systems.

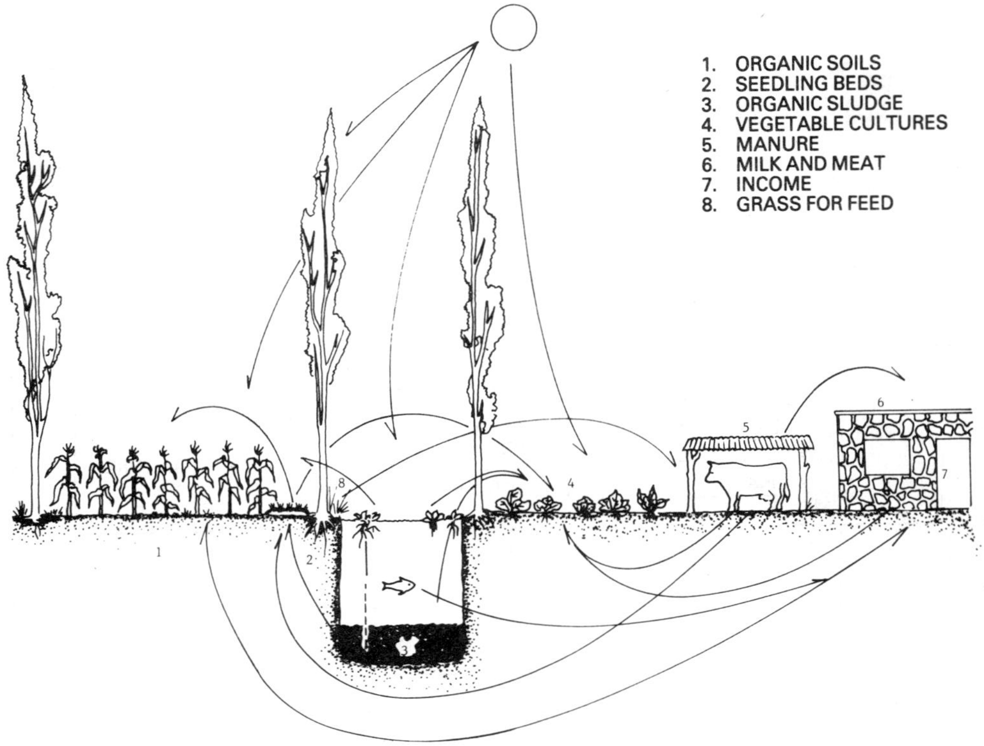


**Figure 15—2.** The energy cycle of a chinampa (after Gómez-Pompa, 1978)

Subsistence farmers possess their own form of logic, which differs from that of scientists and decision-makers (Morales, 1980).

In some Latin American countries, the leadership changes every 6 years. During this term, leaders must set new and precise goals; they determine the work of scientists and its application through the implementation of development programmes. However, continuous failures indicate that written projects never really get applied in the field, and that research results are rarely diffused and applied at the farmers' level. And the fact is that the rural population is worse off each year.

## EDUCATING THE SUBSISTENCE FARMER TO BETTER MANAGE HIS RESOURCES

The experience gained through the "chinampas" and integrated farm projects has shown a need for redefining research strategies to make them applicable to rural development. By this is meant "education" and not new forms of research. In effect, it can be proven that conventional research logic, although intended to be universal for all kinds of actions and circumstances, is actually marked by characteristics which make it most relevant in highly specific development projects and contexts. Another orientation is proposed, which is derived from observations of the differences between conventional science and the traditional, concrete approach of the local practitioner. This system is based on the following ideas:

First, in the process of conventional research, one must distinguish between the methodological interest *per se,* the proof of its value and the interests which intervene in the applicability of its results. In general researchers tend to concentrate on testing the value of the method and on the advancement of abstract knowledge, leaving it to politicians, industrialists or merchants, whether, and how, to apply its results. In consequence, their research results may affect rural and agricultural development adversely, since it disrupts, or even competes with and/or contributes to, the dissolution of the farmers' own economy. This is especially true if there has been no ethnological study of the effects a project may have on the normal life of the community or what will happen when new consumer-based habits replace traditional ways of life. The most notorious effect is the failure to fulfil basic needs. Cash-crops (coffee, cocoa, copra, sugar cane, etc.) destined for outside sale replace the polycultural practices with which farmers formerly satisfied their requirements.

Second, the process of organizing conventional research tends to be a mechanical one — restricted laboratory scale studies giving way to field tests, then pilot scale projects, which eventually lead to the so-called "extension" activities. In the process, some elements of truth may be found, but shortcomings of the procedure should also be recognized. Thus, the farmer is neither an ignorant "empty shell" nor a virgin market for new products. In the farm economy, a local systematized experience is present in which the farmer's repeated perception of this environment as well as his agricultural practices are transmitted from generation to generation. This science of the concrete is expressed both in classifications and names, as well as in rules of thumb, that take into account climatic conditions, predator-prey relationships, and finally a balance between self-sufficiency and the external market. The classical investigator who overlooks this systematized experience is thus denying an essential dimension of the very same environment to which he is trying to apply his results. To this extent the so-called universal value of research becomes a trap rather than a contribution to rural development.

Third, there exists a matter of "scale" in considering the problems of rural development. Research carried out on a type of seed or on a certain technology is justified in itself in terms of the rules of the "scientific" method. But studies and experiments on production systems should not be considered in isolation from the real conditions in which results are to be applied. If production systems are studied in isolation, neither the farmers' experience nor the initiatives of the researcher is valid. On the other hand, if the experimental phase is carried out in close cooperation with local farmers, then this provides a way whereby the farmers' experience can help shape the experiments, and at the same time prepare the ground for the application of research in improving farming practice. It should, however, be noted that not all communities have acquired the same degree of experience and initiative, and, therefore, a method for the analysis of social and technical organization in rural communities becomes necessary, as part of the research effort.

Fourth, many rural development projects, especially where they concern soils and hydraulic schemes, have demonstrated the difficulties involved in trying to manage the natural environment. But farmer-devised systems have often been able to cope successfully with the same problems left unsolved by development projects (Collier, 1981; Ewell and Poleman, 1980; Hanson and Koesoebiono, 1977). It has often been pointed out that large engineering schemes do not duly take into consideration ecological and social aspects (e.g. Barkin, 1977). It is a fact that this experience and tradition are often used in project planning. Perhaps the participation of ecologists and social scientists would help solve the problem, provided that the subsistence farmers' experience as well as laboratory experience were taken into account.

Fifth, the farmer's own knowledge of his environment, along with contributions from scientists should be the basis for field studies and research. The step from individual experiences to more extensive forms of experimentation can develop gradually through education taking place on the farmers' own level. As positive results are achieved, the farmer teaches the farmer and a certain complicity and confidence are developed. In this way, the "chinamperos" of the Valley of Mexico have taught their techniques to farmers in San Pedro (Tabasco), La Mancha (Veracruz) and Tecocomulco (Hidalgo). In a similar way, fishermen from one region can teach other fishermen their techniques.

One example of this transfer of experience is in the northern part of the State of Puebla, where the Indians have constructed and are presently managing a 4 ha integrated farm, which was designed on the basis of experiences obtained in the Valley of Mexico. The farm encompasses vegetable production, pig and fish culture and use of digestors. Groups of farmers from other regions have visited the farm, and on the basis of a mutual interchange or "teaching" of experience, new knowledge is shared. This is accomplished both through work in the field and through the laboratory. A similar experience is being carried out in the community of Buena Vista, Emiliano Zapata, Veracruz.

Sixth, in recognition of the need for subsistence farmers and researchers to have the opportunity to exchange experiences on an informal basis, a series of 10-day courses has been established. Through these courses there has been a mutual opening of frontiers, a confidence gained, and a common terminology established, with the result that a real exchange of field experiences takes place. The first course of this kind was held in 1978 in collaboration with MAB and the State of Veracruz. Thirty-five subsistence farmers from five states of Mexico attended. By 1981, three years later, three relatively advanced projects had been established, with a community "promoter" in each project in charge of activating work schemes.

Seventh, the farmer himself not only raises problems associated with contamination, erosion and aquatic population control, but also points out possible ecological reserve zones or studies of available resources which have not been fully utilized.

In conclusion, the idea of "environmental education" applies as much to the subsistence farmers — and what interchange occurs between them — as it does to those ecologists, scientists and agronomists involved in the planning process. In order to accept this equality of relation, it is necessary for the scientist to adopt new methods and ways of thinking, and perhaps also to acquire a certain humility towards one's fellow beings.

## References


Barkin, D. (1978), *Desarrollo Regional,* Nueva Imágen, México.

Collier, W. L. (1981), *Social and Economic Aspects of Rice Cropping Systems in the Coastal Wetlands of Indonesia,* Paper presented to First International Workshop on Research Priorities in Tidal Swamp Rice in Indonesia.

Ewell, P. T. and P. Poleman (1980), *Uxpanapa: Agricultural Development in the Tropics,* Pergamon Press, New York.

Gómez-Pompa, A. (1978), Vino Nuevo en Odre Viejo, *Mazingira* (Pergamon Press, Oxford), 5, 49-55.

Gómez-Pompa, A., H. L. Morales, E. J. Avilla and J. J. Avilla (1982), Experiences in Traditional Hydraulic Agriculture, In K. V. Flannery (Editor), *Maya Subsistence: Studies in Memory of Dennis E. Puleston,* Academic Press, New York, 327-42.

Gómez-Pompa, A. and R. Venegas (1976), *La Chinampa Tropical,* INIREB Informa 5, Instituto Nacional de Investigaciones sobre Recursos Bioticos, Xalapa.

Hanson, A. J. and Koesoebiono (1977), *Settling Coastal Swamplands in Sumatra: A Case Study for Integrated Resource Management,* PSPL Research Report 004, Center for Natural Resource Management and Environmental Studies, Bogor Agricultural University, Bogor.

Marten, G. G. and L. A. Sancholuz (1982), Ecological Land-Use Planning and Carrying Capacity Evaluation in the Jalapa Region (Veracruz, Mexico), *Agro-Ecosystems,* 8, 83-124.

Morales, H. L. (1980), Rural Development, Science and Political Decision-Making: Diverging or Converging Tendencies? *Impact of Science on Society,* 30 (3), 167-77.

Morales, H. L. (1981), Hacia una Economia del Pantano: Las Granjas Tropicales Integradas y las Chinampas, *Biotica* (INIREB, Xalapa), 5 (3), 103-14.

Venegas, R. and J. Meyer (1981), *The Nitrogen in the Chinampas of Mixquic, Tucta and La Mancha,* INIREB-Bioaqua, INIREB, Xalapa.

CHAPTER 16

# Interpretation of the Mediterranean Environment: a Programme of Cooperation

*Fernando Gonzales Bernaldez, F. El Cadi and Rémi Perelman*

## Abstract

An international cooperative programme launched by the MAB Committees of Tunisia, Spain and France has the aim of sensitizing local populations to their immediate environment and modifying in a positive way their relations to their natural and cultural surroundings. Pluridisciplinary workshops have been organized for presenting information on the environment in a form adapted to different populations. Networks for interpretation of information have been set up, and illustrative information materials prepared on the spot at the local level. The scientific approaches and information of MAB Project Area 13 ("Perception of environmental quality") have proved useful in the development of programme content, but also in providing insights on the way of presenting information and linking such information to topics of direct interest to the public (e.g. topics relevant to their day-to-day activities and concerns).

## *Résumé*

*Un programme international de coopération a été lancé avec l'appui des Comités MAB de la Tunisie, de l'Espagne et de la France dans le but de sensibiliser les populations à leur environnement immédiat et de modifier positivement leur comportement à l'égard du milieu naturel et culturel. Des ateliers pluridisciplinaires sont chargés de présenter un système d'information sur le milieu de manière adaptée aux différentes populations en créant des réseaux d'interprétation et en les illustrant à l'aide de documents élaborés sur place. L'information et les approches scientifiques du thème No. 13 du Programme MAB (Perception de la qualité de l'environnement) ont été utile non seulement pour l'adoption des contenus du programme, mais aussi pour leur présentation et leur liaison avec des sujets d'intérêt pour le public (activités de la vie quotidienne).*

## INTRODUCTION

### Background to the project

In 1975 a Franco-Tunisian cooperation was initiated in Tunisia within the framework of the MAB Programme of Unesco, particularly within MAB Project Area 13 on the perception of environmental quality. The work began with a multidisciplinary Franco-Tunisian group nominated in Tunisia by the Ministers for Agriculture and Equipment and in France by the Minister of the Environment. This group was charged with a study of the evolution of the coastal landscape and the preparation of a description of the way the resident population on one hand and tourists on the other hand perceived two particular sites: at Cap-Bon and on the Isle of Djerba.

During the period 1977-80, this study was at the origin of a pedagogic experiment launched at Djerba under the name of the Pilot Operation for the Interpretation of the Mediterranean Environment (OPIEM). Within this project, a methodology has been developed to sensitize the public and students to their environment. Institutions and experts from Tunisia (from relevant ministries and technical organizations and the Association for the Safeguard of the Isle of Djerba) and from France (from the Ministry of Youth and Sports, Ministry of the Environment, etc. . . .), as well as international experts (from Great Britain, Spain, Sweden), participated at different workshops organized at Djerba.

In 1980, the operation became a tripartite one with the cooperation of the Spanish National Committee for MAB, following a preparatory meeting and seminar held in Madrid respectively in June and December 1980. The Tunisian, French and Spanish participants had the support of the "Diputación" (General Council) of the Municipality of Madrid and of the Municipality of Collado-Mediano in the Sierra de Guadarrama. The OPIEM, initiated at Djerba, can thus be considered as a methodological experiment designed to inspire similar activities in the countries bordering on the Mediterranean.

### Objectives

The general objective of this programme of international cooperation is to sensitize the local populations, notably the public and the school population, concerning their environment so as to modify in a positive way their habits with regard to the natural and cultural environment. The principal aim is to improve the understanding of the environment in all its aspects by active pedagogic methods, based on the interests and sensitivity of the people concerned, and as close as possible to their life styles and their environment. The methods tend to highlight the relation between the different elements of the environment and various human activities.

## Methodologies

The methodology used owes much to the efforts at systematization and projects carried out by the Countryside Commission for Scotland, which initially tried out these methodologies in Djerba. Two essential aspects of the approach should be stressed: first, the importance of setting up a "mother group" which promotes the extension and development of the operation in an autonomous manner; and second, the need to ensure a mixture of specialists from within and outside the country, as well as the presence of photographers, designers, graphic artists, etc.

In each case, the main focus of work is on methodological workshops comprised of specially constituted groups in which there are specialists concerned with different aspects of the natural and human environment, teachers, administrators and experts in environmental education, as well as, most importantly, a large proportion of the residents of the region being considered. This group works in several subgroups, which are given the task of exploring different themes for interpretating the environment. The task consists of presenting the environment in a manner which is adapted to the different populations concerned, in exploiting the natural and cultural potential likely to provide a support to the pedagogic action of interpretation.

For this, networks of pedagogic circuits have been created, linked with a certain number of centres of interest. These networks concern, for example, the discovery of the natural environment (coast, lakes . . .), the rural environment (olive groves, palm groves, water economy . . .) and the urban environment, and give due attention to local cultures and crafts. Information materials are adapted to the different groups of the population concerned: they include leaflets, simple brochures and more elaborate documents, explanations posted up in the most interesting sites, recommendations on the parts of an area to be visited and studied, written guides, audio-visual documents, permanent exhibits, etc.

Experience has shown the interest of combining scientific research undertaken within different themes of the MAB Programme with a pedagogic treatment of the results obtained. In fact, even in areas which are well known from the standpoint of their natural environment, the launching of a research project confers a particular educational interest on the locality concerned. This is the case for research on perception of the environment which has been integrated at Djerba and at Madrid within pilot projects for the interpretation of the environment. The method used within OPIEM has the following phases:

1. Identification of the objectives for each operation and the persons to whom it is directed within the general theme of sensitization of populations to their immediate environment (preparatory seminar, leading to data collection phases 2 and 3);

2. Identification and study of the pedagogic potential (characteristics of the natural, cultural, architectural environment, activities of the population), themes and places of particular interest which can be exploited from a pedagogic viewpoint. The use of specific physical spaces which can be explored and discovered actively by the public is one of the methodological features of great importance in the overall operation;

3. Analysis of the scientific information available: for example the physical, ecological, historical and cultural characteristics of the region and the sociological characteristics (perception, system of values, etc . . .). Selection of content as a function of objectives and target audience and of the characteristics of the localities available;

4. Elaboration of a programme and calendar encompassing such aspects as beginning, functioning and means (existent, to be created, or to be assembled). This is developed during the course of a workshop by: illustration of methods using pedagogic documents elaborated locally and in a limited time (a few days) by the seminar participants; this phase constitutes a key characteristic of the OPIEM. The workshop also examines proposals for administrative, scientific and pedagogic structures necessary for the development of the operation, adapted to each situation under study. Plans for interpretation centres and programmes, linking where possible the natural environment with socio-economic activities, are also considered;

5. Evaluation, continuation and expansion of the operation in order to promote the training of teachers through workshops and periodical seminars and gradual improvement of the pedagogic methods used.

## Case studies

Within this general framework of objectives and methodological approach, specific operations are developed in particular case studies. Three such case studies on environmental interpretation are described in the following paragraphs: one concerned with the Isle of Djerba (Tunisia), the second with a national park in the Western Pyrenees (France), the third with the Collado-Mediano to the north-west of Madrid (Spain).

## ISLE OF DJERBA CASE STUDY

The environmental interpretation programme on the Isle of Djerba (Tunisia) was finalized by a meeting of experts held in 1980 (ASSIDJE, 1980), which was the culmination of several earlier seminars and meetings (Ministry of the Environment, 1980) and which took advantage of studies undertaken since 1975 under the auspices of the National Centre for Landscape Studies and Research and the Association for the Protection of the Isle of Djerba (CNERP, 1976-79; CNERP-ASSIDJE, 1977).

### Objectives

The main objective was to define actions enabling pedagogic use to be made of the findings of MAB studies on perception of environmental quality. The resulting materials were to be geared to sensitizing the population of Djerba, and especially students and schoolchildren, to environmental evolution and degradation, to promoting a policy of preservation by upgrading and renewal of the environment. Djerba is situated in a semi-arid zone, and is subject to the risks of desertification. The protection of the environment is of vital importance and justifies actions for modifying people's behaviour to the environment. We thus hoped to develop links between the natural environment and the daily life, the cultural heritage and the historical roots of the people of Djerba.

### Identification of centres of natural and cultural interest

An important step was to identify centres susceptible of contributing to the network. The group decided to concentrate its activity on several themes of special interest within which certain underlying threads could serve to link up particular locations which are accessible and which would provide a physical stimulus to the senses, to personal exploration and discovery. Three themes and centres of interest were retained.

**Agriculture:** The "olive tree" was chosen as the focal point for project work related to agriculture, with the Roman oil-mill of Oualagh providing a centre of interest which could be visited. The oil-mill is an ancient subterranean installation which used camels for its motive power. It is one of few remaining examples of many similar mills which existed on Djerba until only a few decades ago. Now largely replaced by modern installations, such oil-mills played an important role in the economic life of the island, and were well adapted to the local conditions, especially to the climate. Having provided Rome with oil, such mills have considerable historical interest.

The upkeep and opening to the public of one of the rare oil-mills still in perfect state is intended as a reference point in space and in time, which commemorates the oil-growing past of the island and acts as a sort of yardstick for the study and understanding of the present. In addition to its architectural quality, the mill provides a point of attraction. The continuum of olive tree-olive grove-adjoining land (uncultivated areas, pathways) can be used as tangible spaces open to personal discovery and documented in special publications.

**Handicrafts:** The principal link is that of weaving, from the raw material (wool) to the finished product (carpet), and illustrating such aspects as the techniques in raising sheep, traditional and modern methods of weaving, the economic factors involved in weaving, and the aesthetic concept of the carpet. The "fondouk" Jemni at Houmt-Souk, a traditional market tavern, has been chosen as a physical centre of interest, where authentic and living examples of handicrafts can be demonstrated and a market for wool and carpets held. Other traditional crafts of Djerba could be treated in a similar way, such as pottery (from the extraction of the clay to the finished product) and plaiting and basket work (using the ecology of the plants providing the source of the raw material as a starting point in the interpretation programme).

**Natural environment:** We have selected a number of links between agricultural and artisanal themes in an attempt to illustrate the many interconnections between man and his environment, the ecological effects of people's activities, etc. The topics of water cycle and soil stability have provided a focus for explaining the concept and localization of habitat. Two centres of interest that can be visited have been selected because of their favourable pedagogic characteristics: Point Ras-Er-Rimel, formed by a sand dune and muddy clay banks, characterized by rapid change in littoral shape and dune relief; Lake Guellala, a coastal depression, partly fed with drainage water from an agricultural zone, rich in animals (birds in particular) but subject to considerable human influence. It is noteworthy that the interpretation activities at Lake Guellala have led to proposals for protection of this site, following recognition of its importance by the local and regional authorities.

## Pedagogic approaches

Figure 16—1 provides an illustration of how links can be developed among the three principal themes (agriculture, handicrafts and the natural environment) in a way that can be related to the concerns of daily life and economic well-being. The rectangles refer to concrete exercises of observation that can be carried out in localities chosen as centres of interest. The lines between the rectangles in Figure 16—1 symbolize the existence of conceptual or material links which can act as a basis in explaining the web of relations which underlie the themes proposed. Each rectangle and each line can be exploited pedagogically. In this way the relationships between the natural environment and human activities that are well known to the public and to the students can be illustrated.

The various teaching exercises have been documented and illustrated in "teachers' handbooks" prepared during the OPIEM seminars at Djerba (El Cadi, 1980a,

1980b; Ministère de l'Environnement et du Cadre de Vie, 1980). The booklets designed for teachers contain 92 pages of exercises which offer suggestions for such observations. A good starting point was the general schema of the Djerba landscape which provides a synthesis of some of the principal trends of man-environment interactions on the island, such as:

—the presence of *cuestas* formed by the fracture and the movement of the soil crust;

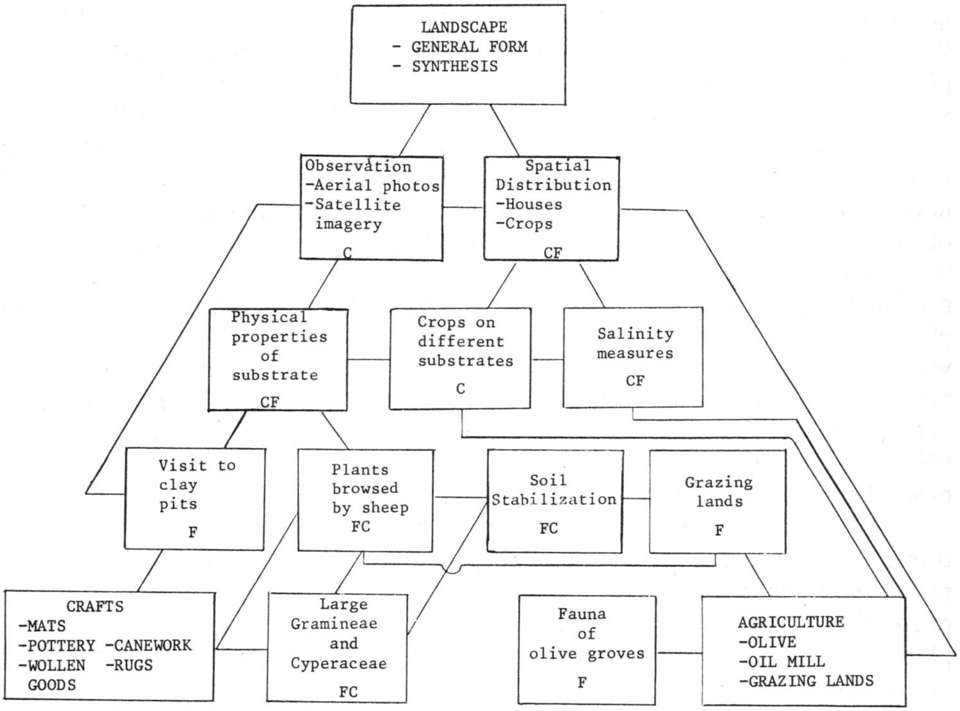


**Figure 16—1.** Conceptual framework linking craftmaking (pottery, canework), agriculture (olive groves, grazing) and the natural environment (general representation of the landscape) at Djerba. The boxes correspond to specific exercises described and illustrated in guides for teachers (C – classroom exercises, F — fieldwork)

—the accumulation of alluvial soil in parts that have subsided, and subsequently the creation of rocky zones with very shallow soils in the raised zones;

—the presence of slightly salty water in the geological structures at levels higher than sea level (water with a low salinity becomes more salty in the lower slopes and towards the sea);

—the dynamics of coastal sands, which drift into the interior.

This schematic representation, or simplified model, of the rural landscape can be used as a reference point for observations of a very different nature, such as those depicted or described in other parts of this contribution.

The accent placed on the salinity of the water and the mesological character of its distribution can be justified by the fact that, because of the aridity of the climate, the level of subterranean water necessary for irrigation is a limiting factor important for agriculture, and that too intensive pumping accentuates this salinity by the influx of salty water. A group of exercises on this topic includes constructing more or less complex pieces of measuring apparatus, sampling well-water, obtaining water with different salt concentrations, germinating seeds and growing different plants with these concentrations of salts. Reference to a map of measured salinities should permit understanding of the part played by salinity in the siting of different crops, vegetation and dwellings and of the need for research into other types of relations.

Observations on different soils and geological substrates — and correlation with crop distributions, land use, salinity, etc. — provide interesting exercises which improve perception and knowledge of the immediate environment, as well as use of land, etc. Together with observations on the salinity, they lead to a better understanding of man-environment interactions. The relations between handicraft production (e.g. pottery) and geological substrate can be demonstrated using aerial photographs and satellite images combined with visits to clay mines and to the workshops of potters at Guellala. Observations of the large Gramineae and Cyperaceae of the sandy dunes and peri-dunes and of the coastal *sebka* can be organized in the form of exercises which highlight the view of: the world of handicraft workers who use these plants in making baskets, mats, etc.; sheep breeding; the phenomena of wind and coastal erosion, which are extremely important on the island.

It is possible to identify subjects of observation in relation to the process of transformation of raw materials (clay, wood, reeds) into traditional handicraft products. These primary materials provide an entry to understanding important phenomena in the natural environment. In the case of the large grasses and other plants used by the artisans, it is easy to study their mechanical properties, root systems and growth, whether in sandy areas or inundated coastal soils and to demonstrate their geomorphological role. For example, the distribution of *Ammophila arenaria* var. *arundinacea, Sporobulus punges, Lygeum spartum, Imperata arundinacea* and *Phragmites australis* correspond to characteristic stages in the succession, humidity and stability of the substrate. The effects of vegetation destruction and sand movement can also be demonstrated. Links can be drawn between such phenomena and sheep raising, and in turn with the raw material (wool) and other handicrafts.

Among the information materials produced is an atlas of plants classified according to palatability and utility for different herds. Brochures destined for teachers are only one part of the documentation produced within the projects, and these are being integrated into a very extensive system of education and communication, which includes leaflets, brochures, films, audio-visual materials and panels. Various centres of interest are directed not only at schoolchildren (the priority target) and to the Djerbian public, but also to tourists. Information on the educational activities at Djerba can be obtained from the Association for the Protection of the Island of Djerba, Houmt-Souk, Djerba, Tunisia.

## WESTERN PYRENEES CASE STUDY

The interpretative plan and teaching programme for the National Park of the Western Pyrenees have been elaborated during a workshop on the interpretation of the Azun Valley held in Val d'Arrens in July 1980, and have been revised during the course of later meetings. Trials on the interpretative plan for the National Park are being carried out with the help of the French Ministry of the Enviroment, particularly the Department for the Protection of Nature.

### Objectives and specific methods

The essential objectives of the programme are: to develop the awareness of the inhabitants of the valley to the richness of their natural and cultural heritage and to help to spread this consciousness among visitors, with a view to promoting positive behaviour with regard to the natural and human environment. It is hoped to reach these objectives by means of a programme of environmental interpretation which seeks to present and diffuse information in a readily understandable form.

A large amount of scientific information is available concerning the natural and cultural environment of the Western Pyrenees National Park and adjacent areas. There is need to compile, systematize and above all to provide a suitable presentation of this information, so as to attain the desired objectives and to reach the target audience. With this in mind, a number of multidisciplinary workshops have been organized. Participants have included members of the staff of the park and of its scientific committee, naturalists, historians and local teachers, graphic artists, specialists in environmental education as well as Spanish counterparts. The task of these groups was to imagine and prepare prototypes of a range of interpretative materials considered to be useful. These materials were in various forms, such as posters, leaflets, audio-visual documentaries, booklets, etc. The end product from each workshop has been presented to the principal local decision-makers, elected representatives, civic associations, etc., so as to obtain their comments and suggestions before finalizing and distributing the interpretative materials.

### Information for the general public

Workshop participants retained three main themes as vehicles for communication in the first phase of the experiment and have produced several types of experimental teaching guides at the methodological level. First, for the natural environment, a draft guide has been prepared in booklet form, whose leitmotiv is the "Path to discovery of nature" of the Plan d'Aste. Second, the topics of history, traditions and handicrafts have been treated in a locally prepared leaflet, whose underlying theme is "A week in the life of a nailsmith of the valley". Third, questions relating to agriculture, landscape and rural architecture have been included in a booklet "In the Azun Valley following the *herechou*". The link here is the ash tree, which is known as the *herechou* in the local dialect.

It is intended that the overall interpretation programme will include the production of films, exhibitions, dioramas, audio-visual materials, publications (lexicons, guides, books). We have also endeavoured to reflect the relation between the natural environment and typical human activities through history. This is notably the case in a series of materials devoted to "ecologically associated information" such as: nailsmith/iron mines; slatemaker/schists; clockmaker/beech; carpenter/pine, etc.

"A hundred years ago . . . a nailsmith in the Azun Valley", the title of one of the three documents produced locally to sensitize the public, illustrates the use of graphic techniques as well as of a local personality in the elaboration of interpretative material. It comprises a leaflet of ten pages. Research on the theme included studies of archives belonging to a member of the group living in the valley. Preparation of the illustrated manuscript ready for the printer was carried out by members of the group during a three-day workshop, and included contributions from a graphic artist, an inhabitant of the valley, a specialist in environmental education, a collaborator from the national park, and a teacher. The leaflet forms part of a set of documents designed to illustrate the theme "History, traditions, handicrafts", and to link this theme to other aspects of the natural and cultural environment of the region. Day-by-day descriptions of the activity of a nailsmith make it possible to evoke briefly certain characteristics of the physical environment, and the use made of it by the village community.

A number of guidelines have been followed in presenting scientific information.

1. The necessity to limit the volume of information presented and to adapt it to the target audience (this explaining one of the crucial roles of the link or leitmotiv in allowing a selection of material while creating a coherent whole);

2. The interest of the historical dimension for the understanding of the environment and the interpretation of a given environment (evolution of the environment in relation to historical events);

3. The importance of basing the interpretation on the cultural heritage as linked to local ecological features, an approach which tends to increase the interest of the public;

4. The need to devote sufficient study time to the form and content of the presentation, and to develop close working relations with graphic artists in the process of preparing an integrated document.

## COLLADO-MEDIANO (MADRID) CASE STUDY

A pilot operation for the interpretation of the environment in the Collado-Mediano, northwest of Madrid, was launched during a preparatory meeting in June 1980, and at a seminar in December 1980, organized by the OPIEM Subcommittee of the Spanish MAB National Committee. The creation of two workshops were proposed, which could be converted into pilot centres on environmental education: the zone of the Sierra del Castillo near the village of Collado-Mediano, an area characteristic of the Sierra de Guadarrama, a favourite place for open air recreation of people from the Madrid urban conglomeration (project supported by the "Diputación" of Madrid and by the Municipality of Collado-Mediano); and the property of Tres-Centos of the Municipality of Madrid, characteristic of the green wooded zone near the city of Madrid. This zone to the northwest of Madrid has been for several years the site of a MAB field project on the perception of the environment, which it was thought desirable to combine with the educational activity.

### Objectives and scientific underpinning of educational programme

The objectives fixed for the whole of the educational operation in the area northwest of Madrid are: to make the population of Madrid agglomeration more aware of the natural environment and cultural heritage of the recreational zones in the region, and to promote enlightened behaviour with regard to the natural and human environment; to develop in the inhabitants of the rural zones affected by the Madrid agglomeration, particularly those of the Sierra de Guadarrama, an increased consciousness of the natural and cultural values of their heritage; to take advantage of the results of the MAB research project on perception of environmental quality in the northwestern zone of Madrid, adapting the scientific information obtained for the purposes of environmental education.

Within the context of these overall objectives, a number of the findings of the environmental perception project have proven useful in the educational operation in the same zone. For example, information on the preferences for different types of countryside and motives of interest of different groups of the public (Gonzales, 1981; Gonzales and Parra, 1980) have been used in drawing up a typology of preferences, using psychometric techniques. These procedures have been used to study changes in perception and in sensitivity as a result of the educational activities.

Work on perception by the local population (Gonzales *et al.*, 1981) has established parallels between the countryside as perceived by different groups of the population and the scientific descriptions of the same areas (Figure 16—2). These results have been used in developing the educational activities (historical reconstruction of changes in land use and their causes, preparation of landscape interpretation documents) and in facilitating communication between different types of "culture" (between the rural population and developers).

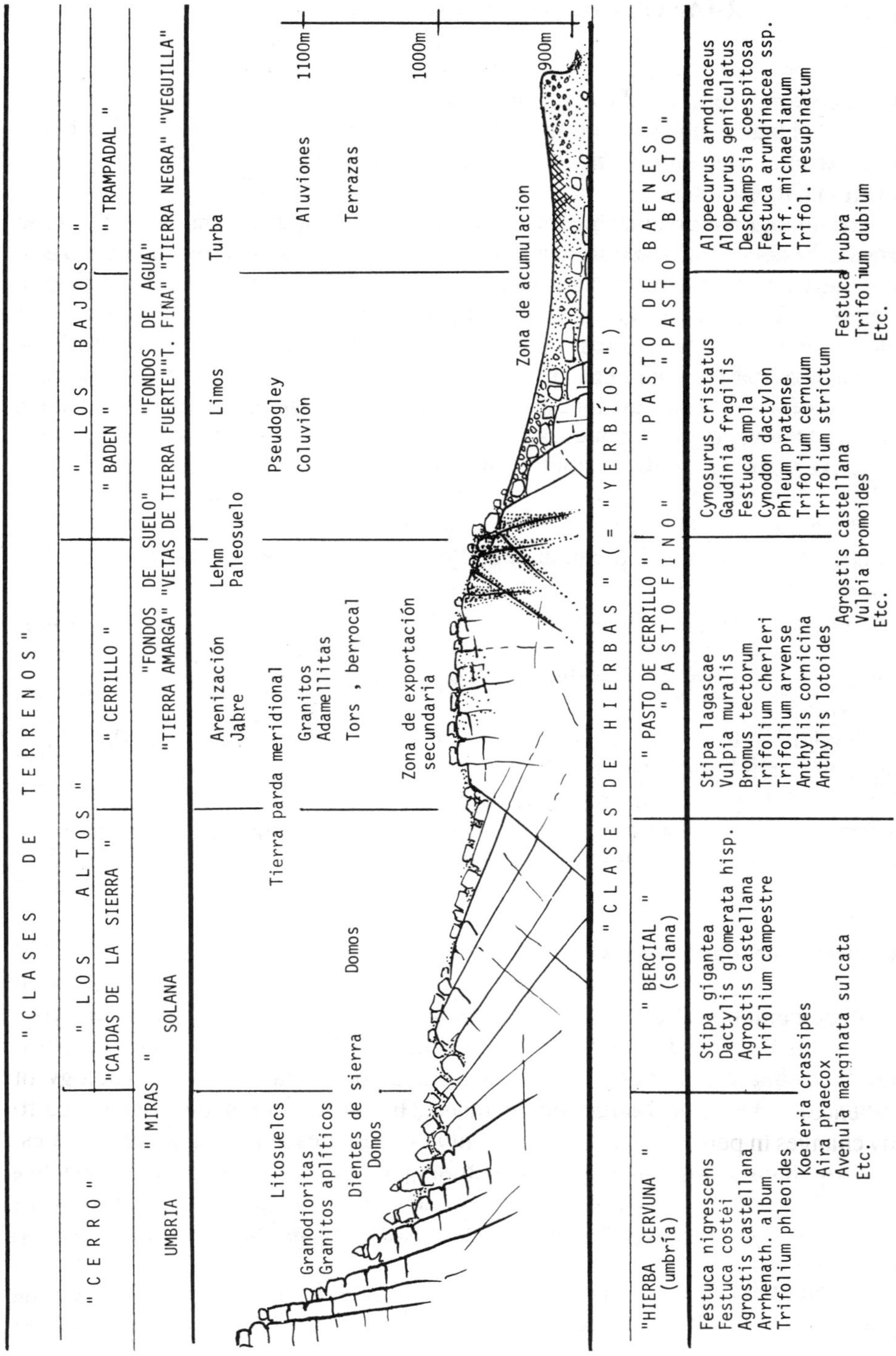


**Figure 16—2.** Comparison of how local people (traditional stock-breeders in the Sierra de Guadarrana) and scientists conceive and describe the same landscape. Vernacular terms in capital letters, scientific terms in small letters

Integrated ecological surveys of the study zone (systematic inventory and syntheses of the geomorphological, pedological, biocenological characteristics and their relation to actual and previous use) carried out within the MAB field project have been very useful for the educational programme, in the preparation of guides and in organizing studies with different groups of the public. Historical changes in the landscape, vegetation and fauna, have also furnished information of wide educational interest.

The adaptation of available scientific information for environmental education purposes has been carried out by an interdisciplinary group comprising ecologists, pedologists and engineers familiar with the problems of the region, as well as local historians. The prime target has been the schoolchildren and the inhabitants of Collado-Mediano, and it is expected that in the final phases of the operation the schoolchildren of the communes in the province will benefit from the installations, trails and centres of interest that are being developed. The municipality has made available to the project a zone in the Sierra del Castillo, where a nature trail is being developed, while the "Diputación" has provided a villa which has been used as a laboratory and teaching room near to the natural environment.

The main educational objective is to increase awareness of the environment by means of interpretation of the natural and human environment (urbanism, agriculture, animal breeding, landscape, ways of life, etc.), with a special emphasis on historical perspectives and changes in the economic and social conditions. Knowledge of these changes makes it possible to explain the presence of very characteristic elements in the natural and urban landscape, whose interpretation is not always very obvious.

In addition to the use of scientific information in elaborating the educational programme, the knowledge of older people in the village has been used to provide details of changes in life style and living conditions and in ways of using the environment. At the same time, this has proved effective in integrating the older people as active participants in the project.

## Interpretative materials on Collado-Mediano

The documentation for the environmental interpretation of Collado-Mediano (Diputación Provincial de Madrid, 1981) is intended for a wide public, since it is considered that information on the pilot zone can be extrapolated to a much wider region in the centre of the Iberian Peninsula. It consists of a booklet of 80 pages, where the historical-ecological interpretation of the village and its neighbourhood is presented in graphical form and with simple text. The historical perspective provides the essential thread of the booklet, in that it links important aspects of the present and past life of the region such as a quarrying, animal breeding, weekend tourism, etc.

Among the central themes is the granite quarry. Empirical knowledge on the quarry and its ecological effects, its traces in architecture, its dependence on economic and industrial factors, etc., can be illustrated in the context of the geological history of the region. The quarry thus serves as an entrée to discovering the geology of the region. Breeding and its important historical changes are also used as one of the links, to reflect changes in land use and landscape transformation (trees, bushes, herbs, problems of fire, etc.). Linked to other land uses, animal breeding constitutes an excellent leitmotiv for presenting history as it can be deduced from careful study of the countryside. Fauna and flora are also presented in a historical perspective and relations drawn with socio-economic factors. Among the aspects treated are changes in plant and animal populations (including decrease in numbers of large predators), consequences of the abandonment of pastures, invasion by *Cistus* and *Stipa*, effects of fire.

## DISCUSSION AND CONCLUSIONS

It is premature to draw up a balance sheet on the influence of OPIEM in changing attitudes and behaviour; such judgement must await the results of evaluations that have already begun for Collado-Mediano. However, some features of the operation seem positive and worthy of retention for use and experiment in the future. For example, multidisciplinary workshops for the conception and preparation of pedagogic materials (films, audio-visual programmes, booklets, brochures, panels), produced locally and within a limited time period, promote the learning of skills and interaction between participants; this is particularly important in respect to international cooperation, given the complementarity of the problems, experience and viewpoints that are brought together. The integration of graphic artists and designers and other specialists in visual and audio-visual presentations within workshop teams has produced better results than designs and materials produced later using sketches. Moreover, an outline or model of an interpretative programme can be produced relatively quickly and cheaply using the workshop approach.

Two main types of scientific information (mainly generated through the MAB Programme) have been used for environmental education — the results of ecological studies (particularly within an historical perspective) specific to the zone concerned, and the results of studies on the environmental perception of the people who are the target of the education programme. The results on perception provide useful clues on how to present environmental phenomena that relate to different cultures and interests, and the most efficient ways of communication (identification of landscape units, use of vernacular terms, especially place names, etc.). The methods and results of local studies of perception are also of enormous interest in the evaluation of attitudes and behaviour. In short, scientific information in the field of perception seems to be absolutely essential if environmental education is to facilitate two-way communication between different groups (for example planners and rural populations). It should also be an obligatory and integral component of development programmes.

Another conclusion concerns the usefulness of links and leitmotivs, both thematic (pottery, weaving, breeding) and physical (specific places or areas to visit, such as a sand dune, olive grove, pond, granite hill, oil-mill . . .). Place examples are particularly important in ensuring that education is active (role of personal discovery) and stimulates the use in environmental interpretation of all the senses (sight, smell, hearing, touch . . .). Concrete links are better than abstract considerations in the discovery of relations and interrelations. Themes from daily life and the nearby environment, as well as the historical interpretation of special circumstances (life style, customs, "roots", etc. . . .), stimulate interest and focus attention, especially when related to information on vegetation, hydrology, fauna, soils, etc.

The participation of local people in the preparation of activities and documents (information materials) is another key ingredient of the approach, for two main reasons. First, interpretation programmes for visitors can benefit from the experience and perception of those who helped shape the local environment. They also benefit from the larger and more vigorous support they receive than if, as is too often the case, they consisted of materials prepared by experts from outside the region, in projects planned outside the region and "parachuted" in. Second, for the residents of the area themselves, the deliberations and concrete achievements that accompany the interpretative process are positive factors which tend to renew and reforge links between the community and its environment. The whole approach is one that complements and completes the more common, the more partial, programmes of agricultural extension work (that aim at increasing productivity, especially through environmental modification) and tourism development (that stress the attractiveness of the environment from the viewpoint of static scenery).

## References

ASSIDJE (Association pour la Sauvegarde de l'île de Djerba) (1980), *Réunion des Experts dans le Cadre de l'OPIEM,* Rapport de Synthèse, Document No. 23, Secrétariat SONNIVAS, Djerba.

CNERP (Centre National d'Études et Recherches sur les Paysages) (1976-79), *Série d'Études Non Publiées et Relatives à Différents Aspects du Paysage de Djerba,* Équipe franco-tunisienne du Paysage littoral, ASSIDJE, Djerba.

CNERP/ASSIDJE (1977), *Recueil Bibliographique de Textes Concernant l'Histoire et le Milieu Naturel et Humain de Djerba,* ASSIDJE, Djerba.

Diputación Provincial de Madrid (1981), *Collado-Mediano. Aspectos Naturales y Culturales,* Servicio Forestal de Medio Ambiente y contra Incendios, Madrid.

El Cadi, N. (1980a), *Textes et Exercices. Commission No. 1: Ecologie,* ASSIDJE, Djerba.

El Cadi, N. (1980b), *Relations Entre les Thèmes: Ecologie, Agriculture et Artisanat,* ASSIDJE, Djerba.

Gonzales Bernaldez, F. (1981), Modelos de Paisaje en Zonas de Montaña, Paper presented to *Coloquio Franco-Español sobre Zonas de Montaña,* Casa de Velasquez, Madrid.

Gonzales Bernaldez, F. and F. Parra (1980), Dimensions of Landscape Preferences from Pairwise Comparisons, In *Proceedings of Conference on Applied Techniques for Analysis and Management of the Visual Resource,* USDA General Report PSW-35, Berkeley, 256-62.

Gonzales Bernaldez, F., F. Parra and M. A. Garcia (1981), Environmental Preferences in Outdoor Recreation Areas in Madrid (Spain), *J. Environmental Management,* 12.

Ministère de l'Environnement et du Cadre de Vie (1980), *Coopération Franco-Tunisienne pour l'Environnement. Une Pédagogie de l'Environnement sur un Paysage,* Rapport OPIEM No. 1, Ministère de l'Environnement et du Cadre de Vie, Paris.

Parc National des Pyrénées Occidentales (1980), *Il y a un Siècle un Cloutier dans le Val d'Azun,* Parc National des Pyrénées Occidentales, Pau.

CHAPTER 17

# The Chilean Altiplano: Natural Environment and Socio-Cultural Traits of the Aymara Population

*Milka Castro, Alberto Veloso and Eduardo Bustos*

Abstract

The Aymara people of the high plateaux of northern Chile have traditionally taken advantage of different microclimates for the production of food and other products, for their own use and for exchange with lower altitudes. Over the centuries, the Aymara have been subject to a series of outside influences, which has transformed their religion, their socio-economic system and patterns of land use. Though it is still too early to assess the impact of modern transportation and spread of basic education, the Altiplano people retain an acute perception of their natural environment, a perception which can have important implications for the development of new productive systems and in helping ensure the balance of man-environment relationships in the region.

*Résumé*

*Le peuple Aymara des hauts plateaux du nord Chili a traditionnellement tiré profit des microclimats pour la production de nourriture et autres produits, pour leur propre usage ou pour des échanges avec les habitants des basses altitudes. A travers les siècles les Aymara ont été soumis à diverses influences extérieures qui ont transformé leur religion, leur système socio-économique et leurs habitudes d'utilisation de la terre. Bien qu'il soit encore trop tôt pour estimer l'impact des transports modernes et de la diffusion de l'éducation de base, les peuples de l'Altiplano gardent une perception aigüe de leur milieu naturel, perception qui peut être porteuse d'importantes répercussions sur le développement de nouveaux systèmes de production et pour aider à assurer l'équilibre des relations homme-milieu dans la région.*

## THE NATURAL ENVIRONMENT

THE northern part of Chile, comprising the provinces of Arica and Parinacota, has a steep relief which extends in a transversal section of about 150 km from sea level to Lake Chungara at 4,600 m. This relief can be subdivided (Veloso, 1981) into: the coastal plains of the Pacific, the coastal ranges (Cordillera de la Costa) and the desert Pampa: the Andean foothills and the Sierra de Huaylillas; and the western chain of the Andes and the Altiplano.

The Altiplano corresponds physiographically to a large plateau varying in width from 20 to 70 km, between 17°30′ S and 28° S. Its average height is 4,200 m. The distinct geomorphological units resulted from intense tectonic movement, particularly during the Pliocene epoch, which modified the existing coastal conditions. The morphology of the Altiplano was affected by intense volcanic activity during the Plio-Pleistocene epoch — activity which still continues today. Post-glacial vulcanism was a determining factor in the creation of a series of volcanos which, like the Payachatas, of the Holocene and recent ages, are over 6,000 m high. Another consequence of this vulcanism was the formation of closed basins with large deposits of water (Lake Chungara, Lake Cotacotani), as well as the formation of marshes from which the local rivers spring (Lauca, Caguena). Glaciation also affected this area, but only up to 4,500 m level, some 1,000 m below the current snowline. The Andean foothills and the coastal regions were not directly affected by glaciation.

Along with this great diversity of relief there is also a great variety of climates and vegetation. These vary according to altitude. Both the human and animal populations are therefore subject to differing atmospheric pressures within adjacent areas. Their responses are correspondingly complex and are geared to the most efficient use of the resources available, with a well-defined pattern of social organization and use of the land.

The region's climate is heterogeneous, with the relief acting as a barrier to climatic influences from the tropical zone. These influences can however be felt on the higher western slopes of the Andes. These different influences have made for sectors which differ in their amounts of rainfall, relative humidity and temperature ranges. Areas above 4,000 m have summer rains (December to March) because of the tropical influences. These rains are torrential and are often accompanied by snow-storms or hail. The year-round temperature is low with the monthly minimum often falling below 0° C, due to considerable daily fluctuations in temperature. In winter, the temperature often drops to 0° C or below, during the night. The region's water resources are provided by rivers such as the Lauca and the Caquena, which flow eastwards towards Bolivia. There are also important and permanent water deposits in lakes and lagoons (Cotacotani and Chungara), which are prevailing features of the Altiplano landscape.

Between the 3,700 and 4,000 m levels, the vegetation is low and shrublike. Above the 4,000 m level, the shrubs are replaced by scrub (*Festuca orthophylla*) and tufts of plants up to the 5,200 m level. This areal distribution of vegetation is the predominant form, and is cut by islands of hygrophilous vegetation located in marshland (*bofedales*). There are also small scattered woods of *Polylepis tarapacana* and formations of *llareta* (*Azorella compacta*). The climatic conditions in the foothills are milder, with a tendancy to desert-like plant life including succulent plants (*Cactaceae*) and low-lying shrubs known as desert "jaral". In the foothills a distinction must be drawn between the conditions prevailing in ravine vegetation areas (*Putre, Zapahuira*) where water is always available and where there is an azonal form of vegetation resulting from introduced species and crops.

Along the coast and in the desert pampa region, there is almost no rainfall. Oceanic influence is responsible for the mists (*camanchacas*) which drift inland towards the desert through the valleys and coastal ravines. These give rise to patches of desert-like vegetation. In the ravines themselves permanent waterways (e.g. the Lluta and Azapa rivers) supplement the effects of the mist and thus favour intensive agriculture. The ocean ensures stable temperatures throughout the year, ranging from 18.7° C (monthly average in July) to a summer high of 22.3° C (February). Daily fluctuations are much narower than further inland.

## THE NATIVE POPULATION

Between the years 1000 and 1500 BC, the pre-Inca Aymara population which was settled along the shores of Lake Titicaca expanded into adjacent lands. The groups situated near the Chilean Altiplano were known as the Kingdoms of Lupacas, Pacajes, Carangas and Lipez. The scarcity of resources, a characteristic of the high plateaux, led these peoples to devise a peculiar system of exploitation so as to make use of the different microclimates. Thus, corn and chili were obtained from the hot valleys, coca from tropical humid forests, potatoes and cereals from higher altitudes, and llamas and alpacas grazed in the highest plateaux. Such a system included a complex network for the interchange of products ensuring access to the resources of lower lands, and thus serving to counteract the limitations of life at high altitudes.

Throughout their history the communities of the Altiplano were also subject to outside influences, such as the Incas, the Spaniards and the Chilean Republic. Contact with Europeans, especially, meant a deep transformation of religion, and of economic and social organization. For one thing, some of the native population, that had developed a relatively productive economic system, were forced to labour in the mines and as servants to the Spaniards; others had to work as muleteers in order to satisfy the transportation needs imposed by the new economic scheme of the Europeans.

The land use changed as a consequence, with some land being devoted to pasturage and some to newly introduced species (wheat, fruit trees, vineyards, ovines, etc.). Because the communities living in higher lands depended on those inhabiting lower valleys (up to 3,500 m), the main result was a complete collapse of the Andean economy. The shortage of workers led to the importation of black slaves, mostly to the lower villages. The autochtonous population either mixed with them or moved to other localities.

In spite of these pressures and the changes they brought about, the Aymara groups even nowadays have some ancient cultural traits. Their beliefs and rituals have been very carefully preserved. In the holidays corresponding to those imposed by Catholic liturgy, a mixture of religion and paganism can be observed; for instance, the coca leaves and the "cocoroco" (cane sugar alcohol, 96 per cent) are ritual elements included in the celebration of the Patron Saints of the community.

The Andean beliefs are a reflection of the dependence of man on supernatural forces. Protection is invoked to avoid natural disasters such as drought, epidemics and disease, storms, predation, etc. The goodwill of the gods is sought by sacrificing animals, by burning ritual objects or even by offering blood to Mother Earth, a goddess known as Pachamama. Most ceremonies are held within the nucleus of the family, since the family is the basic unit of production. There, women and children are in charge of pasturing, an activity essential to the Altiplano economy. The men, although they may play a role in this context, are mainly obliged to seek unskilled work that will be paid in money, so that they are gradually incorporated into the cash economy.

Under the ancient system of land management, high efficiency in grazing was achieved because the family at large owned a rather extensive piece of land with diversified forage resource. Under this system, the family property was not subdivided. However, due to changes in the ownership of land, initiated in the 1930s, and to the emigration (mostly of the younger generations) to the urban centres from 1950 on, the system is no longer general. Livestock are represented by herds of sheep — not always well adapted to the climate and thus, of low fertility and productivity for meat or wool — and South American camelids, llamas and alpacas, better adapted to grazing in poor lands. The alpacas need green, hard pastures however, and thus the mobility of herds and shepherds are a common feature of Andean life. During the rainy season (December to March) in the Altiplano, pasture is abundant and the animals are kept in the *bofedales,* whereas in the dry season (April to November) the llamas are taken to the hills, and the grass in the *bofedales* is reserved for the alpacas.

Families living at high altitude (over 4,300 m) move their animals to places where creeks are found, and the presence of green pastures in these areas leads to the establishment of seasonal villages named *majadas,* strongly dependent on water resources. All parts of the animal are used: wool, skin, meat (either fresh or dried as *charqui*), blood, viscera, fat, etc. Even the foetuses have a magical value and are thus commercial items. The faeces, once dried, are used as fertilizer. Curiously, only the milk is not used. The male llama is commonly used as a transport animal.

The natives of the Altiplano have an accurate and fine perception of this natural environment. Their knowledge of plants and their uses — including medicinal properties — is remarkable, but mostly among the older individuals. Through the grazing activities, they have built up a complete inventory of the plants that may feed or harm animals or humans. Other plants are used for ritual or magical purposes. The diet is broadened by the traditional system of exchange *(trueque)* with the agricultural communities of the pre-Andean range and lower lands. Usually, trips taken for this purpose, which reach even the coastal valleys, are undertaken once a year. Potatoes obtained from the lower lands are frozen and dehydrated, taking advantage of the extremely low temperatures of June and July, thus being transformed into the so-called *papa chuño* that remains consumable for very long periods.

Large groups travel not only for trading, but also for the celebration of religious festivities. There are many traditions related to the protective role of some hills, the most venerated being Tatasavaya, Sajama, Sirijnay and Jachura (in Bolivian territory) and sanctuaries such as Copacabana, close to Lake Titicaca.

With the opening of new roads into the Altiplano great changes have occurred. Trucks loaded with different merchandise and small merchants can today reach even remote Altiplano villages, thus modifying their previous economic status. Followers of some Christian churches have once again evangelized the villagers, hence drastically changing the myths and beliefs of the Andean people.

In the high plateaux of the Andes, men have learned to survive by turning a potentially hostile environment and the scarcity of natural resources to their own advantage. They have developed a kind of historical and cultural fitness based on an autochthonous perception of their surroundings. The question that remains to be answered is: what will be the outcome of the changes imposed on Andean life by the advent of modern transportation and basic education now available to the Aymaras?

## References

Castro, M. (1982), Estrategias Socioculturales de Subsistencia en las Comunidades Ayamaras Altoandinas en el Interior de la Provincia de Arica, In *El Ambiente Natural y las Poblaciones Humanas de los Andes del Norte Grande de Chile (Arica, Lat. 18°28′S),* Volume 2, *Los Poblaciones Humanas del Altiplano Chileno: Aspectos Genéticos, Reproductivos y Socioculturales,* Unesco, Montevideo, 99-132.

Murra, J. (1975), *Formaciones Economicas y Politicas del Mundo Andino,* Instituto de Estudios Peruanos, Lima.

Veloso, A. (1981), Ecological, Biological and Genetic Research Amongst the Vertebrate and Human Populations of the Altiplano (Chile), in *L'Homme et Son Environnement à Haute Altitude, Environmental and Human Population Problems at High Altitude,* CNRS, Paris, 37-43.

CHAPTER 18

# The Use of a "Key Problem Approach" in a Technical Cooperation Project for a Himalayan Hill Area

*Jiro Kawakita*

Abstract

A key problem approach has been refined as part of a technical assistance project for a group of Nepalese hill villages. Various kinds of holistic presentation were effectively used in the elaboration of this approach, notably a system for integrating qualitative data. The use of this sort of approach is increasing rapidly in Japan, including its application in the solution of environmental problems and in environmental education.

*Résumé*

*L'auteur a formulé une "approche des problèmes clefs" lors de l'exécution d'un projet d'assistance technique dans des villages montagnards népalais. Plusieurs méthodes de présentation holistique ont été employées avec succès dans cette approche, notamment un système d'intégration des données qualitatives conçu par l'auteur. Ce type d'approche se répand actuellement très rapidement au Japon, y compris son utilisation pour la solution des problèmes environnementaux et l'éducation relative à l'environnement.*

## CULTURAL ECOSYSTEMS OF NEPAL

THE cultural ecology of Central Nepal can be divided into four major zones, three of which coincide with the various natural ecosystems found at different elevations (Figure 18—1). The Hindu zone is found in the lowlands under *c.* 1,200 m above sea level, the Tibetan zone is found in the highlands above *c.* 3,000 above sea level while between the two (i.e. between c. 1,200 and 3,000 m) are found people who live according to traditional tribal culture at the local level, but who look to either of the adjacent Hindu or Tibetan civilizations for social life outside the community. The fourth culture — that of the Newars — has spread out from the Nepal Valley to the market towns throughout the country. Until the Gurkha Conquest in 1769, many city states were found in the Nepal Valley. The Newar culture has assimilated many traditions of this mode of life. The layering effect of these Himalayan cultures is illustrated schematically in Figure 18—2.

## A HOLISTIC TECHNOLOGICAL ASSISTANCE PROJECT

Since 1963, the author has worked with a technical cooperation project in a place called the Sikha valley, in the foothills southwest of Annapurna South. The main inhabitants are Magar people. The project was formalized in 1974-75 and five villages of the Sikha valley were to participate under ATCHA, the Association for Technical Cooperation to the Himalayan Areas (Kawakita, 1975; Kawakita and Niitsu, 1978; Yamada, 1978).

### Understanding ecosystem condition

Before the implementation of the technical cooperation project, my understanding of the environmental problems of the Sikha valley was as follows:

**Succession of ecosystems:** Some centuries ago, subsistence was based on swidden agriculture and hunting (Figure 18—3). Gradually, the small, semi-migratory settlements were transformed into something more sedentary, so that they formed a stable cluster around a core. Extending outward from the core, in concentric circles were terraced upland fields, a transitional zone called *pakha* (grassland dotted with bushes and trees) and forests called *jangal.* In agriculture, a three-field system was adopted with different animals such as cattle, buffalos, sheep and goats in each. During the summer, the animals were taken from the pastures to the *jangal*, and sheep and goats were grazed in the alpine grasslands called *bugyen*, if these

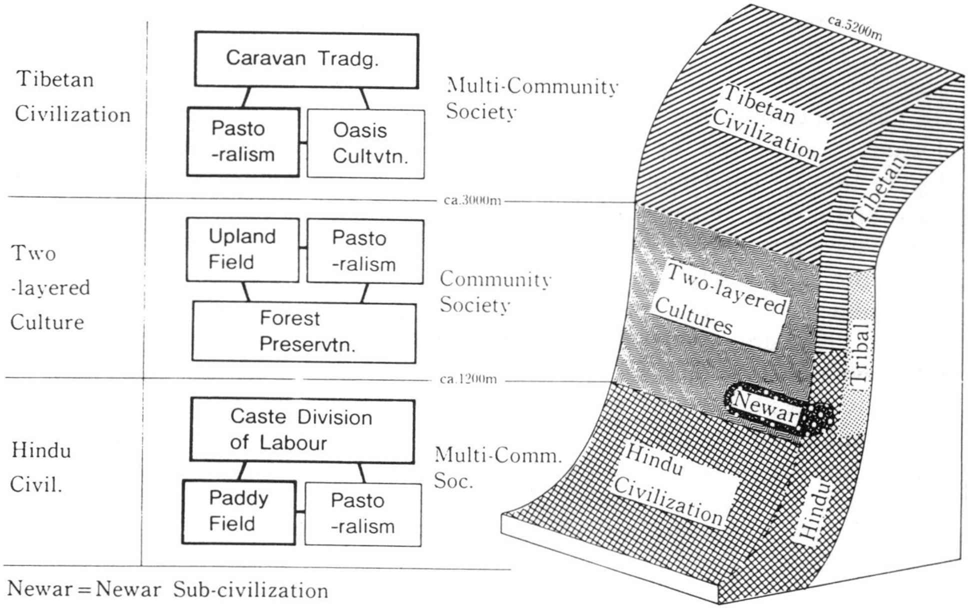


**Figure 18—1.** Four cultural ecosystems in Nepal

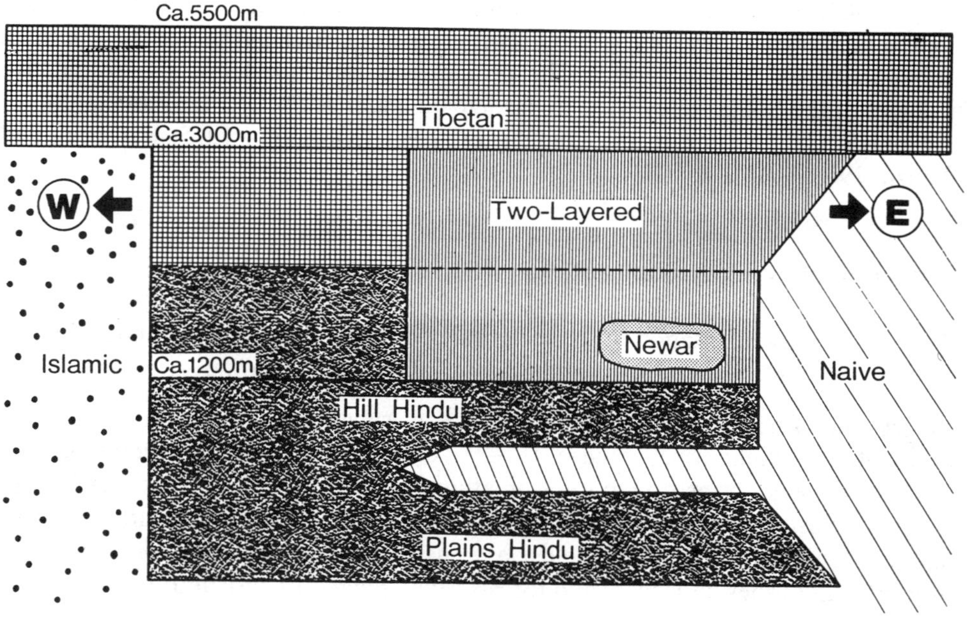


**Figure 18—2.** Arrangement of ecosystem and settlement patterns in the Himalayas

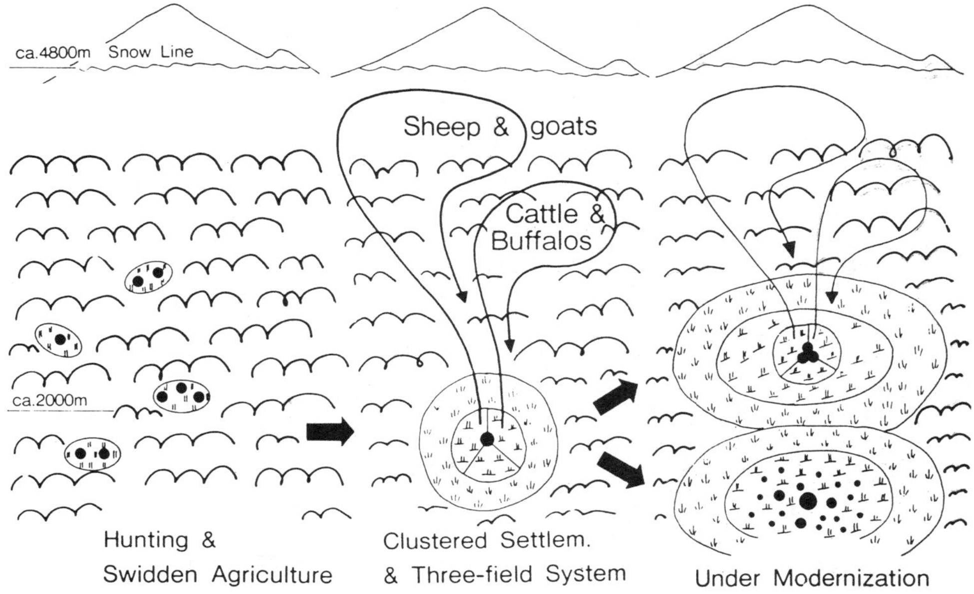


**Figure 18—3.** Three successive stages in ecosystem development in the Sikha valley

were available. Animal wastes were utilized for fertilization of the upland fields after harvests.

During the past 100 years, with increasing population and other pressures, a third stage is underway. In villages with strong social cohesiveness, the concentric structure has simply become enlarged, without changing the pattern. But in areas where social solidarity is weak, the settlements have become dispersed. In both cases, summer pasturage in *jangal* and *bugyen* has declined.

**The structure of the environmental problem:** Serious environmental problems have appeared in conjunction with this third stage (Figure 18—4). Initially, the direct and indirect influence of modernization caused rapid population increase, bringing pressure on the limited resources of the villages, while pushing people to emigrate. Excessive construction of terraced fields and expansion of the *pakha* led to destruction of the forests. In consequence, two major problems have occurred.

As forests became more distant from settlements, the collecting and carrying of grass and firewood became a terrible burden for villagers. Therefore feeding of animals within each settlement only became possible at the lower levels, which meant that there was less manure for the crops and the harvest decreased. Also the water supply from nearby springs was exhausted and people were forced to spend many hours every day in carrying water from more distant springs. In brief, the following process may be observed: shrinkage of forest, shortage of labour to help in daily transport, decrease of agricultural and animal productivity.

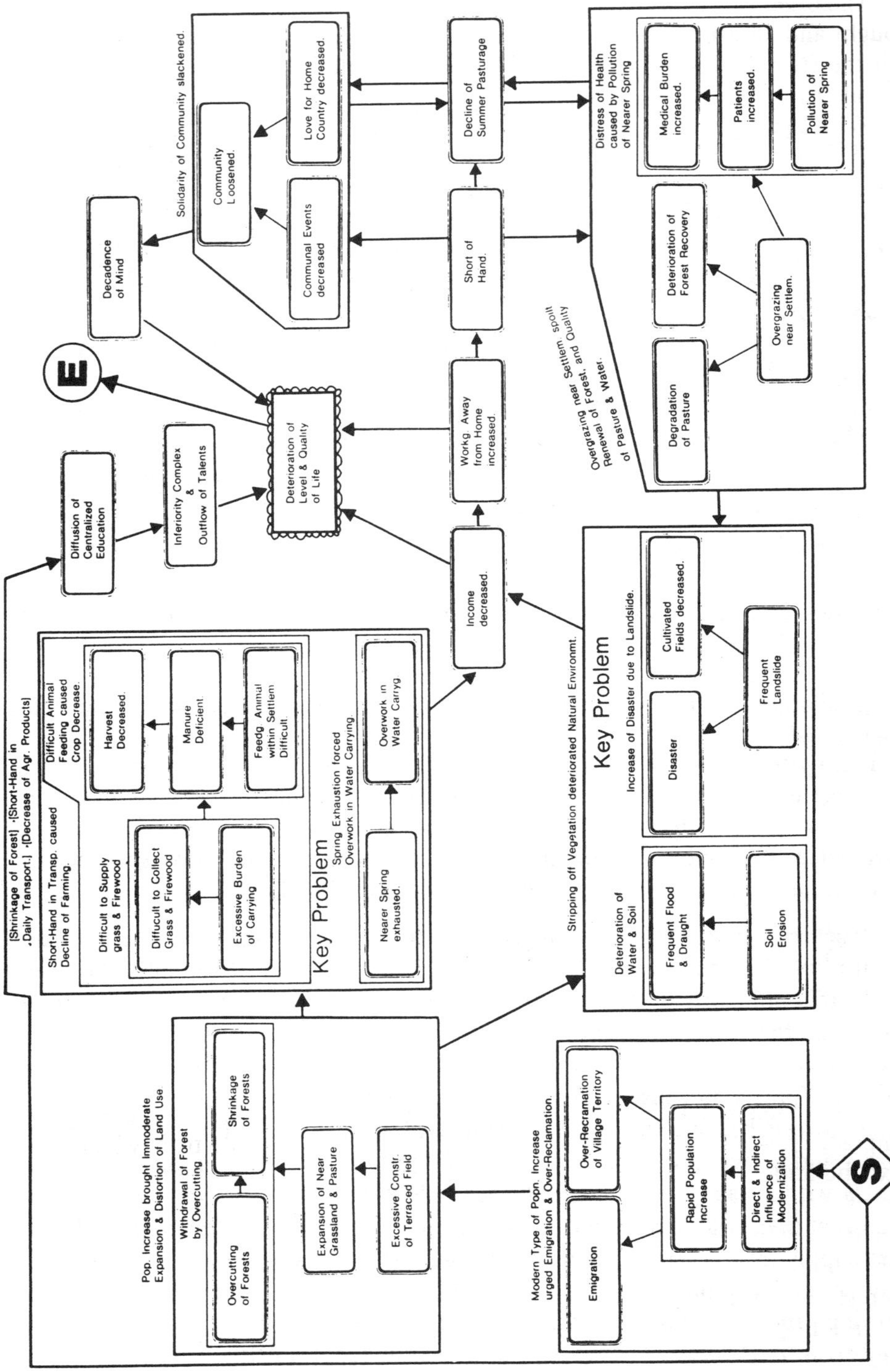


**Figure 18—4.** The process of ecosystem breakdown and degradation in Sikha valley

The other problem stems from soil erosion. Along with it have come flood and drought and landslides, all more frequently than before. These disasters, together with the stripping off of vegetation, have contributed to the deterioration of the natural environment.

With lower productivity, income has dropped, villagers have had to go further away to work, with resulting labour shortages in the village. For this reason, traditional summer pasturage in distant *jangal* and *bugyen* has declined, while overgrazing near the settlement has intensified. Moreover, quality of drinking water in nearer springs has deteriorated because of overgrazing, and sickness has increased due to water pollution.

All these problems have gradually undermined the sense of community solidarity. Communal events including festivals have decreased and with the drop in living standards people have become demoralized.

At the same time, other problems have emerged with the spread of primary schooling. As in many countries, the system is centralized, using standardized textbooks filled with foreign ideas which overwhelm the villagers with a sense of inferiority about their own home land. The result, even in this remote area of a Himalayan country, is a 'brain drain' and a general trend which discourages self-reliance and any efforts to develop and maintain the local environment.

## Appropriate technologies and their implementation

Having singled out the transport of firewood and water as a key problem, some appropriate technologies have been investigated, which might provide assistance: pipeline (PL), ropeline (RL) and natural force pump (NFP) (so-called hydraulic ram). The evaluation of their utility for Sikha is shown schematically in Figure 18—5. On the left is represented the "good old days" when the population was small. The forest was moderately and evenly used, and cultivated fields were fertilized sufficiently. Everything was near at hand. In the middle of the figure is shown the present situation; everything, including tree seedlings in the forest margin, has been cut or grazed. Only a small portion of the upland field zone has received sufficient manure; springs are distant and sometimes located below settlement and are thus polluted. Landslides are frequent near settlements.

On the right is a representation of ATCHA's scheme for redressing the ecological imbalance in Sikha. First a ropeline would be installed to help give access to remoter forests and forest resources (tree leaves and grass for fodder and manure, compost, firewood, etc.) so that they might be collected evenly from wider areas. Next, building a pipeline would permit bringing abundant and better water from distant springs to the settlement, without expanding any labour. If the water source were located below the settlement, it could be raised by natural force pump. With such labour saving devices, time and energy are available for such projects as afforestation and distributing fertilizer to fields.

In Figure 18—6, Note 1 refers to the tasks already implemented by ATCHA. Note 2 refers to future actions to be undertaken by ATCHA. Note 3 refers to actions that were spontaneously initiated by the inhabitants in response to ATCHA's projects.

## FORMULATION OF KEY PROBLEM APPROACH

Figure 18—7 illustrates the different steps involved in our approach to analyzing and dealing with specific environmental problems in a certain area. The first step (1) is to develop awareness in the project participants. The second step (2) consists of surveying and identifying local people and places so that the outside field workers establish a rapport with the new environment. The third step (3) is to learn how the local inhabitants perceive of things. The fourth (4) is to draw up a list of proposed research items and field work activities. The fifth (5) is the integration of the various elements making up the cultural ecosystem. The sixth (6) is for project workers and inhabitants to identify and evaluate the key environmental problems. The last two steps (7 and 8) concern the selection and/or invention and introduction of approp-

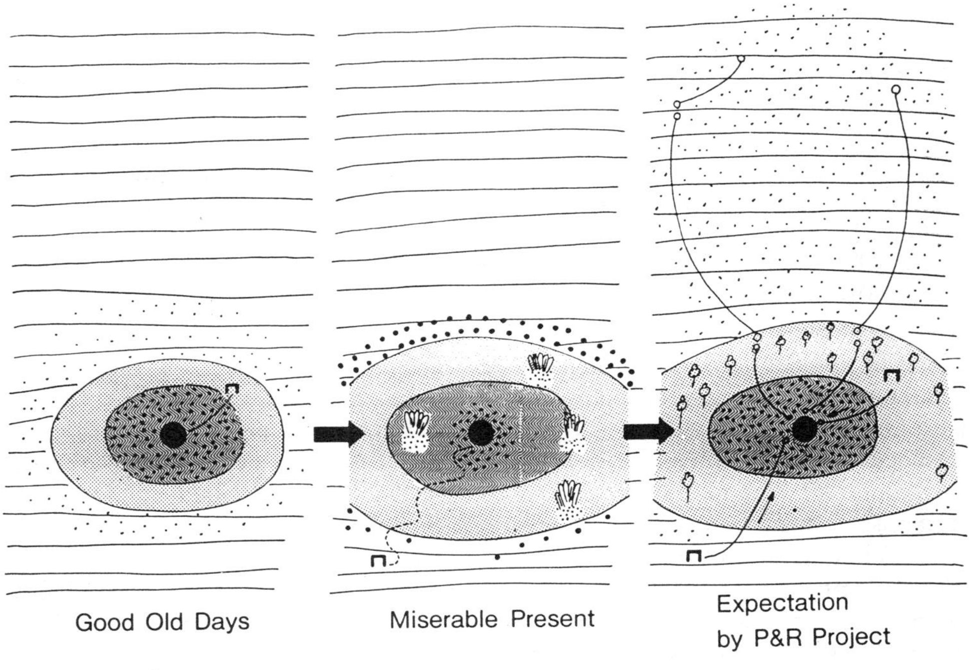


**Figure 18—5.** Schematic representation of ecosystem degradation and rehabilitation

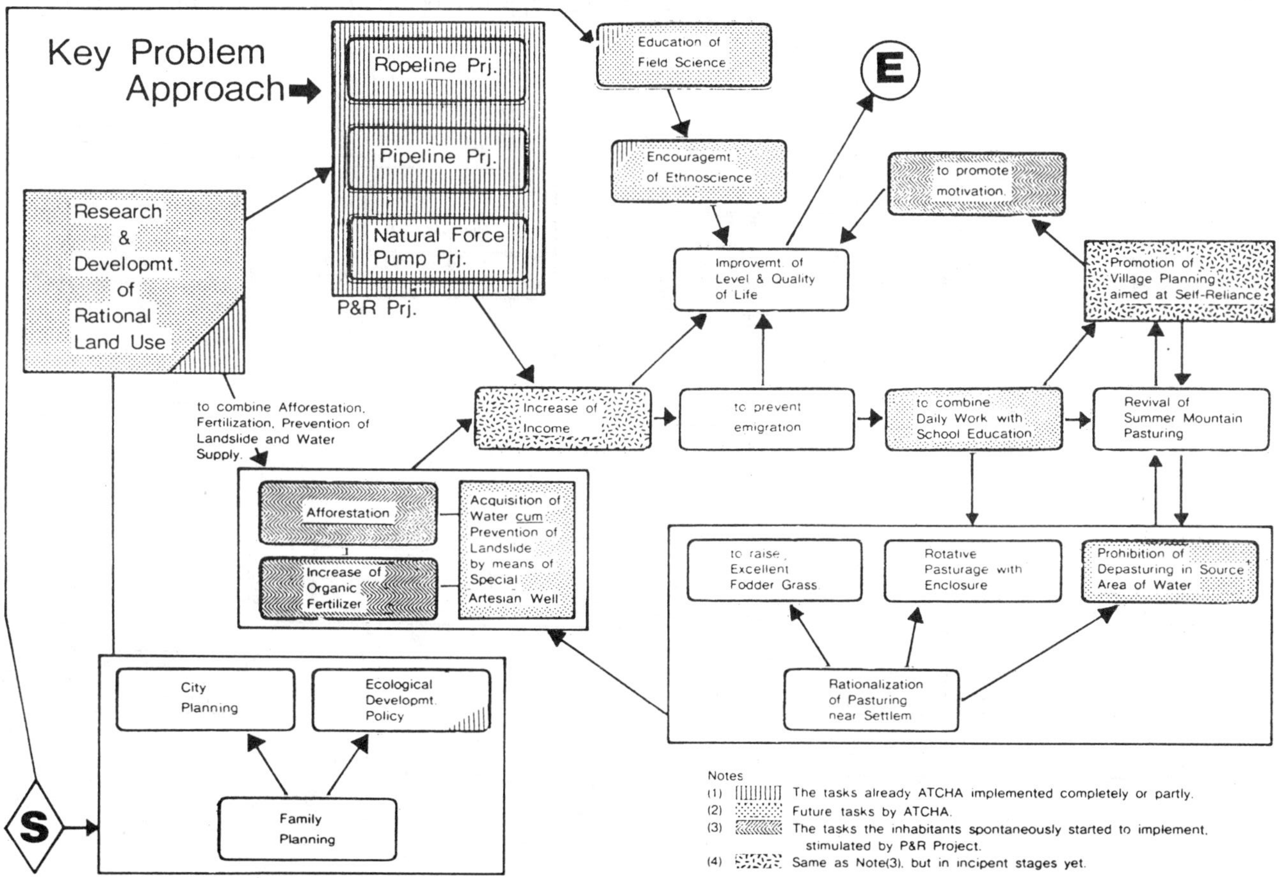


**Figure 18—6.** Technical cooperation scheme for ecosystem rehabilitation in the Himalayas, developed by the Association for Technical Cooperation to Himalayan Areas (ATCHA)

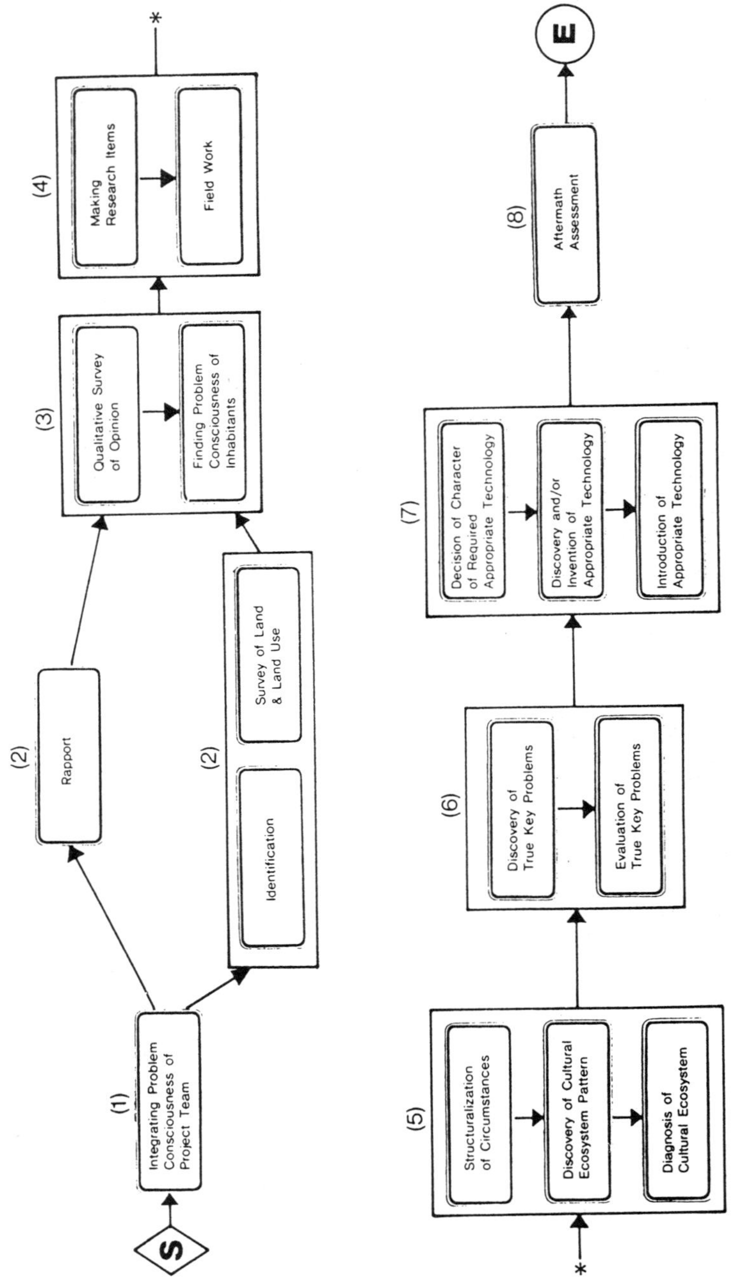


**Figure 18—7. Principal steps in the key problem approach**

riate technologies through a cooperative endeavour and the technological and environmental assessment of the success of the overall exercise.

For the integration of qualitative data, a system of data processing has been devised called the KJ method. The most important principle of this method is that data must be allowed to speak for themselves without being distorted to fit any preconceived theory or hypothesis. The method is speedily being diffused in almost all sectors of Japanese society since 1967, including the field of environmental science. Its value lies in enabling us to get a holistic view of the natural and cultural circumstances surrounding an environmental problem.

## References

Kawakita, J. (1975), Technical Assistance in the Himalayas, *The Wheel Extended*, (Toyota, Tokyo), Special Summer Issue, 26-36.

Kawakita, J. and K. Niitsu (1978), New Possibilities in International Technical Cooperation. *The Wheel Extended*, 7(4), 2-7.

Yamada, K. (1978), Impact of Technical Cooperation in the Himalayan Hill Area, *The Wheel Extended*, 7(4), 15–22.

CHAPTER 19

# Promoting Public Awareness of Enviromental Issues in the German Democratic Republic

*Egon Seidel*

Abstract

In the German Democratic Republic, both renewable and non-renewable natural resources are intensively utilized in the interest of a continuous steady growth of industrial and agricultural production. Within this context, this contribution describes some of the activities for systematic promotion of public environmental awareness and environmental protection in the GDR. The importance of providing the broadest possible scientific-technological information, suited to different groups within the community and to the environmental problem being treated, is explained. Emphasis is also given to the need for close cooperation among all governmental and social institutions dealing with environmental problems and the various media for communicating with the public.

*Résumé*

*En République Démocratique Allemande, les ressources naturelles tant renouvelables que non-renouvelables sont intensivement utilisées en vue d'une croissance ferme et continue de la production industrielle et agricole. Dans ce contexte, cette contribution décrit quelques unes des activités poursuivies en RDA pour la promotion systématique de la prise de conscience par le public de son milieu et de la protection de celui-ci. On y expose combien il est important que l'information scientifico-technologique la plus large possible soit présentée dans une forme adaptée aux différents groupes de la communauté et au problème d'environnement auquel on s'attache. L'accent est mis sur la nécessité d'une étroite coopération entre les institutions gouvernementales et sociales traitant de ces problèmes d'environnement et les différents chenaux de communication avec le public.*

## THE PROMOTION OF ENVIRONMENTAL AWARENESS

THE German Democratic Republic (GDR) has recorded a stable and continuous growth in industry and agriculture since its foundation. This means that renewable and non-renewable resources, which are already intensively used, have to be utilized even more intensively and it places ever higher demands on the economy and technology to solve important problems such as: the development of more efficient technologies so as to ensure a better utilization of wastes; the implementation of low-waste, low-energy and low-water technologies on a broad national economic scale; the disposal of toxic wastes. Their solution will lead to a reduction of environmental pollution and an increase in the availability of the natural resources (particularly considered in terms of air, water, soil, flora, and fauna).

Because of the increasing complexity of environmental problems the solution of the above-mentioned problems is not merely a scientific and technological one. On the contrary, it is necessary to understand the links between economic, social, political, cultural, and natural (including ecological) factors and to establish a dynamic equilibrium between nature and society. The promotion of active environmental awareness of the public is important (Seidel and Protze, 1980), because only with the active participation of the broad public will it be possible to solve environmental problems in the interest of society.

In the GDR, environmental education is a concern of society as a whole and constitutes a general integrated component of the unified socialist system of education and training, including the work among the public. Education is continued for all strata of the population, in all social spheres, within the framework of formal education, through special educational and training institutions (from kindergarten to university) and through social organizations, enterprises, mass media, and other vehicles of informal education (Unesco, 1977). It is characterized by the following general objectives (Protze, 1981):

—Dissemination of fundamental knowledge on environmental protection enabling the citizens to better comprehend the complexity and dynamics of environmental protection and to understand the scientific, socio-economic and technological interrelations in environmental protection;

—Education promoting responsibility of the citizens for their environment, with the aim of encouraging active, creative participation in the solution of problems concerning the natural, working, and living environments and in rational environmental management;

—Establishment of an optimistic state of mind based on a scientific world-outlook recognizing the objective possibility of a long-term, complex solution of environmental problems within the context of an increasing material and technological basis of the national economy;

—Improvement in decision-capabilities at all levels in relation to environmental protection;

—Promotion of education related to the resolute struggle for international peace and disarmament, because of the wasteful exploitation of natural resources resulting from increased armaments.

Thus, environmental education in the GDR provides essential instruments for environmental protection when the ". . . . dialectics of what is feasible today and what will be possible to achieve tomorrow" are mastered and when the unity of education and training, of theory and practice is achieved.

The social and economic development in the next coming years will set new, qualitatively higher standards for the public concerning the problems of environmental protection. It is essential to prepare workers, scientists, engineers, designers and employees so that they understand better how to manage rationally natural resources, e.g. natural water resources.

The basis is the systematic dissemination of general environmental knowledge within the framework of formal environmental education which is carried out in the framework of a number of social organizations and institutions, including: Advisory Council to the Council of Ministers; ministries, territorial governmental organs, enterprises and institutions in spheres of the national economy related to the environment; the National Front; the Society for the Dissemination of Scientific Knowledge ("Urania"); the Society for Nature and Environment at the League of Culture; the engineers' organization "Chamber of Technology"; and scientific societies.

Of particular importance is the need to make available well-founded and comprehensive information about the tasks, results, aims and solutions of environmental problems, through preparation and diffusion of different kinds of publications. To enable working people to counteract environmental dangers and nuisances, it is necessary to inform them about the social and economic causes and effects of such dangers. For example, promoting understanding about the manifold interrelations between water management and other spheres of the national economy, and the functions of numerous social organizations in this field, requires a continuous programme of heightening of environmental awareness.

# EXPERIENCE AND RESULTS IN PROMOTING ENVIRONMENTAL AWARENESS

The basis of activities of the MAB National Committee of the German Democratic Republic to promote environmental awareness is the long-term plan, the five-year plan and the annual national economic plan for scientific-technological and economic development of the GDR, considered in relation to international programmes such as Unesco-MAB and UNEP. In this way, it is possible to ensure that scientific and technological developments are put to use at national and international levels, and that knowledge of the scientific and technological advances necessary for environmental protection are developed in a systematic and planned way. The latter entails making optimal use in planning future studies of the experiences gained by individual scientists and institutions taking part in MAB projects, for example by discussing programmes and results at colloquia. It also calls for rapid dissemination of new knowledge in various ways, and in particular through education measures at the non-formal level. The population as a whole is invited to contribute their own proposals for the better utilization of natural resources, for instance concerning the recycling of wastes and the introduction and use of low-waste technologies for reducing emissions and preventing environmental pollution.

## The public and MAB projects

The MAB National Committee cooperates closely with different educational bodies and programmes in relation to MAB themes and projects. As illustration, mention can be made of various activities related to freshwater ecosystems. In this field, good results have been achieved in working with the public on problems of eutrophication and restoration of surface waters. One key element in this work has been coordinated planning in respect to public information materials (articles, reports, radio and TV programmes, lecture series, exhibitions and shows, films and slide-tape lectures, printed publications, wall newspapers and posters) produced by different organizations.

Planning has also helped to ensure a continuity in work for promoting public interest and involvement. This is reflected in the great variety and volume of information on eutrophication and restoration of surface waters in the press, radio and television. Numerous publications deal with important problems of mass growth of algae in lakes and ponds, of deep-water aeration in reservoirs and of deep-water removal from lakes. The conditions for the discharge of communal and industrial waste water into water bodies are explained, with a view to preventing the mass development of algae and of water weeds. Information is made widely available on the beneficial effects of applying laws, standards and other administrative measures to nutrient input into surface waters.

Much effort is also devoted to explaining essential preventive measures against eutrophication in a manner understandable to the public. Specific activities have included the organization of press visits to the Lake Müritz region (to demonstrate results of MAB studies on multiple resource use) and the preparation of two films on possibilities and methods of protecting lakes against excessive nutrient input. Other films have been produced in the past three years and shown during numerous training programmes, dealing with such topics as water balance, automatic control of flowing waters and measures to counter oil spills.

In a similar way to these activities based on MAB work on freshwater ecosystems, environmental education has also been promoted in other problem areas dealt with by MAB, including temperate forest ecosystems, and the environmental effects of pollution and large engineering works, and nature conservation. For instance, within the overall scheme for nature protection in the country (Table 19—1) the mass media have provided extensive coverage to results obtained at the Steckby and Vessertal biosphere reserves, which have been used as part of teachers training programmes. A special issue of the journal "Landschaftspflege und Naturschutz in Thüringen" (Landscape Conservation and Nature Protection in Thuringia) has been devoted to the Vessertal biosphere reserve, and a similar publication is projected for the Steckby biosphere reserve. And considerable public attention has been expressed in the 10-month post-graduate training course on ecosystem management at Dresden, designed for participants from developing countries in Asia, Africa and Latin America and supported since 1977 by UNEP, Unesco and a number of MAB national committees.

At a somewhat different level, the outcome of MAB projects has been published in a special exhibition "Environmental Protection", that is open each year from May to October in the grounds of the International Gardening Show at Erfurt. This is visited each year by some 600,000 people, and in one year an estimated 12,000 visitors saw projections of seven films related to MAB projects. The results of MAB activities are also publicized through the GDR agricultural exhibition "Agra".

The significance that is attached to environmental protection in the GDR is also reflected in the variety of publications on environmental matters issued by publishing houses in the country. The range of specialized and popular scientific literature includes all spheres of environmental protection. Out of a total of 78 publishing houses, 19 have activities in this field. More than 2,000 public libraries of a general or specialized character, in all the larger towns and villages, offer these publications to the population. Furthermore, scientists involved in MAB are regularly invited to comment on topical problems of environmental protection, in daily newspapers, on the radio or in TV programmes. In the first 9 months of 1981 alone, a total of 34 programmes of 3 to 30 minutes duration on environmental questions were produced by the central national radio in the categories "science", "economy" and "school broadcasting". A similar range of programmes was aired by the 12 regional stations and studios at the country level.

**Table 19–1.** Different types of protected areas and landscapes in the German Democratic Republic, and their role in environmental protection and education

| Type Characteristics | Natural monuments | Nature reserves | Protected landscapes | Biosphere reserves | Nature parks (proposed) |
|---|---|---|---|---|---|
| Number and surface area | More than 9,100 monuments | —735 reserves<br>—90,700 ha<br>—0.84 per cent of the GDR territory | —410 protected landscapes<br>—1.9 million ha<br>—18 per cent of the GDR territory | —Steckby-Lödderritzer Forst (2,113 ha)<br>—Vessertal (1,284 ha) | Müritz lake region (5,300 ha) |
| Key features and examples | Single objects or areas up to 3 ha having scientific-cultural-historical value | —Protected forest<br>—Protected water or fen<br>—Geological reserve<br>—Zoological reserve<br>—Botanical reserve<br>—Nature reserve with diversified and complex character | —Area with beautiful scenery<br>—Management according to special landscape protection plans | —Representative areas with various types of ecosystems or unique biotope characteristics, divided into core region and buffer zone towards the neighbouring area<br>—Confirmed by Unesco in the framework of the MAB Programme | —Large landscapes with multiple functions where all natural resources are managed in a central, planned manner |
| Functions | —Preservation and maintenance of these objects<br>—Scientific studies<br>—Education purposes | —Preservation of biotopes and species<br>—Scientific research<br>—Genetic reserve<br>—Environmental education<br>—Culture | —Recreation<br>—Contribution to the creation of stable ecosystems and landscapes<br>—Economic use while preserving landscape characteristics<br>—Examples of optimum landscape preservation | —Protection of ecosystems and species<br>—Long-term scientific studies on structure and dynamics of ecosystems<br>—Environmental monitoring<br>—Excursions and education | —Multiple use of stable landscapes<br>—Recreation<br>—Research<br>—Bio-monitoring<br>—Environmental monitoring<br>—Preservation of flora and fauna<br>—Education and culture<br>—Representation |

## Environmental problems and non-governmental organizations

A number of social organizations play an important role in working among the public on matters of environmental protection. For example, the National Front, an association of some 330,000 volunteer citizens of all political forces, has as one of its major aims the improvement of working and living conditions of the overall population. Through its branches in various districts, towns, villages, etc., the Front promotes activities on the theme "More beautiful towns and villages — Join in", designed to encourage citizens to participate in improving their living and working environments (Lüer, 1981). As a result new facilities have been developed in living quarters, considerable national economic benefit has accrued and above all the responsibility of citizens for their environment has been raised. Voluntary work campaigns for spring and autumn clean-ups are held in enterprises, institutions and neighbourhoods, and involve more than 2.5 million people. As part of these work campaigns, wastes are collected as a contribution to pollution reduction. For the maintenance of green spaces in residential areas, 85,700 maintenance contracts were concluded in 1981 with house communities and families, and more than 2 million citizens regularly take care of 343,000 ha of green areas.

The Society for the Dissemination of Scientific Knowledge (Urania) plays a key role in work on environmental protection among the public, through popular presentations of the latest scientific and technological findings. About 700 out of a total of 45,000 scientists, social scientists and technologists, cultural experts and educationalists give public lectures in enterprises and institutions, in all towns and larger villages. Urania also sponsors special functions for young people in schools and youth clubs, under its programme "Jugend-Urania" (Merzdord, 1981).

The Society for Nature and Environment, founded in 1980 by the GDR League of Culture, has 40,000 members who work in expert and special interest groups at central, county and district levels. Examples of groups and activities include "park groups" (who make an important contribution to the reconstruction of parks and landscape gardens), the Society for Nature and Environment, excursions and field projects (e.g. on problems of use of former mining areas in the Leipzig area) and landscape days. The scientific work of the specialized groups on subjects such as recording and mapping of endangered animal and plant species and bird ringing, is evaluated at conferences on environmental protection that that have been held in the last two years in various counties and districts of the country. Provision is made for the outcome of these conferences to enter into the planning process.

Technological aspects of environmental protection are treated by the Commission of Environmental Protection, at the presidium of the GDR engineers' organization Chamber of Technology, which concentrates mainly on technological solutions for environmental problems related to industrial production processes. It is also involved in measures for in-service training of environment commissioners, technologists, designers, constructors and staff members of environmental monitoring and protection bodies. "Environmental Protection" sections of the Chamber of Technology have been set up in some 200 enterprises, within which some 4,000 technicians seek concrete solutions for environmental problems associated with their enterprises.

## CONCLUSIONS

The experience of the MAB National Committee in the German Democratic Republic indicates that for environmental education of the public it is necessary to undertake such actions as:

1. Informing the editors of publications about important issues, problems, correlations and remedial measures in the field of environmental protection;
2. Disseminating news about various environmental initiatives and activities, with expert advice so as to provide high-quality treatment of key problems and appropriate articles, publications, radio and television shows, etc.;
3. Coordinating important projects planned by enterprises, institutions, the Chamber of Technology, Urania and other institutions having their own publications;
4. Planning, organizing and coordinating series of lectures, exhibitions, films, slide-tape shows, printed matter, wall newspapers and posters.

The appropriate bodies mentioned earlier ensure that these tasks are carried out to obtain the optimum results. In close cooperation with publishers, results and experience are regularly exchanged and generalized. The overall target is to promote environmental education among the public in such a way that the immediate benefits of wise management of natural resources are visible to everyone.

### References

Lüer, O. (1981), *Masseninitiative der Nationalen Front in Form des "Mach-mit-Wettbewerbs"*, Unpublished manuscript.

Merzdord, D. (1981), *Beitrag der Urania fur die Umwelterziehung in der DDR,* Unpublished manuscript.

Protze, B. (1981), *Bedeutung und Aufgaben der Umwelterziehung,* Unpublished manuscript.

Seidel, E., and B. Protze (1980), *Erziehung und Ausbildung im Umweltschutz und einige Schlussfolgerungen für die Aus- und Weiterbildung in den Universitäten und Hochschulen der Deutschen Demokratischen Republik,* Institut für Wasserwirtschaft, Berlin.

Unesco (1977), *Intergovernmental Conference on Environmental Education,* Tbilisi, Final Report, Unesco, Paris.

CHAPTER 20

# The Scientific Community and the Development of Environmental Education in Indonesia

*Indonesian MAB National Committee*[1]

Abstract

An overview is given of the role of the scientific community and the Indonesian MAB National Committee in initiating, developing and carrying out a programme of formal and non-formal education in Jakarta, Indonesia. The activities, experience and achievements of the teams charged with the task of implementing this programme are also discussed.

*Résumé*

*Une vue générale est donnée sur le rôle de la communauté scientifique et du Comité National Indonésien MAB dans le démarrage, le développement et la mise en oeuvre à Jakarta, Indonésie, du programme d'éducation scolaire et extra-scolaire relative à l'environnement. Les activités, l'expérience et les résultats obtenus par les équipes chargées de l'exécution de ce programme sont également discutés.*

[1]This paper was prepared collectively by those involved in the Environmental Education Programme sponsored by the Indonesian MAB Committee. Participants in this programme agreed that the Indonesian MAB Committee should be credited with the authorship of this paper.

## INTRODUCTION

As part of the celebrations in 1977 of the 450th Anniversary of the founding of the city of Jakarta, the municipality of the capital city of Indonesia convened in June 1977 a three-day workshop on environmental education at the instigation of the Indonesian National Committee for the Man and the Biosphere (MAB) Programme. The workshop was attended by many prominent Indonesian scientists and educators, and represented the culmination of a series of scientific gatherings on environmental problems, conservation of natural resources, environmental information systems and other related themes sponsored by the Indonesian MAB National Committee and other agencies. International events, such as the UN Conference on the Human Environment held in Stockholm in 1972, had a marked influence on many Indonesian intellectuals. But at the same time, the rural people already possessed considerable environmental wisdom and tradition. The scientific community in Indonesia sought to take advantage of this knowledge in the convening of the workshop, for presenting a feasible programme on environmental education which could be carried out at formal and non-formal levels.

## THE ROLE OF THE SCIENTIFIC COMMUNITY

In the past two or three decades, ecological studies by biologists have addressed some of the multitude of environmental problems faced by the country. Like many other developing nations, high population, uneven distribution of people and backward technology have contributed to the poverty of the majority of Indonesians. In attempts to fulfil basic human needs, it was natural that many people turned to nature. But unfortunately their activities in exploiting and manipulating natural resources led, more often than not, to the creation of serious environmental problems. Shifting cultivation, opening of poorly planned transmigration areas, large-scale timber exploitation, indiscriminate use of pesticides, etc., all provided examples of poor management. In developing the social economy of the country, an increase in the number of endangered species and the destruction of the indigenous culture that had helped maintain ecosystem balance had been considered unavoidable or ignored.

In recognizing these various problems, the Indonesian scientific community suggested that appropriate action should be taken for promoting change in attitudes to man-environment relationships, from one where man has only privileges and no obligations to one where man has privileges but also obligations to ensure a high quality living environment. Until now, most people have subscribed to the view that education is the most powerful force to bringing about the desired changes. Previous studies, in many schools, revealed that the ages of 9 to 13 were the most crucial, in that children of these ages were particularly responsive towards change, and once the idea of living harmoniously with their environment had been implanted it remained. These ages coincide with grades 4, 5 and 6 in the elementary schools of Indonesia, and in rural areas they represent the ages when children are likely to leave school. This is why it was decided to initiate environmental education at this age group level, both through formal as well as non-formal channels.

The environmental education workshop held in June 1977 underlined the need for experiments in the field of environmental education. As a result of this recommendation, four principal agencies — the Indonesian MAB National Committee, the Indonesian Institute of Sciences, the Municipal Government of Jakarta and the Agency for Education and Culture — agreed to plan and manage a series of such experiments on a joint basis. For this purpose, the Chairman of the Indonesian Institute of Sciences set up two working committees, with members from the National Biological Institute, the services of the Municipal Government of Jakarta, and education specialists and other experts from the Agency for Education and Cultural Research and Development of the Ministry of Education and Culture, as well as from the Institutes of Teacher Training of Bandung and Jakarta. One team concentrated its efforts on formal education, the other on non-formal.

## ENVIRONMENTAL EDUCATION AT THE FORMAL LEVEL

As might be expected, environment is not treated as an independent subject or taught as such in the existing curricula. Though there are instructional materials on the environment in the syllabi of several school subjects, environmental aspects are not emphasized or presented in a comprehensive manner. As a result, teaching tends not to motivate learners to appreciate their interdependence with the environment.

A study has therefore been made of existing school curricula, in order to determine the best methods of teaching both for school children and for teachers, to develop educational modules and to organize refresher courses for teachers. The outcome of the studies indicated that:

1. It was not necessary to introduce environmental education as a new discipline or new subject area in the school curricula;
2. The existing curricula contained many entries for environmental education, if the relevance of environmental problems could be demonstrated;
3. Environmental education could be conveniently "grafted" onto several subject areas such as religion, Pancasila morality, physical training and health, social sciences, and natural sciences;

4. The integrated approach should be an intrinsic characteristic of environmental education.

## Environmental education at the primary level

A teacher's guide on environmental education and teaching modules for fourth, fifth and sixth grade students have been developed. Before implementing this environmental education programme on a pilot basis in 15 selected schools, the teachers were given special training in the content and concepts of environmental education. In addition, they were trained to teach environmental education using the subjects already taught at school, and to encourage affective learning in their students. Affective learning is a very important aspect in teaching about the environment, in that it helps students develop an emotional basis from which an enduring dedication and commitment to the preservation of a high quality environment might develop. The role of a teacher dedicated to the development of a high quality environment is very important.

The results of the project in pilot schools has so far been successful. Evaluation of the results obtained shows that there was an heightening of environmental awareness in both teachers and students. School children showed a change in attitude towards the school environment. To produce a lasting impact, the within-school learning sessions should be linked to informal activities at home and within the community. A non-formal environmental education programme for parents and other members of the community is therefore coordinated with the formal one.

Environmental education activities in the pilot project schools is still being monitored in order to assess the constraints met by the teachers in implementing the programme, and to find ways to overcome them. Given that environmental education is a new activity in education, and that numerous internal as well as external constraints can therefore be expected, the teachers are encouraged to discuss their efforts with others concerned with the development of positive attitudes toward the environment.

## Environmental education at the secondary level

A comprehensive environmental education programme at the secondary level has yet to be developed. Environmental topics are included in existing curricula, but not in a way particularly relevant for the development of positive attitudes towards the environment. We believe that it is not necessary to develop a separate environmental education subject at the secondary level. Rather, environmental concepts or themes should be integrated within the existing school programme, and appropriate materials are being developed at the present time.

The teacher holds the key to the success of environmental education, and we are now monitoring the implementation of an environmental education programme that has been carried out for trainee teachers at the teachers' training school. This programme has aimed at developing the trainee's competence to teach about the environment, for the environment.

### Environmental education at the tertiary level

Environmental science is being offered as a discipline or separate subject in Indonesian universities, especially in the schools of sciences and technology. Ecology is offered as a subject in the departments of biology and of geography at the undergraduate and graduate levels. In several universities and other tertiary institutions, environmental science is offered at the post-graduate level. In addition, centres of environmental studies have been established in several Indonesian universities and an environmental education course for undergraduates is available at the Institute of Teachers' Training and Education in Jakarta.

## NON-FORMAL ENVIRONMENTAL EDUCATION

The target audience for non-formal environmental education within the urban system is very broad, and includes school-age children at home, housewives, labourers, white collar workers, shopkeepers and others. Given the limited means available, we have concentrated on programmes designed to have a far-reaching influence, including posters and brochures, parental education activities, slide and film programmes, briefing of school teachers, writing and drawing contests.

Sets of posters and brochures have been published. Six posters on the natural environment of Indonesia and three brochures on environmental problems are available. These are made available free-of-charge to grade schools in Indonesia and distributed through the Department of Education, public libraries, youth science clubs, boy-scouts, etc.

In cooperation with the team working on formal education, parents of school children have been provided with information about the environment in which they live. For this purpose a weekly visit is made to a community unit in Jakarta, consisting mainly of about 250 households. Discussions with groups of about 50 people focus on neighbourhood problems such as clean water, garbage, sewage, sanitation, greenery, etc.

Slide and film shows are presented on a twice-weekly basis in selected grade schools, with a view to providing information on the natural resources available in Indonesia for supporting basic human needs, how these resources are used, the effects of human activities on these resources and the requirements for their sustained use. Following the presentation of the audio-visual programme, evaluation sheets are completed by the students enabling us to evaluate whether the students have understood the message or not. Through such evaluations, we are able to improve our presentation.

In addition to school visits, we have invited school teachers to attend one-day seminars on the Indonesian environment, with particular emphasis on natural resources, in which a minimum of four speakers give their views on the subject at hand. The practical examples given are those readily available in the environs of the school, so that the teachers may use such examples to illustrate their courses. Following a discussion session, we invite the teachers to express their views on whether or not such a seminar is useful, and suggestions are solicited as to how we can be of further help. This kind of seminar is held on a monthly basis, and is attended by approximately 60 teachers from 30 different schools.

To supplement these various activities which are currently limited to the Jakarta area, every six months a writing contest is sponsored which is open to school children between the ages of 7 and 13 from all over Indonesia. Various environmental themes are selected, and the children are free to choose the title for their articles. The committee also work with one of the children's magazines. Each quarter, this magazine provides a series of drawings depicting environmental problems induced by human activity (e.g. erosion and flooding resulting from deforestation), which the children can look at and then compose a story. In addition to this, once every six months, a drawing contest is organised for school children, open to any grade school children in Indonesia. Usually about 700 to 800 entries are received. Several newspapers publish the drawings of the winning entries.

## CONCLUSION

The Indonesian MAB National Comittee considers environmental education as one of its main activities. The success achieved in Jakarta seems to presage the spread of the environmental education programme to other areas of Indonesia much more rapidly than expected. In fact a plan is now being worked out to set up roving instructors to train teachers at teachers' training schools all over Indonesia. The teaching materials, hand-outs, posters and brochures published by the formal and non-formal education teams are in such great demand that the future prospects for this programme look most promising.

SECTION 3

# Providing the Types of Information Needed for Decision-Making on Land Management

CHAPTER 21

# Introduction

*Francesco di Castri, F. W. G. Baker and Malcolm Hadley*

It is a misconception that scientific research is primarily concerned with the solution of practical problems: in reality, scientific research is a self-motivating activity with its own carefully-structured reward system.

We do not need any more research, we cannot apply more than a fraction of the research information that has been produced.

THE two perceptions of scientific research above shed light on the continuing debate on the "relevance" of science. Unless scientific data are obtained with a specific end in view, they may not be applicable, or if applied lead to ambiguous or inadequate results. Much ecological research is indeed of a basic generic character, and to be useful, needs to be supplemented by site specific studies. And the variety of climatic, biological and socio-cultural systems found in the world means that specific land-management programmes just cannot be blindly transferred from one place to another, as one might transfer a piece of chemical technology for example.

But even when geared to practical ends, research may not produce applicable results, because it is addressing what the managers consider to be the wrong questions, because it does not take sufficient account of the needs and perceptions of the potential users of research findings, because the solutions proposed by scientists may be too naive, because evaluation and feedback procedures are not adequate for shaping future research.

Even when sound data exist, they are often not applied. There may be lack of political interest in receiving objective advice on the ecological consequences of environmental policies. The social and economic costs of application may be too great. Many data remain inaccessible and are not presented in a form that can be readily used by planners and decision-makers. Or when they are presented, they are no longer in phase with the "moving targets" of land use problems.

The eight contributions to this section examine various dimensions of the problem of providing ecological information for decision-making. Topics taken up include: the information needs for decision-making processes; the participation of scientists, planners and local populations in planning and execution of research;

approaches to research design, to the communication of results, and to the evaluation of research; integration of natural and social sciences, and of non-governmental and government approaches, within research efforts.

In Chapter 22, Martin Holdgate discusses the environmental information needs of the decision-maker, drawing on his experience as a research biologist and university lecturer who has in recent years become chief scientist within a large government department. Greater attention needs to be given in environment research to the different needs of decision-makers at various levels of the social hierarchy within each country. For example, farmers, foresters, fishermen and others who manage the environment directly, need information about how biological production is likely to be affected by major external environmental variables and by the control of management methods — and at what cost; these needs are, in principle, already tolerably well served by ecological research. At another level, that of national or regional policy-makers, the contribution of ecologists may be more methodological or conceptual — in helping ensure that the major variables in complex systems are identified, that uncertainties are indicated, and that the effects of decisions on the behaviour of systems are monitored.

In the following contribution (Chapter 23), Serge Antoine addresses some of the ways in which thinking on relations between environment and development have evolved during the last fifteen or so years. There has been a change in vocabulary and concepts, with the introduction of such concepts as "polluter pays" and recognition that separation of environment and development is not a realistic approach to the problem of combining economic growth with improvement of environmental quality. Indications of changing attitudes to development-environment relations include the increased importance accorded to ecological equilibria, to geographical and cultural identity, to the elaboration of resource use strategies, to questions of time scale. Antoine presents some conditions for the introduction of environmental perspectives in development programmes, such as a firmer scientific basis for management and the testing and adoption of environmentally sound technologies. Other conditions include the need for change in institutions, in financial accounting procedures, in the time horizons of the planning process, and in international economic relations.

Participation of research workers, local people and decision-makers in the planning and carrying out of research is the theme of a case study from Switzerland presented by Paul Messerli, Charly Darbellay, Erwin Stucki and Albert Chapalay (Chapter 24). The Pays d'Enhaut is one of four MAB study areas in Switzeralnd, where the general problem is how to reconcile new and often conflicting demands for winter and summer tourism with traditional forms of land use based on agriculture and forestry. A special association has been formed in the Pays d'Enhaut to facilitate exchange of information among the different parties concerned in the research effort. The local people are resolved not to be "guinea pigs" for scientific research. It is hoped that the sort of relations that have been initiated between researchers, community leaders and individual citizens will lead to a fruitful conclusion for all concerned. Of particular note is the participation by local people in the research, contributing their own knowledge and skills to the overall scientific effort.

Integration is also the theme of the next two contributions, particularly integration among different scientific disciplines. In Chapter 25 Anne Whyte describes some of the difficulties encountered in bringing together the social and natural sciences in problem-oriented research on the environment. Barriers to integration include historical, institutional and perceptual constraints, as well as problems at the conceptual level. In tracing changes in attitudes to social sciences within the MAB Programme, Whyte describes some of the approaches to integration that have been tried out, such as human ecology, human use systems, perception studies and integrated models. Ingredients for fruitful cooperation are suggested, including the orientation of projects towards management decisions or policy alternatives, the involvement of local people, the adoption of a strong modelling focus, the key role of internal communication and the project leader in the integration process, the importance of small-scale in project design. A major need is for working models which can integrate natural and social science data in a form that can be incorporated into management decisions.

Brian Spooner discusses the objectives sought through integration from a slightly different angle (Chapter 26). He argues that the social sciences (and indeed the natural sciences) encompass a variety of approaches, all equally valid, leading to different but complementary results. There is no single approach capable of giving full weight to each of the major relevant variables in the analysis of relationships and interfaces between everyday life and natural processes. Any approach is by definition selective. No approach can comprehend everything. Any approach is likely to have something to learn from others. And therefore, what is required is an exchange between people with different approaches to a central ecological problem. By viewing different approaches as complementary and not competing, the ground is prepared for effective collaboration among disciplines as well as the more important goals of improved communication and application.

Communication is a central question in considerations on the use of ecological information in decision-making processes. John Jeffers prefaces his contribution by highlighting two common misconceptions — alluded to above — which lie at the root of difficulties of communication between research scientists and the wider community which might make use of the results of research. The first misconception is that scientific research is primarily concerned with the solution of practical problems. The second misconception is that decision-makers, administrators, and politicians are eager to receive objective advice about the ecological consequences of their policies: in reality, their response is closer to that of "Don't confuse me with facts — just confirm my preconceptions." Jeffers identifies the main actors or organizations involved in the communication of research results, and identifies the objectives of these actors and the end-products of their activities. In recent years procedures have been refined for encouraging resource managers and administrators to participate in the informal interactive communication which normally takes place only between research scientists. New technology can play a key role in this process of establishing an effective interface between researchers and managers, with a second interface to the general public.

Evaluation and feedback should be integral parts of a research effort. They provide a means whereby research can be modified to take account of changing circumstances, both within and outside the research project. Research coordinators and science planners are among those seeking and testing new criteria for evaluating environmental research, and in his contribution, Bengt Lundholm presents a personal and necessarily preliminary assessment of a large-scale ecosystem study on coniferous forests in his country Sweden. Lundholm traces some of the main features in the planning and development of the project, including the role played by the international advisory council and team coordinator. Positive aspects of the project include its role in education and training, in promoting the country's international scientific reputation in specific technical fields, and in contributing tangibly to the country's most critical environmental problem (that of "acid rain"). Perceived shortcomings include those of information handling, of inclusion of socio-economic aspects. The underlying message is the need for more sensitive measuring sticks for assessing multidisciplinary research on environmental problems.

Finally, in Chapter 29, Jim Dooge and Mike Baker trace aspects of the development of international scientific cooperation over the last two hundred years, drawing on examples in such fields as polar research, geophysics, biological productivity and human adaptability, oceanography, environment. Cooperation between non-governmental and intergovernmental organizations has proven effective in a number of joint ventures, including the Study of the Feasibility of a World Scientific Information System, the International Geological Correlation Programme, the Global Atmospheric Research Programme and the Global Environmental Monitoring System. Combining the strengths of the two types of organization can bring benefits to both.

CHAPTER 22

# Information Needs for the Decision-Making Process

*M. W. Holdgate*[1]

Abstract

Requirements for ecological information depend on the people taking the decision, the kinds of question to be answered, and the social, economic and administrative context of the decision-making process and time-scale. Ecological research can contribute detailed information to management decisions, and the concepts of ecology also have something to contribute to the process of analysis and decision-making. But much ecological research is, quite properly, generic in type, dealing with broad processes and issues, while decisions are often site specific. This in part contributes to the so-called "application gap" in ecological information and calls for supplementation of generic research by site specific studies. Another factor is that of communication, and the need for simple summarization of ecological information without distortion, presented in a way suited to the particular type of decision-maker and the level of detail required.

*Résumé*

*La demande d'informations écologiques est fonction de ceux qui auront la responsabilité de la décision et de la nature des questions auxquelles ils devront répondre,*

[1]The views expressed in this paper are those of the author and not necessarily those of the Department of the Environment and Transport.

*du contexte social, économique et administratif dans lequel se présente le processus de prise de décision et de son échelle de temps. La recherche écologique peut apporter des informations détaillées pour les décisions d'utilisation des terres et les concepts de l'écologie sont à même également de contribuer aux processus d'analyse et de décision. Une grande partie de la recherche écologique est tout à fait justement d'un caractère générique puisqu'elle traite des processus et des problèmes en général, tandis que les décisions sont le plus souvent spécifiques à un site en particulier. Cela constitue un des éléments de ce qu'on appelle "la lacune dans l'application" de l'information écologique et demande l'approfondissement de la recherche générique par l'entremise d'études spécifiques de sites. Un autre facteur en est la communication et le besoin de résumer simplement l'information écologique sans la déformer et de la présenter de façon à l'adapter à chaque type particulier de décideur avec la somme de détails convenant à chacun.*

## INTRODUCTION

How can ecological information be used as a basis for decisions about land management? There is a logical proposition in this objective. It is that science — especially ecological science — is potentially important as a foundation for environmental management policy, and that ecological information should be collected for this purpose.

Many readers would take this proposition as a statement of the obvious. Ecology includes an understanding of the factors that determine the composition, stability or rates of change of natural environmental systems, and ways in which their productivity may vary in response to natural factors and human manipulation. There are many examples of environmental damage and reduced productivity that have occurred because available ecological knowledge has not been applied in management. To quote two only: the causes of desertification, and many practical remedies for it, are well established and the Plan against Desertification (UN, 1978) incorporates ecological insights. But the problems arise in ensuring the effective implementation of this plan. The World Conservation Strategy (IUCN, 1980) argues that in 1980 world fishery landings were less by some 15 to 20 million tonnes than they could have been had available scientific knowledge been applied in better management. Other contributions to this book provide further illustrations of how the investment made by people and communities in ecological research is being wasted because the information resulting from this research is not being supplied effectively. Nonetheless, it remains a fact that many people in governmental agencies and other authorities who take decisions about land use planning do not appreciate the potential contribution of the professional ecologist — and I believe that ecological scientists must bear a considerable share of the responsibility for this state of affairs.

I suggest that in considering the needs for decision-making, two issues need to be addressed. Firstly, can the types of decision that should draw on ecological knowledge be identified? Secondly, can some properties of the system be discovered by which ecological information is translated into management — or fails to be used? I propose to address, especially, this question of the "application gap" — which in many ways parallels the "pre-development gap" that lies between research and exploitation in the engineering and industrial world. By analogy with that problem area, it is possible that the imperfections in the system with which we are concerned could have three causes:

1. Research is not producing applicable results because it is addressing the wrong questions;
2. The results of research are not reaching those who could apply them because the communication system is not correct — potential users for example simply being unaware of what ecologists could offer;
3. Research results are not being applied because the costs of application are too great.

The remainder of this paper focusses on the first two of these questions, but in ignoring the third it must not be forgotten that to be acceptable, especially in developing countries with scarce resources, the need for land management to be cost effective is paramount.

## STRUCTURE OF THE DECISION-MAKING PROCESS IN ENVIRONMENTAL MANAGEMENT

It is helpful to approach these questions by attempting to analyse the kinds of information needed by those who are managing the environment in various ways. This is so because the requirement for information must depend on:

1. The people taking the decision;
2. The kinds of question they have to answer;
3. The social, economic and administrative context of the decision-making process and its time-scale.

Throughout the analysis it is important to remember that very few major policy decisions, even in the environmental field, are made by working scientists and scarcely any are made on the basis of scientific data and scientific analysis alone. Economic and social factors, historical, legal and administrative tradition, and the general attitude of a non-scientific public inevitably predominate at many points in the decision process.

It is also essential to recognize that decisions about management of the environment are made at many levels in the social hierarchy. Figure 22—1 attempts to illustrate this fact. It shows a series of groups — which in real life are of course both diverse and overlapping. In this model the *operators* are taken to be people who actually manage the environment directly. They are farmers, foresters, fishermen, conservationists looking after nature reserves and also urban and industrial developers and those who release potentially polluting effluents to the environment. These people, by their actions, affect the structure, direction and rate of change of ecosystems, the productivity of the environment, and the extent to which environmental resources are safeguarded or consumed. The *advisers* in this system are people who guide the operators so as to enhance the benefits they derive from their

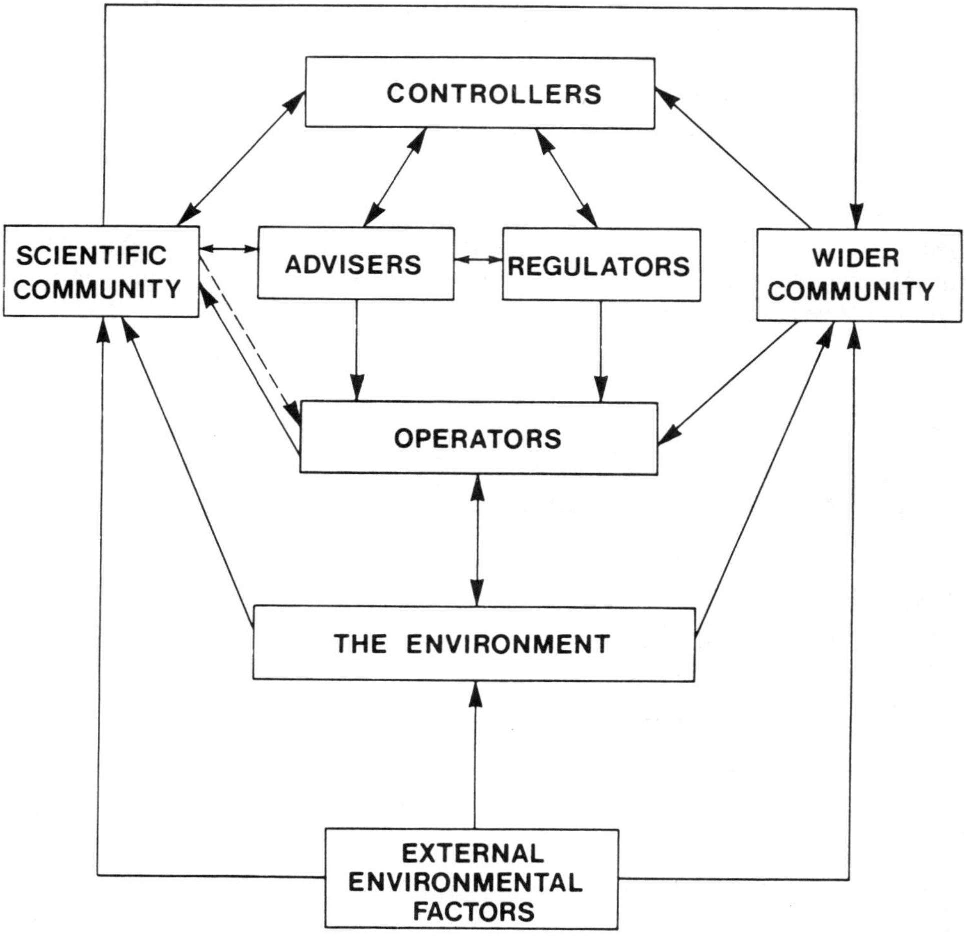


**Figure 22—1.** The interacting groups within the community who determine how the environment is managed. For definitions see text

activities or to avoid damaging or illegal acts. The *regulators* also interact directly with operators: they are individuals in control agencies who provide instructions on standards to be met, or actions to be avoided, but generally leave the advisers to provide guidance on the best means of achieving these requirements. The two roles are evidently complementary and interlocking, and in some national agencies they are effectively combined.

Behind these groups there are three more. The *controllers* in the diagram are national or regional policy-makers in government, or senior managers in regulatory agencies, who determine what the standards imposed by the regulators should be or what production goals the advisers should encourage the operators towards. The group embraces ministers in central or provincial government, and also their senior policy-advisers.

Outside this direct interactive system, the *wider community* embraces people at many levels: the peer group in a village that influences operators because it is a repository of custom; the consumers whose tastes and wealth influence operators through the market; environmental groups demanding higher standards of environmental protection from the controllers; and all those other elements whose interacting perceptions and demands influence policy. It could be argued that the *scientific community* also forms a part of this wider social grouping, but they are shown separately in the diagram because they have a distinctive role in the context of this contribution. They are the source of basic information derived from research. Some of this is conveyed directly to operators, some filtered through advisers, some injected directly into policy and some has a more diffuse influence on policy through the wider community. All these factors may be influenced by *external environmental factors* such as climate, whose fluctuations affect operators directly, the wider community through the market and when extreme events worry people and cause them to demand new social goals, and the scientific community because these external environmental variables are an essential element in research.

This model is evidently crude and is only one of many that could be constructed. Nonetheless it provides a useful framework against which to consider the information needs and kinds of decision taken by the different elements in a complex society. It is however important to recognize that these elements all work within a common system, represented in Figure 22—2, but with different emphases in terms of the kinds of decision they take and the consequent information needs.

One major group of operators, for example, is especially concerned with an environmental system and its biological production. Farmers, foresters and wildlife conservationists have as their chief goal the control, and in many cases enhancement, of biological production or the maintenance of particular ecosystems. They need information about how these are likely to be affected by major external environmental variables and by the controllable variables that we call management methods. They need to know the cost of creating a particular change through management. They also need some information about social needs, as expressed through the market, preferably on a time-scale of a decade or more. But they do not need a great deal of the underlying ecological theory that is the concern of the scientific community. This information commonly needs to be interpreted to the operators

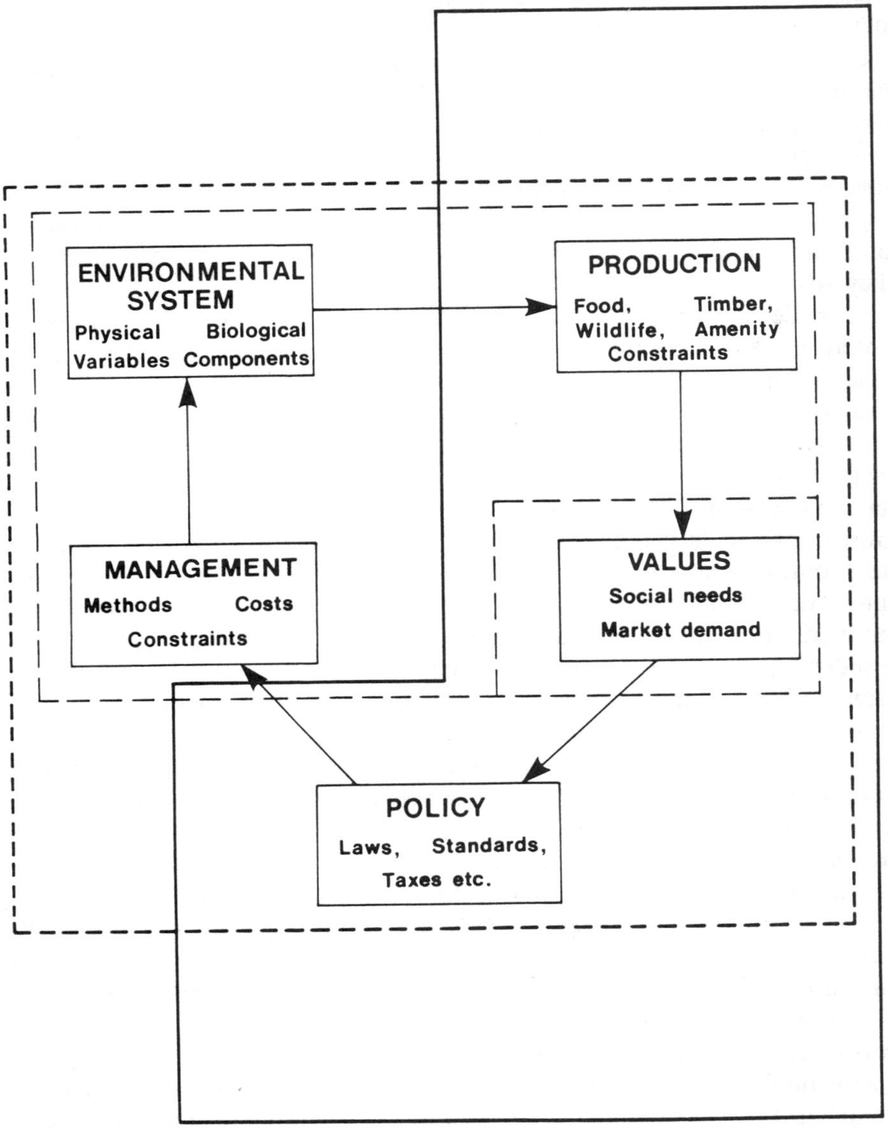


**Figure 22—2.** The environmental decision-making system and the differing information needs of three groups of actors. The light broken line surrounds the areas of principal concern to the operators, the heavy broken line the interests of their advisers, and the continuous line the parts of the system of chief concern to controllers

through the advisers (even though some erudite farmers or landowners may be interested in basic science and be their own advisers in this sense).

The adviser needs much closer touch with the details of scientific research. It is unlikely that he will be an active working member of the research community, but he may well work for a national advisory service or agency that has its own research branch pursuing projects designed to meet the adviser's needs, these being in turn determined by the adviser's perception of the requirements of operators. The strength of this advisory service clearly varies from community to community, being most advanced in developed countries although it is expanding today in many parts of the developing world. Indeed it can be demonstrated that one accompaniment of development is an increasing role for the advisers and a diminishing role for that part of the wider community that is the repository of tradition and influences the operator directly through the village or social peer group. A conflict situation thus tends to develop between the operator's peer groups in the wider community and the groups of advisers. This is especially so when the policy-makers — the controllers — higher up in the system have decided to steer management practices away from traditional modes. For example, shifting cultivation systems in many tropical areas achieved a considerable degree of ecological efficiency when the rotation times were long and population densities low, but these traditional management practices broke down when populations became denser and fallow periods became shorter. The repository of collective wisdom in the community is likely to hamper rather than assist advance in such a situation. Similarly, the success of the Green Revolution in many areas depended on the degree to which the introduction of higher yielding strains was accompanied by education in their use — and especially in the application of fertilizers and the more intensive cultivation practices that many of the new cultivars demanded. Here again, advisers (educators) had the task of persuading operators to turn away from social custom.

For all these reasons the adviser needs a wider information frame than the operator, also encompassing the market system but embracing wider social goals determined by the controllers, details of the constraints imposed by regulators and the wider community and an awareness of the new insights offered by the scientific community. But advisers also need to be on their guard against misleading information — whether from external scientific experts or fellow advisers in other countries — for the uncritical international transfer of ecological technology can be just as damaging as in other human environmental fields. To quote one simple example, when tea growing was transferred from Assam to Kenya shade trees were planted. In Assam these had been found to be beneficial because they helped to prevent high leaf temperatures. In Kenya, at higher, cooler altitudes, this shade was not needed and the trees reduced production by competing for water with the crop.

The above analysis deals with the needs of operators concerned to enhance biological productivity, and with their advisers. But much the same pattern holds for urban and industrial developers. As operators, these need to know how the environmental system will respond to their actions. If the regulators have laid down standards of environmental quality to be preserved, the operators need to know about the rates at which the materials they release will disperse, be diluted or be transformed in the environment. If the quality objectives are set in terms of potential impact on a target, the operators need to know about the dose/effect relationships for the substances they may discharge. Their advisers also need to know about these matters and to be abreast of developing scientific knowledge since this will enable them to alert operators to potential new concerns that may inspire fresh regulations. The advisers may also have a task of weaning operators away from traditional practices that are needlessly disruptive of the environment, and informing them of new technologies that will allow reduced environmental impact at no greater cost. It is common place for industrial groups to have advisers within their own firms because the application of scientific understanding to achieve regulations and standards by the most cost-effective route is important to the efficient and productive operation of the firms concerned.

The regulators in this area need less knowledge of detailed ecological parameters. They need to have more cognizance of the socio-economic system including the demand for environmental products or industrial expansion. Generally they have the task of applying policy determined by the controllers, but guiding the latter on the practicability of achieving new social objectives. The regulators need to interlink with advisers (or have their own independent advisers) so as to judge the impact of their regulations on the productivity of operators and to ensure that regulations can readily be transferred into practical measures.

The controllers — the senior policy-makers — need to know still less about ecological detail but they must be guided about what can be achieved through changes in management and regulation, what the nature and scale of impact of new developments are likely to be, and about the costs of attaining new social goals. Their concern is therefore all embracing but the information they require often needs to be presented as balanced summaries of issues about which a great deal of detail may be available elsewhere in the system, and about which preliminary debate has also gone on elsewhere. For example, a decision on the degree to which a pollutant should be regulated is likely to be made by ministers on the basis of brief position papers. However they need to be satisfied that their professional advisers have genuinely reviewed the wider range of available data. It is therefore customary for the factual analyses presented to them to be accompanied by a second kind of summary detailing the steps by which that analysis has been generated and the groups that have been consulted in its formulation. This is an important point in decision-taking: it is not sufficient simply to be provided with information: it is essential to know how that information has been put together and where it comes from.

Policy-makers also need to have machinery for reconciling information of radically different types. For example, it is commonly agreed that pollutants capable of damaging human health, livestock, the productivity of ecosystems or the beauty of the landscape need to be controlled. It is commonly argued that such control must be based on science, in that it should be founded on an understanding of the relationship between the exposure of a target to a substance in the environment and the resulting effect. The worse the damage and the more valuable the target the stronger the case for control would be. This process is essentially one of risk *estimation*. But the controller — the policy-maker — has also to pay heed to the public reaction to the hazard. People are well known to rank targets and types of environmental damage in an order which does not always correspond to the scientifically assessed statement of risk, and the demand for a control measure may well change with variations in public perception. For example, DDT and other organochlorine pesticides were initially hailed as a boon to agriculture and to public health alike, and combined effectiveness with the great merits of cheapness in production and safety to operators. The new generation of organophosphorus pesticides that replaced them were undoubtedly potentially more hazardous to human operators. The transition was brought about by the recognition of the unwelcome side-effects of the organochlorines on wildlife — a development that would have been improbable before the modern growth of conservation as a social concern. For this kind of reason the policy-maker also has to conduct a process of risk *evaluation* and take these concerns into his final decision — alongside the evaluation of the relative costs of damage and control. The scientist should not be too critical of this process: one major duty of government policy-makers is to satisfy their electorate and tax payers that they are properly embodying the wishes of the wider community in the laws and regulations they impose. If there is scientific evidence that those wishes are based on an imperfect perception of the facts, the role of the scientist, in partnership with the senior policy-makers, is to ensure that the wider public is better informed so that the gap between objective risk estimation and public risk evaluation narrows and the system becomes more efficient in its use of resources.

Another important dimension of variation arises because these various groups of actors often work on different time-scales. Operators commonly have a time perspective of a few years — although wildlife conservationists work to longer time horizons. But national policy-makers ought generally to be thinking in decades. Given that applied ecology is about prediction, this also creates different demands on the scientific community and the advisers interpreting their results at different levels in the decision system.

## THE APPLICATION GAP

Against this background, answers can be sought to the first two questions about the application gap. The first is whether ecological results are not being applied because in fact they are not applicable — whether the research, judged from this standpoint, is addressing the wrong questions.

Clearly this is too simple a proposition because it is now clear that the different components in the decision-making process have different information needs. For example, one major group of operators is concerned to enhance biological production, for instance relating fishery intensity or the cropping of wild animal populations or natural forests to sustainable yield or optimizing production from managed crops or plantations, or sustaining a pattern of wild ecosystems for purposes of genetic and ecosystem conservation. This group can be regarded as tolerably well served in principle by ecological research. This is because ecosystem and population dynamics, the relationship between physio-chemical and edaphic factors and production, predator-prey and host-pest interactions and similar topics are at the heart of much ecological research. It may be that some people will argue that ecologists spend too much time on rare species and exotic ecosystems and too little on situations directly relevant to human need, but it is hard to argue that the research is inherently wrong. We might discuss however how far it could be made more relevant. Is it helpful, for example, to talk of agricultural, fishery or ecological research as distinct when the same principles clearly underlie them all and when the great challenges to the world so clearly lie in ecologically wise development that will provide food, clean water and other natural resources for the growing populations of the developing world?

In contrast, the ecologists' contribution is less obviously and directly useful to the operator, adviser or regulator concerned with natural trends or human developments that do not lie internally to bioproductive systems. In this class I include such impacts as that of atmospheric carbon dioxide on the hydrological cycle (via global temperature) as well as human impacts through pollution, industrialization and urbanization. Going further, I would argue that ecologists are even less geared to contribute to the highest levels of national or regional decision-making. This is partly because the questions in these areas are less familiar to research ecologists. They include (for example):

Given that a change in extrinsic environmental factors or in management practices will change the environmental system and its product;

1. Will it operate in a desirable way? If so, what is the scale of the benefit? If not — does it matter? What is the scale of the disbenefit? By whom is the benefit or disbenefit perceived?

2. Is the benefit sustainable? At what cost?

3. Is the change adaptable to new social needs or effectively irreversible? If a disbenefit, can it be alleviated thus optimizing the situation? At what cost, and by what means?

4. Are there conflicts between social groups (for example when an industrial development may benefit the operator but not the wider community)? If so, should the polluter be made to pay and how much?

5. Are there other conflicts based on inadequate information so that data susceptible of a wide range of interpretations are being used differently? For example, is an optimistic prospective developer minimizing the risk of his actions while concerned social groups are maximizing them? Is the imprecision too great for an acceptable decision?

6. Are there conflicts based on different approaches to the future (e.g. for the use of different test discount rates and other economic conventions)?

It is obvious that only some ecological research will be pertinent to such questions. The same holds for economic research, for environmental economics is far from being an advanced skill. Another problem arises because such questions are commonly asked in a site specific way, about particular developments that will change the environment at particular points. In contrast, much ecological research is generic in type, concerned with broad processes or issues that may be studied in particular chosen localities but rightly aim to provide knowledge of wider application. The balance between generic and site specific research is not easy, but most funding agencies rightly prefer the former. The result is that when questions related to an industrial development arise, the ecological research results available are unlikely to be sufficiently precise to answer all the detailed questions that may then need to be addressed by the policy-maker. In this kind of area it may therefore be contended that the application of knowledge to sound decision may be hampered by a lack of "the right kind of study". The paradox in this assertion is however now obvious: what is right research from the point of view of science may be wrong research for those taking a site specific decision.

So far the discussion has focussed on the detailed information derived from ecological research. But the concepts of ecology also have something to contribute to the process of analysis and decision-making. One reason why this is so is that ecologists are familiar with the need to select relevant information from large quantities of data of varying quality. Management decisions about the environment will always be based on partial information if only because it would clearly be impossible for the environmental impact assessment to embrace all possible effects on all species in an area. The correct selection of data can only be made if the system within which the decision has to be taken has first been defined correctly. And it is important to stress that in these terms we are not speaking of the ecological system but of the wider management system. I suggest that this is one major problem in considering the issue of ecological information for decision-making. Ecologists design their research, as has already been noted, in a fashion that reflects their perception of ecosystems or of ecosystem components and interactions such as populations or predator-prey subsystems. This makes application straightforward where the policy decision is within a system that is strongly ecological in character — as in agriculture or forestry. But few people set out to design ecological research about the interplay of ecological and socio-economic systems, and those that do often run into trouble, as the literature on biological indicators of pollution, the valuation of different ecosystems and landscape types in order that priority land use zones can be established, or the development of indices of environmental quality demonstrates.

Nonetheless, ecologists are particularly well qualified to explore questions about the behaviour of complex systems. For example, ecologists appreciate that uncertainty is inevitable: that, as Holling (1982) has put it, we must "expect the unexpected". Ecologists can assist decision-making by ensuring that the major variables in environmental systems are identified and uncertainties indicated. The question for the senior policy-maker is then whether the margins of uncertainty matter. Such an evaluation is often closely linked to the scale and reversibility of the components about which there is an uncertainty. If those components are obviously of great importance, while the available information does not permit their likely response to management changes to be predicted with any reasonable degree of confidence, the pressures for fresh research to reduce the uncertainty will clearly be great. If the components are minor and the changes clearly susceptible to adjustment within the context of management, then significant uncertainty may be acceptable. Either way the existence of uncertainty makes it essential that there is monitoring of the systems about which decisions are taken. Indeed a change in the management of the environment or even a new environmental law can be thought of in scientific terms as a hypothesis which requires testing by the observation of its actual effects. Monitoring in this way will not only contribute to the quality of subsequent decisions in a broadly similar area elsewhere, but will allow further adjustment of management of the particular system in the light of experience. This is one reason why standards set for discharges of pollutants to the environment need to be kept under continuing review. At the same time, it must be made clear that a wildly experimental approach to environmental regulation is not being advocated. This is not the way to inspire public confidence or use the resources of the community wisely. What is important is that ecologists with systems understanding can contribute to the design of monitoring and feedback machinery that will continue over time to improve our management and environmental protection practices.

It is therefore clear that in this kind of decision-making process the "right" ecological research may be methodological. But the second question was whether the application gap is also one of communication. This, it is believed, is the case. Many people involved in the decision-making machinery think of ecologists as people only concerned with wildlife conservation and need persuading that ecological philosophy has more to contribute than arguments for and against the establishment of nature reserves. Because of this misunderstanding — perhaps because agriculture, forestry and fisheries management are not widely understood to be fields of applied ecology — the decision-takers and their advisers tend to ask the scientific community of ecologists the wrong questions — or rather fail to ask them questions to which they could contribute useful answers. There is need to give further consideration to how this communication gap can be bridged. Clearly one essential requirement is for the true nature of ecological science to be recognized more broadly. Professional ecologists must present themselves as people concerned with improving the use of renewable living environmental resources for human benefit in many different ways, and while it is entirely legitimate that one of those applications of ecology is in protecting the diversity of species and natural ecosystems and genetic material on this planet, it is essential to stress that this is only one of the important applications.

Another need in spanning the communications gap is a system for setting scientific data alongside social, economic and other kinds of information. Bringing these to a common base — or at least setting them out in a form that allows intercomparison — is no easy task. Commonly, money values are employed, partly because so much national policy decision has an economic base. But ecologists and others concerned with the environment have often criticized this use of the economic yardstick: well known problems indeed surround valuation of human life, genetic diversity, rare species or natural beauty. This is an issue that needs to be given more thought. Thirdly, the communication gap will not be bridged so long as professional ecologists regard the scientific paper as their principal organ of communication. Such papers are totally unsuited to conveying the thinking and judgements of ecology to most of those engaged in decisions — and I include in this number the advisers of most kinds of operator as well as the advisers and participants in the higher levels of the decision-taking process. If the communication gap is to be spanned ecologists need to put their scientific results in an understandable plain language form readily converted into comprehensible, balanced summaries of the major scientific issues underlying a management decision.

But the interactive process cannot just be done on paper. Where decisions involve a range of groups in the community it is very often best to involve them in direct discussion at the outset, in order to agree on the system about which the decision is to be taken. This may not be so essential where the issues lie within a fairly narrow area — as when a farmer changes a crop or converts pasture to arable — but with decisions of wider implication, omitting a significant group of actors can lead to judgements of doubtful validity that are almost certain to be challenged. The use of workshops, bringing the main interests together in order to define the system requiring analysis, and to decide whether the available data are adequate for a responsible decision, is one approach whose value has been demonstrated. Such workshops can also decide on whether simulation models would be useful, and if so how they should be structured. It is unlikely that this kind of workshop can take place at the highest level of decision-taking: it can however involve the advisers within the appropriate government or local administration, thereby enabling them in turn to satisfy ministers or other senior decision-takers that the issues presented to them have been thoroughly explored. This kind of approach offers one very promising way of ensuring that the communication gap is not a severe barrier to the application of ecological data and philosophy in decision-taking. However it is clearly a prerequisite for such communication that the administrative system of decision-making is structured to allow the necessary interactions, and at present it very often is not. The potential users of the information in the decision-making process need therefore to be convinced that the investment of time and effort in this kind of interactive process is worthwhile from their own standpoint and will not simply mean unnecessary expense and delay.

## CONCLUSIONS

Three questions are often posed in respect to the theme of this contribution, the ecological information needs for decision-making.

1. Has ecological research been sufficiently problem-oriented?
2. How can its results be presented in a form more useful to decision-makers?
3. How can ecologists help the wider public to understand ecological matters better?

The first question is almost unanswerable unless the decision-making system is defined. However the preceding analysis suggests that the research has been sufficiently applicable in areas where the ecological understanding has a high degree of relevance to the problems addressed, and where the concern is fairly directly with living environmental systems — as in forestry, fishery, wildlife conservation and agriculture. The generic types of ecological research that have been pursued have also been relevant to many decisions about the management of particular areas of land, the impact of development and the control of pollution — but here generic research of a basic character inevitably needs supplementation by site specific studies, and the questions are over the parameters about which this extra information is essential and whether adequate time is available. If, for example, these variables are likely to show cyclical change extending over decades, or trends which will not be adequately defined in less than five years of research, then the time-scale for the decision-making process may simply be incompatible with the requirement to reduce uncertainty by further research.

Methodological research — especially across disciplinary frontiers — has grown substantially during the past ten years particularly under the impetus of the International Biological Programme (IBP) and MAB and it remains crucial if ecological concepts are to be developed in a fashion suitable to decision-making. At present this kind of research may not have been sufficiently geared to the needs of the decision-making process partly because of difficulties at the interface between ecological and socio-economic systems.

So far as presentation goes, again much depends on the particular type of decision-maker and the level of detail required. The communication of scientific results via advisers to people such as the agricultural community presents few problems, but communication to the more complex decision-making systems concerned with major developments demands interdisciplinary contact and a workshop-type framework to decide on the crucial factors and on how to bring data to a common base — perhaps through a simulation model: this offers one of the most promising ways forward.

Communication to the more general levels of policy-making demands abstraction — clear, simple summarization without distortion. It also demands, on the policy-makers' side, an administrative system into which the ecological data can be received. There is a circularity here. The structure is unlikely to be adjusted unless the benefits of an ecological input are established. But without a compatible decision-making structure these benefits will not be evident. There is no simple way forward but ecologists must always be careful not to over-sell what their discipline has to offer. It is essential that ecological inputs are clear, with margins of uncertainty indicated, and with the extent of supporting research evident.

Communication to the wider public needs to proceed in two ways. One is through the growing area of environmental education, which can make the concepts and uses of environmental science obvious to a wider circle of people. It is particularly essential that training programmes for those who can guide operators on the ground in the developing countries should receive high priority. But there is also a need for more direct communication, especially to counter pseudo-scientific assertions that claim to be made in the name of ecology, but actually select those data that support a predetermined thesis to which a pressure group is committed Professional ecologists — if they want to advance their science — may need to be more positive in presenting clear, balanced evaluations of complex issues before the public. In that way too, they will influence the decision-makers — both because such statements will be directly useful and because they will help to mould public opinion. But to do this effectively, ecologists may need to learn a new language.

## References

Holling, C. S. (1982), Predicting the Unpredictable, *The Unesco Courier,* August-September 1982, 60-3.

IUCN (1980), *World Conservation Strategy,* IUCN, Gland.

United Nations (1978), *Report on the United Nations Conference on Desertification,* Nairobi, August 29 to September 9, 1977, Report A/CONF.74/36, United Nations, New York.

CHAPTER 23

# Environment and Development: Contribution to the Debate in the Developed Countries

*Serge Antoine*

Abstract

This contribution outlines some ways in which thinking on relations between environment and development has evolved during the last fifteen years or so, particularly in industrialized countries. There has been a change in vocabulary and concepts, with the introduction of such concepts as "polluter pays" and recognition that separation of environment and development is not a realistic approach to the problem of combining economic growth with improvement of environmental quality. Some indications of changing attitudes to development-environment relations are described, as are some conditions for the introduction of environmental perspectives in development programmes.

*Résumé*

*Cette contribution donne quelques aperçus de l'évolution de la pensée sur les relations entre l'environnement et le développement au cours des quinze derniéres années, en particulier dans les pays industrialisés. Le vocabulaire et les concepts ont changé, avec l'introduction de notions telle que "le pollueur paie" et la reconnaissance que la séparation de l'environnement d'avec le développement ne constitue pas un abord réaliste du problème que présente la conjugaison de la croissance économique avec l'amélioration de l'environnement. Elle décrit quelques signes du changement d'attitude vis-à-vis des relations développement-environnement, et elle fait ressortir quelques unes des conditions nécessaires pour l'introduction des perspectives de l'environnement dans les programmes de développement.*

THE reconciling of environmental policy with development is an important but often overlooked problem. It is incorrect to make believe that the environment is linked with rich countries having zero growth or that it is difficult to include environmental concerns within the needs of developing countries. Three aspects are examined: vocabulary, changes in growth rates in developed countries, and the means available to reconcile development and the environment.

## THE VOCABULARY

The environment is a dimension of society, of culture, of the economy, of space, of life. It is a prospect of the world, an indication of where we may be going, such as those developed by the Club of Rome. The focus will be on personal observations of changes in the quality of life in countries which have a history of development, not on figures or quantitative comparisons.

There is a growing awareness of the need for a more sober development paying greater attention to waste, to resources, to ecology, to the environment, and in the recently developed countries a feeling for "ecodevelopment". In the 1965-70 period, particularly in the Anglo-Saxon countries, concern was concentrated on the fight against pollution and on attempts to repair the effects of industries and of urbanization. There was a tendency to "externalize" the environment in order to identify it more readily. As a result there developed a basic contradiction between growth and a satisfactory environment. This "individualization" of the environment made it easier to analyse the societal situation, and with the industrial era the cost of pollution was transmitted to the consumer and to communities, whether in free market or planned economy countries. This individualization made possible a new industry: that of restoring the environment, and with time its growth.

Attempts were also made to provide an indication of the effort and the cost required to fight against environmental degradation. Leontief, and in 1975 the US Environmental Protection Agency, measured it as a percentage of Gross National Product. In the 1970s, the principle of "polluter pays" was born and brought into the public domain thanks to the OECD, but the separation of environment and development is not realistic, and this was soon realized.

First, the definition of the charges imputable to the environment evolve with time: those which may be considered at one moment as an expense may in the future be considered as a profit.

Second, allocating expenses either to the environment or to growth is a function of the degree of activity. What part should one impute to the accounts for environmental improvement such as the costs of rural reconstruction, the conservation of wetlands and of hedges, for example?

Third, to isolate specifically expenses for protecting the environment is to ignore the costs of non-protection of the environment: the costs of destruction or pollution of the natural environment; the extra costs to a town of treatment of water from a river polluted by towns and industries nearer the source; the costs of building and maintaining a swimming pool because the river is too polluted to swim in.

Fourth, to isolate environmental expenses without taking into consideration compensation, or health, in its widest sense is pointless. The migration each weekend of the inhabitants of large towns, eager to satisfy their desires for pure air, space and silence, is an expense linked to the poor quality of the urban environment and is costly to the community in time, energy, material, investments and in deaths.

Thus the relation between classical economic accounting and the environment becomes more and more complicated when one takes into account expenses of a social or a psychological character. This problem of selection causes us not to separate so precisely expenses for the environment from those for development although their character can be quite different. For example, with regard to inflation the effects are not the same: in the short term environmental expenses have a deflationary effect with a deduction from the money available, in the middle term they tend to reduce consumption.

Contemporary economists and a large part of the "ecology movement" are reviewing, in the light of the environment, concepts and practices of economic growth in order to redefine them so that they satisfy social objectives and those of protection of physical environment, both quantitatively and qualitatively. To a large extent because of the environment our societies have developed "another growth" and abandoned the temptation towards "zero growth".

Developing countries have also participated in this debate and have contributed to the creation of a new concept: that of the "New International Economic Order". This idea had just been put forward, when one of the first North-South meetings on development and environment was held at Founex in Switzerland in 1971 (Anon., 1972), as part of the preparations for the UN Conference of the Human Environment held in Stockholm in 1972. At Stockholm, economists and politicians from both North and South began to develop the concepts of a "New World Ecological Order" and "ecodevelopment". Economists of the Third World continue to remind us that with the social inequality between rich and poor it is impossible to think about a more equitable distribution of wealth without postulating continued economic growth.

At the same time the West, while recognizing the contradictions between growth and environmental protection, had begun to study the conditions of growth. It was realized that it was not worthwhile to quantify systematically the qualitative and to include it in cost-benefit calculations. It was even felt that this type of calculation brought dangers, as Gruson and others have indicated. The economic evaluation of the environment opened a new way of decision-making — a kind of ecological policy — integrated into the needs and aspirations of society. It was no longer sufficient "to put flowers in a picture of Leontieff".

## SIGNS AND SUPPORTS

### Increasing focus on ecological equilibria

The environment is no longer split into subunits treated separately: water, air, noise, waste products. Today attention is concentrated more on large-scale equilibria, on ecological laws and on effects which have a wide range of interactions. This is not only the case with ecologists. Economists such as Lesourne (1976) are insisting on the need for global analyses where the relations between economic development and other factors, such as climate, are taken into consideration. Here attention must be drawn to the debt owed to Unesco for the stimulation of this new global approach to the relations between people, activities, resources and the environment. This tendency to global thinking is not incompatible with the limited scale of action: "think globally, act locally" in the words of René Dubois.

### Reintroduction of the artefact, pre-eminence of social factors

In developed societies for more than a century there has been a tendency to produce and focus on objects without taking sufficient account of the problems that can be created. The car is a case in point. One should not forget that a very large part of the population of the world does not have the elementary objects necessary for anything other than survival. The tenants of the "qualitative" should not ignore the fact that many populations have need of the quantitative even in some countries with a long history of development. But a change is already taking place.

The development of the tertiary and quaternary sectors is significant since they "produce" goods that are not traumatic for the environment. The characteristics of new manufactured goods will probably evolve. For example the durability and the ease of maintenance are likely to improve. There has been a tendency to a "throw-away society" with ephemeral objects, but it seems likely that the need for a better management of resources and a revision of the social factors will lead to a review of the durability of objects.

The expression "quality of life", which has become worldwide in the last decade, has resulted from the preoccupation with "another growth". As to the word "environment" it resulted from a worry about growth in the American and the Japanese societies. Then Europe in the 1970s, and then thanks to the Stockholm Conference the Third World in 1972 adopted it as part of their aspirations. The quality of life is a subconscious repulsion from the mechanistic tendency to linear growth. A society should always question its existence. One should not believe in the stability of a quality of life, nor yet in that of a standard of living. People can not say to the State and to the community: construct for us a good urban environment, a good habitat and we will do the rest. Even the quality of life must be prepared as a function of the needs and psychic make-up of people.

## Development of geographical and cultural identity

The concern about the quality of life has caused societies to rethink their objectives. These must be linked to the cultural, national, regional, local desires and to the geographical situation of a country. It is not possible to produce and adopt a stereotype. The quality of life must be related to the environment in which a particular society lives, or, as L. Armand has put it, on the necessity for societies to learn again the rules of the "grammar between man and the environment in which he lives".

During the UN Conference on Population held in Bucharest in 1974 it was agreed that there is no reason why demographic growth should be uniform in a continent or even in a country. But in a world in which the media tend to generalize it is difficult to relearn diversity and to redefine strategies so that they take full account of the geographical, economic and societal factors.

## Increasing importance of resource use strategies

To make the importance of the object more relative, to define better the quality of life and of its social ends, to rediscover the importance of geography are three important aspects of the "new growth". Still more important is the notion of "resources" and the development of better management practices to avoid waste. Over the last 30 years or so, several groups and studies, such as Resources for the Future, the US National Academy of Sciences and the Meadows Report, have drawn attention to this problem and to the limits to available resources.

Should one limit this concern to mineral and energy resources? Space is a rare resource which has been given us once and for all. More and more countries, even those with lots of space such as the Soviet Union, are drawing attention to the increased scarcity of usable space. France still does not heed this because its territory is less encumbered than those of most of its neighbours, but the creation of "spaces for tomorrow", worries about the conservation of the coastal zone, and thoughts about the heritage to be passed to future generations, indicate a growing movement in the right direction.

More careful management of resources — especially non-renewable ones — of town and country planning, recycling, etc., are signs of a change in thinking and of policies. There is, however, much still to be done to develop and apply appropriate policies nationally and internationally in using natural resources while respecting ecological interests. The scientific community can help to accelerate processes of natural regeneration of renewable resources, particularly in plants.

## Ideas of durability and long-time scale

It has always been difficult to introduce the time factor in economic thought and calculations. Considerations of issues in terms of long periods of time have been allowed to atrophy. Classical political economics has always been linked to the ephemeral, the profit margin, the operating costs and these have all been modified by the increased speed of change in the "ephemeral" society. Too little notice has been taken of long-range thinking, or of the fact that our successors may have new factors at their disposal which are not available to us. There is a decrease in the importance of the future the further it is away: there is a greater degree of indifference in respect of what may happen in 30 to 50 years time than in 2 to 3 years time. Can one imagine a more egotistical solution in relation to our descendants, especially with regard to decisions that are likely to undermine non-renewable resources, and to have an indelible effect on our planet and theirs?

Up to the present this short-sightedness has had few long-lasting consequences because human effects on the environment and nature were reversible: this is no longer the case. In spite of the speed of new technological advances, the construction of large dams and of nuclear power stations, for example, may have agricultural, social and economic consequences which last for several decades. Societies can not reason, as in the past, *res sic stantibus*. The increasing rates of change, when the only thing that is certain is the depreciation of money, no longer permit this.

To these two reasons must be added a third: the slow changes or mutations that are occurring in the environment. Social strategies should take account of the risks of modifications (e.g. in carbon dioxide levels, in the troposphere) which might have unforeseeable consequences. The irruption of these ecological phenomena should teach us to pay attention to biological time and to come back into phase with the rhythms of life. Do questions of current non-profitability for new technologies seem adequate if this type of accounting is built into a calculation involving the future? Would the important factory in Athabasca for tar sands, based on an inconceivable doubling of the cost of energy, have been built under such circumstances? If we listened to such factors what would be the margin for manoeuvre of the people concerned with renewable energy who are preparing the way for the introduction of solar energy?

The reintroduction of the idea of durability and reassessment of the importance accorded to long-term considerations will bring back a number of economic ideas that have been left unexploited: capital accounts, energy accounts, material accounts, resources, heritage.[1] More emphasis will have to be given to the unexpected, and to the unpredictable (see, for example: Holling, 1982; OECD, 1979), in spite of the difficulties that such thinking entails. The next 30 years, and perhaps still further in the future, will be more concerned with economy in resource use and will give greater attention to the national heritage, including values such as space, time and milieu.

There is already a strong reaction of populations in developing countries to the implantation of new activities especially when these have, or seem to have an unknown possibility of strange illnesses, discomforts or insidious disbenefits. Ques-

[1]A ministerial group chaired by R. Toulemon has been created in France to study in a few selected sectors the costs of the maintenance of our heritage (Toulemon, 1979).

tions of jobs created or increased incomes are not sufficient to overcome this reaction.

The protection of the environment, the question of the quality of life will cause our societies to think more in terms of centuries. Management of our resources, in the widest sense, will be based on a 50 or 100 year perspective, taking into account increased rationality in relation to obsolescence and a better management of resources, especially those that are not renewable or which are particularly rare. The first priority will be that of having to consider the irreversible.

## INTRODUCTION OF ENVIRONMENTAL DIMENSIONS IN GROWTH-DEVELOPMENT CONSIDERATIONS

There are numerous reflections on the relation of growth to development, in those countries that have a history of development, but those which are concerned with the introduction of a change in the type of growth are rare. It is believed there are six conditions for this new type of growth: a more scientifically based management; ecotechnologies; inherited institutions; financial imagination and innovation; change in planning-horizons, to a 15 year term; a new international deal. Such concerns are a recurring theme in many recent studies on environmental issues (e.g. Barney, 1980; Holdgate and White, 1977; Holdgate *et al.* 1982).

### A more scientific management

It is generally admitted that nothing stops the progress of science. Is this strictly true? The end products of scientific progress certainly end up in a finished industrial or military form. But what of the sciences of comprehension that are grouped under the term "ecology". These are not limited to the study of natural phenomena and objects but include the presence and the role of man in the biosphere and must take account of social demands and practices.

More account is now being taken of the need for sound management of the limited resources of the planet, while the increased emphasis that is being given to the long term requires improved knowledge of the mechanisms and the functioning of the world in which we live. There is however a communication gap between the specialists and the politicians, because the needs of politics do not always correspond with scientific concepts. There is a divided responsibility because the users request help in decision-making and call for results that are immediately applicable, while the scientists tend towards disinterested and exhaustive studies. To overcome this, it is necessary to bring research down to earth by increasing contacts with those who actually manage the land, whose experience and knowledge are implacable.

I believe it is essential to facilitate the dialogue between research workers and people who work on the land, who manage resources on a day-to-day basis. Here, there is a leading role for the professional ecologists, who by their daily activities can introduce into practice more scientific rigour and at the same time (together with the more practical workers) deepen understanding on the need for management of our environmental heritage.

## Ecotechnologies

We are prisoners of an old misunderstanding that technology is in opposition to nature. On the contrary the environment can not do without technology because paradoxically its field of concern is the same as that of technology, that of the interactions between man and his milieu. Ecodevelopment therefore makes use of technology — appropriate technology.

The wise management of resources and of the environment requires a society to seek renewal of materials and make maximum use of solar energy and promising forms of biomass, as well as of recycling. Apart from the removal of pollution, greater efforts are needed to avoid pollution at source so as to escape from the see-saw of pollution and depollution. The search for such "ecotechnologies" will lead us to rethink industrial production systems using as a model the basic principles of the ecological cycles. On the other hand, a wiser use of the "living factory" of the natural environment will lead to an economy in costly and aggressive technologies.

The arrival of ecotechnologies will bring with it "ecological engineering" which will improve our capacity to interact with ecosystems by better applying research in the decision-making process and then to control the functioning and the evolution of biological processes. Ecotechnology will become a not inconsiderable factor of growth and employment in relation to production and services, provided that appropriate resources are devoted to it. It would be unwise to imagine that a few subventions or a little financial support will suffice: the whole research-innovation apparatus must be mobilized, putting special emphasis on technological and economic cooperation.

## Inherited institutions

The third condition for launching new approaches to growth requires a new look at institutions. In view of the growing complexity of systems and equilibria, a larger role must be given to decentralized management of both human and natural milieux. Local communities and associations of citizens can play a decisive role in the future of social structures. In many countries, local communities have modified their attitudes towards the natural environment and to socio-economic activities. In addition the importance of developing new institutions should not be overlooked, such as those in France, in such fields as water management, natural parks, coastal conservation. Decentralization should not be limited to public institutions, commercial firms also have a role to play, with new structures and new concerns for the environment.

This process of decentralization implies that activities should proceed in two directions. First, in the underestimated sector of training. There is a need for informed people in specialized institutions as well as within the general public. Second, an important function can be served by public service agreements, established for a period of several years, that allow the introduction of changes that are economically acceptable and financially possible. This system, know in industry as "branch agreements", could be used for relations between state and local communities.

## Financial imagination and innovation

Ecodevelopment can not be synonymous with extra charges. The problem is not to seek new sources of funds but reorientate charges: to make external costs chargeable, and instead of separating such costs to integrate them. This is one of the keys to ecodevelopment. It can be done by authoritative means or mobilizing effectively the people involved. The former requires new legislation, various constraints, new organizational structures and never works very well. Mobilization can involve economic and pedagogic factors. It should be based on a new system of prices which would fix the imputations, the level of reinternalization, and the deductions to be used to improve the quality of life. Obviously the prices for such a reinternalization can not, in a country with a free market economy, be fixed by the state, but it should, nonetheless, be possible to make progress.

On the one hand, it would be useful to have a real patrimonial accounting — national, public, private — and certain number of efforts have been made in this direction in France following the reports of Gruson (1974, 1979) and Toulemon (1979). The budgetary way is certainly the best: to use taxes to pay for public services for the maintenance of satisfactory environmental relations. But experience shows that there is little room for such manoeuvres.

On the other hand, it should be possible for the state without fixing the price levels to determine how they should be fixed, especially by public enterprises. In spite of the freedom that such enterprises should have, even those with a mission to provide a public service should take account of the costs of decreasing environmental nuisances and of the fight against pollution. Another possibility is a system of contributions. Thanks to the introduction of these in 1964, France has today an operational system for managing water. Other contributions based on the "polluter pays" system could help offset the costs of disposal of wastes and of reducing air pollution.

Finally, with respect to certain environmental nuisances that are tolerable, would it be a disadvantage if the state and the local authorities taxed certain products so as to pay part of the inconvieniences to which the local communities are subjected? Higher prices might slow down production or stimulate research or substitution of products. The taxes could be used to improve the quality of life, promote new techniques or be used for projects of local interest. All of the activities that can contribute to growth linked to an improvement in the quality of life necessitate voluntary action by states[2] regardless of their different socio-political or economic systems.

## Change in planning-horizons

Introduction of ecodevelopment necessitates a reduction in the inconsistencies caused by the immediate and long-term needs. There is a big temptation to mortgage the future for immediate gain, at the expense of the environment. Respect for the environment and careful consideration of resources follow from an ethical viewpoint: diachronic solidarity with future generations. The problems of present as opposed to future choices must be carefully arbitrated and explained.

It is of course not easy to arbitrate such choices. The benefits and disbenefits of both have to be clarified and it would be useful to explain more precisely the conflict between a long-term global rationale and short-term, more fragmented, but more acceptable action. It would be good if the systems based on current knowledge were not systematically favoured in relation to "non-knowledge", worries and questions. Involuntary modifications, of the climate, for example, should be taken into consideration even when unknown. We are not used to making an inventory of the gaps in our knowledge, which is as necessary as knowledge for an understanding of the world and the preparation of future activities.

But this concern for future generations and long-term planning should not lead us to impose our ideas of the quality of life on our descendants. We must avoid the irreversible and leave as much choice as possible. We should leave a "made-to-measure" future not a "ready-made" one. Our descendants should be able to decide the quality of their lives as a function of the time in which they live: they should not be able to say as Musset did in 1836: "Our towns are nothing but left-overs. We live in the debris of past societies. We have not been capable of imposing on our towns the mark of our time".

This does not mean, however, that we have no responsibilities with regard to the future. A policy for new growth will be facilitated by the elaboration and the adoption of plans for a conversion to mid-term planning. With energy, for example, this should mean not a one- to two-year policy but a policy for 15 years: France benefitted from such a change in planning horizons a few years ago in relation to nuclear energy and information.

Such plans should fix objectives, provide rules, make experiments, as far as possible on an international scale. The conversion of economies as a function of the environment or of the quality of life will be a long process. If the changes take place

[2]The creation of new rules by the State depends largely on the legislator and is already being done in some countries. In France laws concerning water (1964), wastes (1975), protection of nature and impact assessment (1976) have provided important benchmarks. For the future priority should be given to international cooperation.

in a series of steps, it will help disarm the opponents. There are few decision-makers who are opposed to a change in development provided this is decided sufficiently far in advance and in a multi-national framework.

Such activities necessitate both a prospective outlook and prospective action. If the change requires 30 years to take effect we should not wait 30 years before we start. The future takes place from this moment on, and can be modified by the actions we take now. This prospective action must include scenarios, not any kind of scenarios but ones of transition. Too much weight is given to exogenous, normative, logical scenarios, too little to those that facilitate evolution. The plans for change should be accompanied by a series of systematic large-scale experiments. We must bring in social innovations and the right to experiment and help groups which are making future-oriented studies. To return to the analogy with tailoring — the suit should be tried before it is worn.

## A new international deal

In order to allow ecodevelopment to be born and take root, one of the first things required is a minimum of international cooperation. In a world where competition is so strong, it would be irrealistic to expect that this take place country by country. Such changes can take place, as has been the case in my own country, by branch agreements that seek agreements branch by branch, for example among car manufacturers, paper makers, cement works, etc., to reduce pollution and nuisances, improve the quality of the products (durability, maintenance, etc.) and of conditions of work. Such agreements for 5 to 10 years should be discussed by trade unions and by states in the framework of international realities and of a "new world order".

Another possibility is to follow geographical zones. To launch planetary activities is not easy — witness the problems over the last 30 years in relation to the oceans *res nullius* of the world. The global scale is large. This is why maximum use of neighbouring groups would facilitate the task. Eco-regional groups are another possibility as the Mediterranean region has shown. The Blue Plan for example: will this prove a stimulus to the 17 Mediterranean states concerned to work together on a series of ecodevelopment projects on a grand scale?

### References

Anon. (1972), *Development and Environment,* Report and Working Papers of a Panel of Experts Convened by the Secretary-General of the United Nations Conference on the Human Environment (Founex, Switzerland, June 4-12, 1971), Mouton, Paris, The Hague.

Barney, G. O. (Editor) (1980), *The Global 2000 Report to the President,* A report prepared by the Council of Environmental Quality and the Department of State, Three volumes, US Government Printing Office, Washington, D.C.

Gruson, C. (1974). *Rapport sur la Lutte contre le Gaspillage,* Documentation Française, Paris.

Gruson, C. (1978). *Rapport du Groupe d'Evaluation de l'Environnement,* Documentation Française, Paris.

Holdgate, M. W., M. Kassas and G. F. White (Editors) (1982), *The World Environment 1972-1982,* A Report by the United Nations Environment Programme, Natural Resources and The Environment Series 8, Tycooly International, Dublin.

Holdgate, M. W. and G. F. White (1977), *Environment Issues,* SCOPE 10, Wiley, Chichester.

Holling, C. S. (1982), Predicting the Unpredictable, *The Unesco Courier,* August-September 1982, 60-3.

Lesourne, J. (1976), *Les Systèmes du Destin,* Projet Interfuturs, OECD, Paris.

Toulemon, R. (1979), *Rapport de la Commission Interministérielle des Comptes du Patrimoine Naturel,* Ministère de l'Environnment, Paris.

OECD (Organisation for Economic Cooperation and Development) (1979), *Interfutures. Facing the Future: Mastering the Probable and Managing the Unpredictable,* OECD, Paris.

CHAPTER 24

# Participation of Scientists, Planners and Local Populations in the Planning and Execution of Research: Some Insights from the Swiss-MAB Experience

Abstract

The contribution comprises three parts. In a first part, the general background and design of MAB work on integrated mountain management in Switzerland are outlined, and some of the similarities and differences of studies in four test areas are discussed. The second, longest part of the contribution deals with one of these study areas, the Pays d'Enhaut region, where research workers, local decision-makers and local populations are collaborating in a programme of research and where a special association has been set up to promote such collaboration. The third and final section presents the viewpoint of the local population. The proposed studies have been received with a certain degree of scepticism by the population, who are not prepared to be reduced to the role of guinea pigs. But with the contacts that have been established, and the consultations and discussions that are in course, a fruitful collaboration between scientists and the local people can be envisaged.

*Résumé*

*La contribution comprend trois parties. Dans la première partie, l'historique et le cadre conceptuel du travail MAB sur l'aménagement intégré des montagnes en Suisse sont présentés et quelques unes des ressemblances et dissemblances dans les études des quatre régions témoins sont discutées. La seconde, la plus longue partie de la contribution, traite de l'étude d'une de ces régions en particulier, le Pays d'Enhaut, où chercheurs, décideurs locaux et population locale tentent une expérience de collaboration dans la réalisation du programme de recherche et où une association a été créée pour promouvoir cette collaboration. La troisième et dernière partie présente le point de vue de la population locale. Cette étude a été accueillie avec quelque méfiance par la population qui n'est pas prête à jouer le rôle de cobayes. Mais grâce aux contacts qui ont été établis et aux consultations et discussions en cours, une collaboration fructueuse peut être envisagée entre les scientifiques et la population locale.*

# 1. Design and Implementation of Swiss Research on Integrated Mountain Management

*Paul Messerli*

## THE CONTEXT

In the late 1950s, in many parts of the mountainous regions of Switzerland, rural emigration was halted by the development of tourism. Tourism was and remains for most parts of the country the only realistic means of stopping this demographic trend and of balancing the economic disparity between populations in urban areas and in rural mountainous zones. The development of tourism has brought about a remarkable change in land use practices, not so much in a direct as in an indirect way, by the complete transformation of economic structures and, with a time lag of some years, of social and cultural systems, as illustrated by changes in social norms and value systems. Within this context, the very precise question that has to be answered for the Swiss government is whether this development can be absorbed without provoking serious disruption of the ecological systems and without compromising the recreational potential of these sites. And if it can not be absorbed as such, what kind of intervention or control measures should be envisaged.

"The modern threat to the Alps" is the challenging title of the map, reproduced as Figure 24—1, which shows the quantitative development of population and tourism around and within the alpine region. The different impacts of this development were discussed during the Council of Europe seminar held in 1978 in Grindelwald. But discussions at Grindelwald, and elsewhere have highlighted the lack of knowledge and experience in integrated mountain management — an

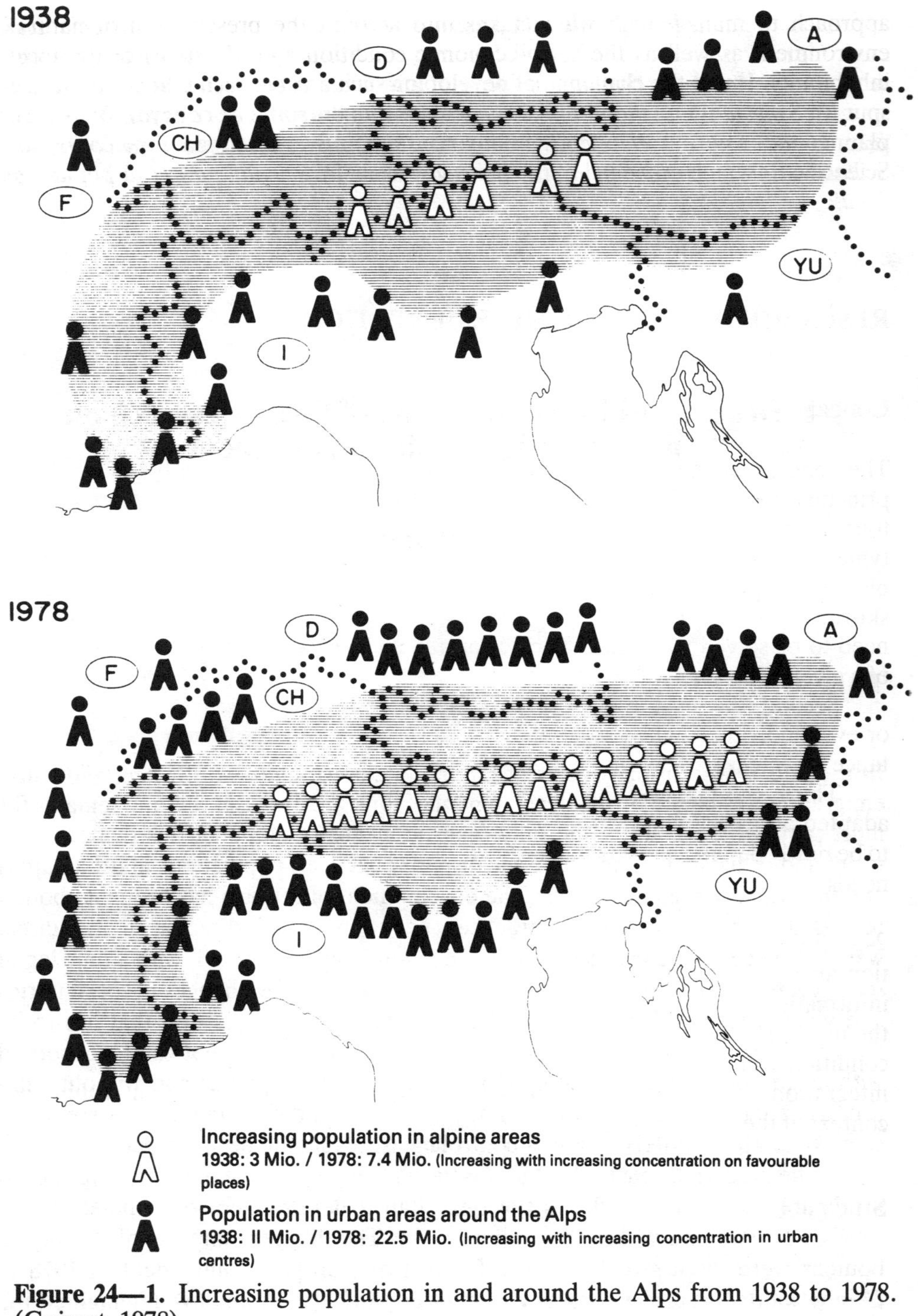


**Figure 24—1.** Increasing population in and around the Alps from 1938 to 1978. (Guizert, 1978)

approach to management which takes into account the preservation of natural environment as well as the socio-economic condition and life styles of the local inhabitants. It was the challenge of developing such an approach that provided the spur for Swiss work on mountain ecosystems within the framework of MAB — work planned in the period 1976-1978 and financed by the Swiss National Foundation for Scientific Research on behalf of the Swiss government.

## RESEARCH DESIGN AND IMPLEMENTATION

### Concepts

The conceptual framework of the Swiss MAB activities has been based on three principal considerations. First, although the history and state of development of tourism in the Swiss mountain regions are quite varied, there exist a number of typical problems common to all regions: for instance, the existence of underexploited and unstable forest areas, the despoliation of scenery due to construction of ski-runs, and use of good agricultural land for tourist settlements. These problems need to be solved within the specific natural, socio-economic and cultural contexts of a given community.

Second, the level of decision-making relevant to ecological conditions is local or even individual. The local and individual levels are therefore of greatest importance for integrated mountain management.

Third, the preservation of the natural environment is no longer ensured by adapted agricultural systems. The types and intensities of land use have therefore to be newly defined in respect to ecological constraints and changing socio-economic needs.

In line with these considerations, a research model was developed (Figure 24—2) showing the interdependences at the local or regional level, whose description and analysis within the time-scale of the last few decades would provide the information required for two main tasks: first, to achieve a better understanding of the interrelationship between socio-economic development, land use practices and conditions, and availability of natural resources; and second, to attain a better integration of individual problems and approaches to their solution within the context of the concrete system under review.

### Study areas

Four test areas were selected as sites for putting this concept into practice (Figure 24—3). Each area has its own characteristics and problems. In Grindelwald, there exist crucial economic and social disparities between the part of the population

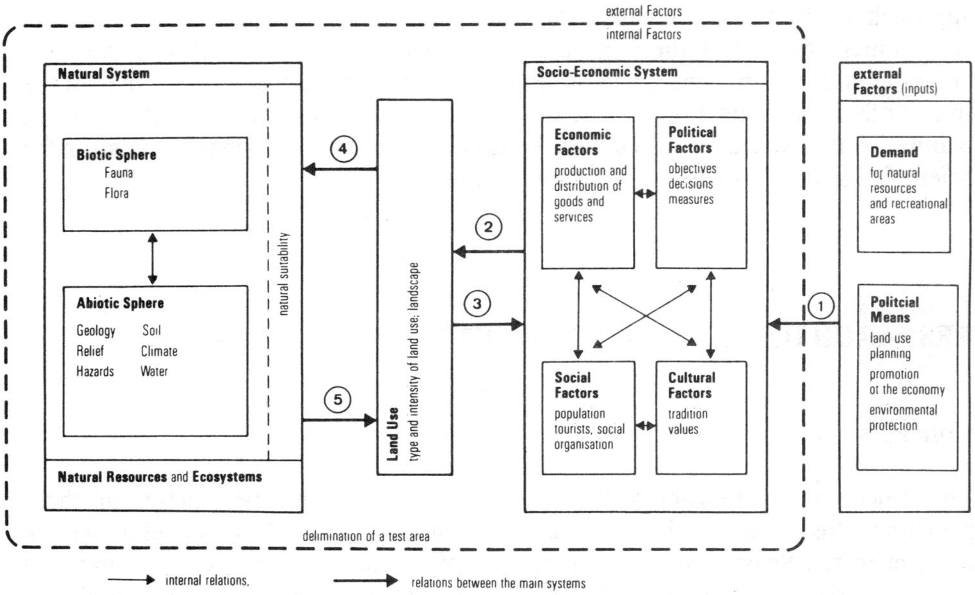


**Figure 24—2.** Schematic presentation of a regional socio-economic ecological system, used as a basis for planning MAB research on integrated mountain development in Switzerland. (1) Influences of the external factors; (2) Socio-economic development and connected change of land use; (3) Feedback of change in land use on population, economy and society; (4) Effects of change in land use on the natural system; (5) Feedback of the change in the natural system on the (human) conditions of living. (Figure published in *Geographica Helvetica,* 1978, vol. 4)

involved in tourism in the centre of the settlement and the marginal agricultural inhabitants. In the Pays d'Enhaut — a favourable area for mountain agriculture — there still exists the problem of depopulation, lack of jobs and the problem of integration of tourism within schemes for rural development. In the Aletsch region, tourist settlements have been created above the timber line during the last two decades, and the predominant problem is one of integration of agriculture with tourism. In Davos, one of the most highly developed tourist resorts in Switzerland, there is a problem of urbanization in a mountainous environment.

## Similarities and differences in approach

Under the general direction of a programme director and a project director for each test area, multidisciplinary research teams have been set up for the four test areas, each comprising 10 to 15 participants. Within the overall conceptual framework, no attempt has been made to standardize approach. Indeed, the scientific approach adopted by the four multidisciplinary groups is quite different, ranging from a systematic approach, as in Grindelwald, to a problem-oriented approach as

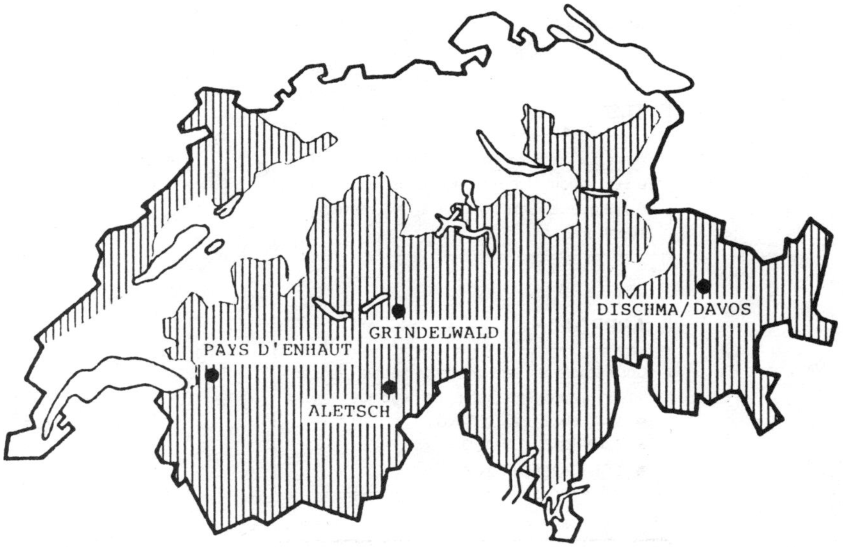


**Figure 24—3.** Map showing the mountain areas of Switzerland (vertical hatching) and the location of the four MAB test areas. The mountain region comprises: 40 per cent of the communes, 24 per cent of the inhabitants and 70 per cent of the total area of the country

in the Pays d'Enhaut. The level and means of integration of local populations and decision-makers within the process of implementation are also quite different from one test area to another, reflecting differences in the practical experience and make-up of the research groups, but also in the mentality of the inhabitants and the political situation in the communities concerned.

Though there is this variation in approach among the four test areas, some well-defined levels of coordination have been identified, for example so as to obtain comparable results at the thematic level. There is also synthesis of information designed to answer practical questions at different levels of decision-making, and coordination of actions for the dissemination and implementation of results, at the individual, local, regional and even national level.

The procedure for implementation of MAB activities on mountain management in Switzerland is illustrated schematically in Figure 24—4. The analysis model shown in this figure enables various scenarios to be elaborated and tested. The "trend" scenario is based on growth of tourism and regression of agriculture. A second scenario explores alternative forms of tourist development, including the integration of tourism into an agriculturally based economy, and the inverse, of integrating agriculture into tourism. A third scenario is concerned with strong ecological constraints and the repercussions on economic activities and social organization.

The 1979-84 period of execution of the programme of field work is expected to provide some important feedbacks, at various levels: at the methodological level;

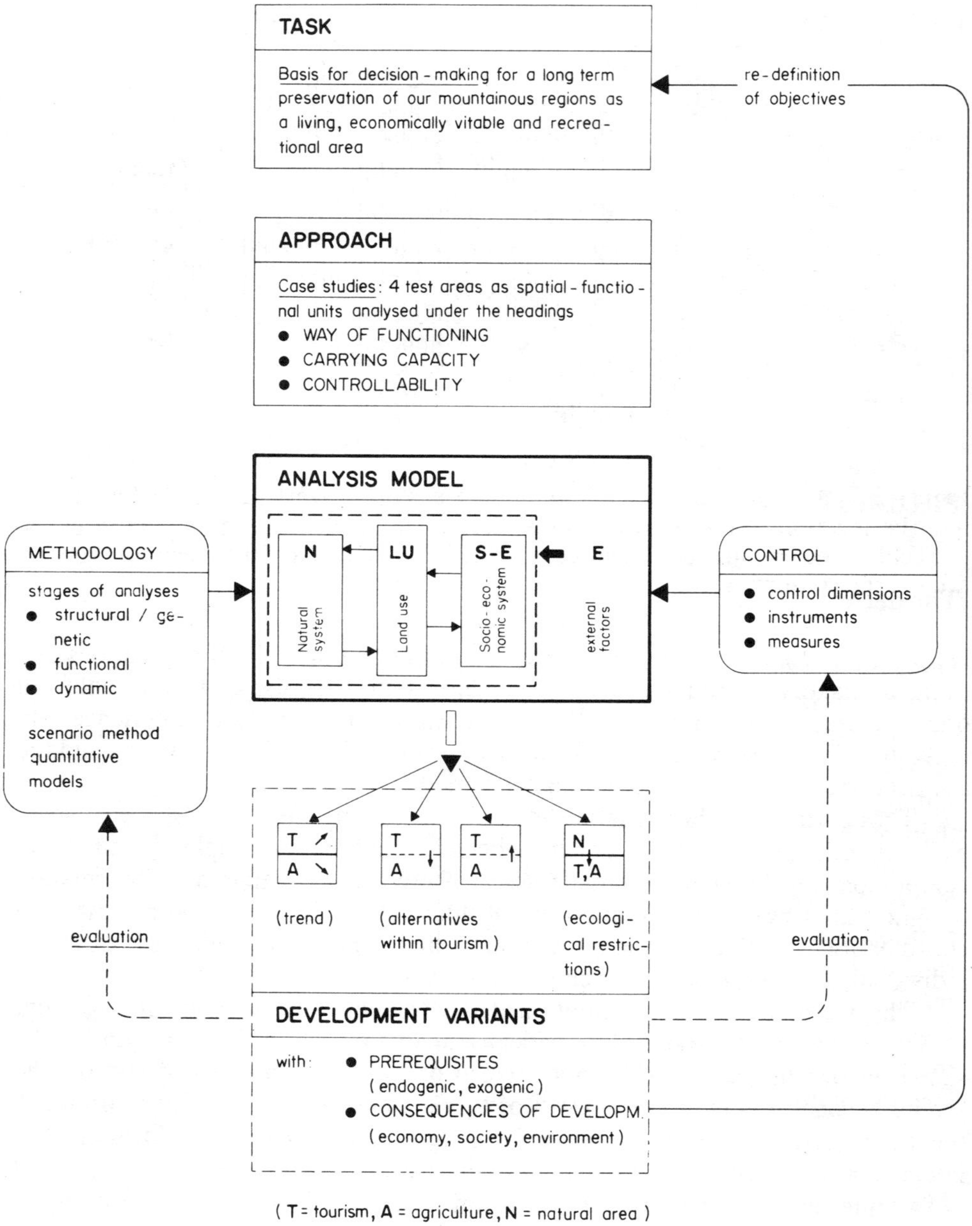


**Figure 24—4.** Implementation procedure of the Swiss MAB Programme

at the political level, by means of measures and control activities; and at the level of local and regional development aims, through the redefinition of objectives. In part 2 of this contribution, an account is given of the early experience in putting the programme into practice at one of the test areas, the Pays d'Enhaut.

# 2. Glimpses from the MAB-Pays D'Enhaut Project

*Charly Darbellay and Erwin Stucki*

## PROLOGUE

### The decor

ONCE upon a time in the Alps not far from the Pays-d'Enhaut, there was a village with a name full of promise: Belavenir. The houses idyllically grouped around the ancient church were homes for some 400 inhabitants. The stables and the barns provided a pleasant ensemble with the houses in stone and wood. The people lived simply, earning their living often outside of the commune, while continuing to work their piece of land. The large number of plots of land resulted from a process of division which has been repeated a thousand times. One result was a marvellous harvest-time picture of the gold of the wheat standing out against the green of the prairies. The beauty of the scene did not reflect the hardness of the life. Work was rare, and year after year the young people left their country. The few families which remained behind did not stay in the old village; they built their comfortable villas several hundred of metres from it, near the new church whose architecture enshrined the break with the past and whose position drew attention to the dynamic new location of the buildings in the territory.

### The projects

In order to be able to work their land more rationally, the people of Belavenir decided in the early 1970s to reorganize and regroup their land. The young people particularly interested in the future of their village proposed to use the occasion to create a community tourist development zone. In this way was formed the Mottaz zone which belongs to the cooperative created by the farmers who made available some of their plots to the project, with a view to increasing local revenue by supplements from the tourists. The idea was interesting, courageous and had the merit of reaching a successful conclusion.

But it was still necessary to do something. For the inhabitants of Belavenir, the solution was simple: several kilometers from the village there was a flourishing tourist spot with some 20,000 beds and this seemed to indicate the way to take. It would be sufficient to divide the land available to the cooperative into small squares of about 500 $m^2$ to make building plots which would easily find buyers. The place is pleasant, communications are quite good, and there should be no problems in selling. The more daring imagined the possibilities of going a little further; after several plots had been sold, it would be possible to invest the proceeds in new chalets which it would be possible to sell later, thus realizing a much larger profit. The plan appeared simple, a helping hand from an urbanist or an economist to orchestrate the whole thing seemed useful but not absolutely indispensable. As a result of personal contacts, help was requested from an institute for rural economy which in turn sought the collaboration of an institute of architecture and urbanism. These institutes, which carried the colours of a prestigious school of learning and expertise, were considered capable of providing the support required.

## The intervention of specialists

From this point, things took another turn. The economists immediately drew attention to the danger of the policy of selling buildings which, although lucrative in appearance in the short term, would not lead to the creation of long-term help. The loss of power over the development by the cooperative, the proliferation of secondary homes which are endemically underoccupied, the charges that the infrastructure would impose on the local population, provided enough inconveniences to make the value of the whole initial project doubtful. For the economist, it seemed much more important to obtain an improvement in the income to the village over a long term rather than a short and transitory increase.

The urbanist refused to consider the zone of La Mottaz in isolation. He studied first the local context and oriented his interest on the old village and drew attention to the following surprising things: the local architecture had unexpected resources and qualities, and could inspire a number of guidelines for construction of a new ensemble at La Mottaz. Paradoxically, the most deserted parts, abandoned to the less well-off, to the old and the unmarried were those that presented the greatest architectural interest. A certain potential for development was thus found in the heart of the village which would necessitate a policy of improving the old buildings. The urbanist then examined the possibilities of a local development plan — perhaps a weighty title for a plan which was essentially limited to cutting up the land available. But this outline was not based on a sound concept of development. In a region where every plan must be accepted by the citizens, the authorities in the commune had a difficult equation to balance if they wished to insure the success of a vote — because on one hand it would be necessary for each proprietor to have at least one plot in the building zone — and on the other hand to try to restrict the dimensions of the building zone so as to avoid imperiling local finances by the expenses which they were likely to necessitate. The result of this perilous balancing act was that the building zones became over large, without coherence and with no real urban policy.

Taking into account all these problems, the urbanist and the economist suggested another possibility: to begin by modernizing the traditional buildings which were being abandoned. These would then be managed by the people of the village and this would provide a bait for the tourists. After this had been carried out, the local development plan would be looked at, particularly from the viewpoint of the needs of the young who wished to stay in the region. Finally, several solutions were envisaged if it proved necessary to develop the zone of La Mottaz, not by building chalets close together on small plots but in creating an organic whole which would function like a village and be able to receive convalescents or classes coming from the lowlands. The management of this new ensemble would stay in the hands of the local people, thus creating work for the young.

## The complexity of problems

For several years things have been at a virtual standstill. The people of Belavenir are no longer quite sure what to do with their zone of La Mottaz. Their initial project seems to them to be pathetic. The many discussions with specialists have taught them the beauty of certain parts of their village. The president of the commune did not appreciate remarks made on the subject of local development, and has discreetly retired from discussions. The projects submitted by the specialists, although attractive, are not sufficiently convincing for the villagers who are absorbed in their traditional tasks, dispossessed of their leaders by an earlier exodus and who no longer have the ambition, nor the means to complete such projects. The specialists have finished their deliberations and discussions and have left. For the present, La Mottaz is rented to a farmer and perhaps one day it will be possible to organize a cooperative small fruit project (the economist has also studied this possibility), but the long-held dream for development is for later . . .

This story, with the exception of the names of the places is true. It illustrates in all its complexity the enigmas that must be resolved in order to ensure the participation of the local populations, authorities and specialists in the development of mountains.

First, the problem of the local populations, grown poorer both materially and humanly and who consider it a return to justice to be able to make a bargain once in their life by selling a part of their patrimony. As a result, any restriction in the field of ownership and in the free disposal of their holdings appears to be an unjust and unacceptable constraint. This gives rise to a strong opposition to anything related to regional development.

Second, the problem of the local authorities, struggling for the survival of their villages and as a result often prepared to accept any form of development, even at the price of a desecration of the natural environment. These authorities are in a difficult position; they are often placed between the hammer and the anvil, realizing the risks of development, but rarely having the means of mastering it.

Third, the problem of the specialists, who are never short of good ideas but who are often far from the realities of daily life for the people of the mountains and who are rarely effective in transmitting their knowledge to the local populations because of a breakdown in communications. This type of situation adds weight to the well established idea in rural populations of an almost insurmountable gap between theory and practice.

This is the challenge taken up within MAB, of finding ways "to work with local populations in identifying the priority problems and in facilitating the application of research results" (di Castri and Glaser, 1979). There are thousands of townships such as Belavenir in the Alps. The research workers and specialists from the various disciplines will be undertaking sterile studies unless they manage to open up a real dialogue with the populations concerned. In order to move in this direction, the people responsible for the MAB operation in the Pays d'Enhaut have initiated an organic collaboration with the local population. This is ambitious, difficult and tricky but also promising, and some features of the operation are outlined in the following paragraphs.

## EXPERIENCE IN THE PAYS D'ENHAUT

### Brief introduction to the region

The Pays d'Enhaut is a district in the Vaud canton of Switzerland. It is composed of three communes: Château-d'Oex, Rossinière and Rougemont. The total population in 1980 was 4,180 inhabitants. The area covered is about 18,000 ha. The regional economy is based on two principal activities: agriculture and tourism. Services are well developed: schools, hospital, public transport, shops and banks. Since 1930 the population has been decreasing constantly. The demographic structure shows acute signs of ageing. The Pays-d'Enhaut has a remarkable environment and up till now the touristic development has been well integrated into the natural environment.

The basic preoccupations of the region correspond with the worries of the proponents of the MAB project, that is to say: the desire for a development which would face up to the problem of depopulation but a desire to avoid the destruction of an environment whose quality is appreciated both by inhabitants and tourists.

## The questions raised

The basic questions which are raised are above all perceived very differently by the various groups concerned, by the local population, the local decision-makers, and the research workers. The local population has to resolve the problems of daily existence. What is best to ensure that my company continues? What dose of fertilizers should I apply to my land? Do I have the possibility of cutting some wood? Would it be possible to obtain authorization to build following the sale of a plot? Where can I build a handicraft workshop? Is it necessary to cultivate such and such a field on a slope or would it be better to leave it fallow? The problems which are faced by the population are therefore of a punctual nature, precise, requiring a rapid decision. On the other hand, the appraisal of the overall effect of the sum of thousands of similar decisions often escapes the local people.

It is up to the decision-makers to evaluate the possible effects of the sum of the individual decisions. What will happen if building continues at the present rhythm? Will there be sufficient free capacity for the future? Does the abandonment of the land constitute a danger for the environment? Should the development of tourist installations be well coordinated or even limited? Is the access to isolated farms sufficient? What should one do to retain the young people in the country? The decision-makers very often find they are no longer in a position to make a choice on a purely empirical basis. After having noted the abuses and the undesirable developments, they recognize the need for a better basis for decision-making in the medium term. This becomes necessary in order to avoid an anarchic occupation of the land, to adapt desired changes to the possibilities of the communes, to coordinate the use of equipment, and to avoid errors.

The specialists in development are called in to collaborate in all of these tasks. But the knowledge necessary for the long-term development of mountains remains superficial. And the research worker, before giving solutions would like to know more, very much more. This is where the "problematique" of MAB comes into play, so as to make more precise the bases which are necessary for the rational utilization and conservation of the resources, to improve the relations between man and his environment, to foresee the effects of current actions on the world of tomorrow, to make man more capable of managing natural resources.

A first task of MAB research is that of providing the knowledge necessary for improved land management, in this case of mountain ecosystems. The essential is to discover the dynamics of mountain systems, to simulate interaction between the activities of man and the natural environment and to study the limits of the carrying capacity of the area.

But the knowledge generated by such studies must not remain the private property of the research workers. There is therefore a second objective in MAB activities — that of developing the principles for development, of explaining the criteria which are applicable and of furnishing usable information materials to local decision-makers in order to improve their knowledge base for decision-making. Such an operation should be considered as a joint undertaking between the research workers and the local decision-makers, who should be considered not as the end point of the information but as partners in the research effort.

But research of the type advocated within MAB should not in any case be limited in its impact to the local decision-makers. It must include another dimension, that of allowing the local population to learn about the research by providing information appropriate to all stages of the project; of improving the overall view of the people concerned on the possible repercussions of different types of development; and of giving them a better basis for choosing the ways and means of managing their region.

## Some landmarks in the history of research in the Pays d'Enhaut

**1969-75: the initial studies:** From 1969 to 1975 the Pays d'Enhaut passed through a phase in which there was an increase in building for tourists. Shortcomings in the use of the land became obvious. The farmers became concerned about the effects of tourism on agricultural land. The proliferation of mechanical lifts and of projects for their installation did not correspond to a well studied programme.

The need was felt to do something but exactly what was not known. The municipality of Château-d'Oex decided to carry out a series of studies. It entrusted M. Chevallaz, a person with good knowledge of the region and good links with it, to undertake a series of studies which would allow the development of the commune. The "Chevallaz Commissions" started working. They undertook three special studies: an agro-sylvo-pastoral study, a study of the skiable area and a study of a centre for secondary education.

The agro-sylvo-pastoral study carried out by the Service Romand of Agricultural Popularization and by the Station for Agronomic Research led to the preparation of a map of the different types of soils which would serve to define the priority zones for agriculture. An agricultural balance sheet was established, with an inventory of the improvements in land use which should be carried out and with forecasts for the best use of each farm. The local farming community and the farmers themselves collaborated closely in these studies and discussed the different options during the course of evening discussions.

The skiable regions were mapped out by a private body, the SGI. This provided an inventory of the potential for tourism, and served as a basis for the coordination of the introduction of tourist installations.

The study for the centre for secondary education was carried out with the help of the architectural office in the region. The objective was to increase secondary education for students of the Pays d'Enhaut and to act as a centre for the creation of qualified employment by functioning as a centre of information on mountains for students coming from outside the region. This was to be taken up later in the regional programme.

**1975-79: the regional development programme:** In this period, innovations were introduced into the Swiss legislation concerning help for investments in the mountainous regions. One condition was introduced, that the communes must organize themselves within a regional framework and elaborate a programme of regional development, otherwise no assistance would be given.

In order to benefit from this new possibility, the three communes of the Pays d'Enhaut joined together in an Association for the Development of the Pays d'Enhaut (ADPE), a private association also open to individuals. The ADPE took over from the Chevallaz Commission, and asked the Institute of Rural Economy of the EPFZ, the Service Romand of Agricultural Popularization and the Service for Development of the Canton of Vaud to carry out the studies necessary for the preparation of a programme of development. Working groups were created in seven different fields: agriculture and forestry; arts and trades; tourism; equipment and services; transport and communications; technical installations and the environment; district financing. With the technical collaboration of the institutions mentioned above, more than seventy people from the region took part in about sixty meetings of commissions, charged with the analysis of the situation and defining the policy for development of the region. The analysis of the situation and of its potentialities (ADPE, 1977), then the objectives and the measures to be taken for development (ADPE, 1978), constituted reference documents for the regional programme, elaborated jointly by specialists and representatives of the local population.

A wide campaign of information was organized, thanks to the initiative of a group of young people from the region. An audio-visual demonstration on mountains followed by in-depth discussion allowed the people of the region to introduce their own points of view. The local newspaper published several articles concerning the work, including a special number on the analysis of the situation and the concepts of development.

On July 13, 1979, the Federal Department of Public Economy gave its official approval to the programme presented by the region, and the programme thus entered its operational phase.

**1979: the MAB research project:** In 1979 the Swiss National Fund for Scientific Research decided to launch a national programme within the framework of MAB. From the numerous projects presented, the MAB Committee retained four, including that of the Pays d'Enhaut. The reasons for the choice of this project were notably the existence of several preliminary studies, the geographical homogeneity of the region, the presence of a dynamic mountain agriculture coexisting with an important touristic development.

The National Fund allocated research to the following: the Institute of Rural Economy of the Federal Polytechnic School with the Service Romand of Agricultural Popularization, the Department of Botany of the University of Geneva, the Forestry Service of the Canton de Vaud, the Faculty of Social and Political Sciences of the University of Lausanne, the Federal Station of Agronomic Research of Changins. This research therefore has a multidisciplinary character; each institution is charged with a certain number of units of research (Darbellay, 1980). The object of the programme was specifically to study alternative solutions for the development of the region, taking into account the ecological, economic, demographic and social conditions and considering the needs of the population.

## From the regional programme to the MAB project

It is interesting to note that several of the institutions charged with research in the MAB project have been working together for some time in the region. This is the case with the Service Romand of Agricultural Popularization, which already collaborated in the first studies undertaken by the Chevallaz Commission in 1969. This is also the case with the Institute for Rural Economy of the Federal Polytechnic School, which from 1976 was charged with the responsibility for studies in the framework of the programme of regional development.

This continuity presents certain advantages: the experience acquired in the region during the earlier studies, the confidence gained with the local population, the habit of collaborating with representatives of the population, the abundance of accumulated data, etc. But the main contribution of these earlier studies to MAB lies without doubt in the identification of the regional problems. The many sessions of the Commission of the ADPE as well as the public debates allowed the specialists involved in the regional studies to make a complete inventory of the problems of the Pays-d'Enhaut. When they were asked to prepare the MAB project these scientists relied on the many observations made by people in the region. We can therefore say that through this approach the population of the Pays d'Enhaut has contributed to the identification of the priority problems in the research. The MAB-Pays d'Enhaut team has adopted an approach which involves research "by problem" and which entails the creation of "units of research"; this approach can be considered a response to resolving the questions which remained unanswered during work on the regional programme.

This continuity also involves risks linked with the ambiguity of the situation. The MAB scientific studies are different to the regional programme, and this difference is not always obvious to the local population. A regional programme is designed and financed by the regional authorities and is related directly to the needs of the region. On the other hand a MAB study is designed and financed by the National Fund for Research. It is not restricted just to the region of study because it is also used to satisfy the interests of scientists in their specific fields and also to provide information useful for other regions which are faced with analogous problems both in the country and abroad.

The MAB Pays d'Enhaut studies are spread over 5 years, from 1980 to 1984. They are integrated into the national Swiss MAB Programme with the three other test areas of Grindelwald, Aletsch and Davos.

## Problems relative to the participation of local populations in research

If in regional planning the specialists and the decision-makers go hand in hand, it is not the same in basic research. A regional programme is an eminently political act; the objectives of development are fixed there and adequate measures are decided. As a result it depends, in the last resport, on the people with regional responsibility to take the decisions on the subject. The role of the planner is in providing a service — that of analysing the situation, proposing possible variants for development, comparing the financial and economic consequences of the choices suggested. But in the end it is the people of the region who decide upon the option which seems to them to be the best. Regional planning is thus, by its very nature, an undertaking which requires close collaboration between specialists and decision-makers.

MAB-type research presents another kind of problem. Here, the researchers are called upon to fix the objectives of the studies and to decide on their content. It is not automatically certain that the local decision-makers and the population of the region to be examined will be associated with such research. Hence the originality and importance of the MAB approach in seeking to associate local people in the definition and conduct of research and in informing them of its results. The inhabitants of the region being studied are therefore considered not as objects of study but as partners.

Nevertheless the collaboration between researchers and the local population involves difficulties, occasionally even risks. In the Pays d'Enhaut there exists a real desire to move from theory into action, and this raises a certain number of problems.

There is, in effect, a profound difference in approaches to the problems of the mountain environment, depending on whether the viewpoint is that of scientist, decision-maker or local population. By scientists we mean the people in charge of the research, specialists in a particular field; by decision-makers the ensemble of regional and local authorities as well as the people who hold influence in the local society and local economy; by local population the inhabitants of the Pays d'Enhaut.

Among the problems which make a dialogue between scientists and population difficult are the varying hopes of the different groups. This situation came to light during the course of a meeting of the Swiss National Commission for Unesco, when a working group addressed the theme "Scientific research for whom?" (Bridel, 1980). In responding to the question of basic motivation for research, the agronomist suggested that it was necessary to save humanity from hunger; from this point of view the mountainous regions must not be abandoned but managed as a precious resource. The research in the Pays d'Enhaut for him therefore is part of a global long-term perspective. On the other hand for the Damounais farmer the research should be useful in the short term: what fertilizer should he put the following day on the meadow to be used for hay? How can he resolve a financial problem connected with his farm? On these examples and many others it is possible to differentiate the spatial and temporal horizon of the scientist and of the working farmer. It is not possible to meet perfectly both their desires. The art of bringing together the two groups resides certainly in the need not to restrict the researcher to the immediate needs of the practical farmer. The researcher must have sufficient autonomy to carry out long-term research; but he should also be constantly aware of the useful — perhaps even immediately useful — fallout of his acquired knowledge. It is in ways like this that effective participation of the different groups can be secured.

Another source of difficulty is the fact that scientific method is conceptual and is always linked to a theoretical base; the language of the specialist tends to be abstract, sometimes even hermetic. The scientist should abstain from value judgements, keep his objectivity, express himself in careful hypotheses which are never easy to verify. He should always maintain a certain detachment to the problems which he is tackling. He has the time at his disposal to go deeper into the fundamental questions. In contrast, the decision-maker — municipal counsellor, business man or political leader — usually has to make decisions rapidly. He does not have the time to wait for the results of academic studies before acting. In addition he must convince, have his ideas adopted — which is not always possible without certain simplifications. For someone who is elected it is at the end of his four years of office that he is called upon to present his accounts: he must therefore make rapid and pragmatic decisions. The scientific studies in his region can raise problems. For sociological studies especially, the conclusions can have acute political implications; there is the risk of "being judged". Such fears, although not often expressed explicitly, also affect the population being studied, which sometimes has the feeling of being a simple object of research. The local people may be reticent because they imagine that certain conclusions from research can be turned against them: fear of new restrictions of the type "sheep must be forbidden in certain zones", fear of sociological studies giving a poor image of the local mentality and social life, etc.

The MAB study was not requested by the region. The test area Pays-d'Enhaut was selected in Berne. The ecological connotation of the project is such as to delay the activities of those who wish to "develop" at any price.

There is therefore both a real interest on the part of the region to be chosen as a study area and also questions about what such a project will really contribute. This is why the participation of the population, and particularly of those interested in the research effort, is an important postulate of the MAB process. On their part, the researchers of the Pays d'Enhaut intend to promote this participation loyally and in a continuing fashion.

## An instrument for participation: the Association MAB-Pays d'Enhaut

A MAB-Pays d'Enhaut Association has been set up to provide a legal framework for the participation of the population in the MAB research work. The statutes of the association were officially adopted by a constitutive assembly on January 16, 1981, and article 3 of the statutes sets out the purpose of the association, which is "to improve the exchange of information between research workers and the population".

The society is open to all the researchers as well as all the persons or groups who are interested in the work being carried out within the framework of MAB. A nine-member committee (six people from the Pays d'Enhaut and three research workers) has the responsibility for ensuring liaison between the specialists and the population; it organizes exchanges of information through the local press or through colloquia and study sessions.

Even before the formal adoption of the statutes, contacts with the population had been established. Prior to the elaboration of the programme of research, the ADPE, which was given power by the communes to coordinate with MAB, had been informed of the projects and had been regularly invited to participate in working meetings; it had responded by nominating a representative. On the other hand, during the regional assemblies, the people responsible for MAB work were asked on several occasions to present the objectives of the research programme and describe the way in which these studies would be carried out.

The real beginning of local participation can be traced to a seminar of the Swiss National Commission for Unesco which took place in the Pays d'Enhaut in June 1980. The place of the meeting had been chosen because of the MAB Programme. The people of the Pays d'Enhaut had been invited to participate in this seminar, and turned up in large numbers.

They were able to learn about the MAB projects presented during the seminar, and their participation in the different working groups gave them the chance to express their views and to contribute their rich experience of the region. The contacts between the specialists and the local practitioners were appreciated by both sides, and gave these meetings their special stamp of originality. In this context, the young people have not been overlooked. The National Commission for Unesco has organized a competition open to all the students of the region on the theme "Pays d'Enhaut in the year 2000". Fantasies, dreams or ideas expressed in words, in images or in models give useful ideas on the fears and the hopes of the young people for the future.

The seminars have also provided a pretext for a wide campaign of public information in Switzerland on the MAB project; television, radio, newspapers have all provided information about the studies. The people from the Pays d'Enhaut who have indicated their interest in MAB are kept regularly up-to-date on its activities through a newsletter which constitutes a link between the partners. This newsletter includes a progress report on the programme, information about provisional results of the research work, news about other test areas, reports of seminars and practical details such as the addresses of participants, etc. The communes are represented in the Association and its Committee, but the Association is not only for officials. It is open to all those who are interested in the research being undertaken, and members of the Association are invited to study sessions organized by the research workers.

Initial assessment indicates a desire for genuine participation. The positive response of a certain number of persons from the Pays d'Enhaut would seem to confirm that the approach was well founded. Should one conclude that the problem has been solved? Such a claim would be rapidly proved wrong by the facts. In spite of the efforts for providing information there are still many people in the Pays d'Enhaut who do not know what MAB is and who have only a vague idea of what the MAB project entails.

The association's committee has accordingly outlined a practical scheme for improving the regular flow of concise and understandable information on the research work. Certainly not all the people are interested in all of the scientific work being undertaken, but many would like to know what the research workers who wander around the country are doing. A scheme has also been launched to enable the participation of individual members of the public in MAB research, according to their interests. A number of other actions are planned, such as: the possibility of starting a MAB chronicle in the Journal of the Château-d'Oex (a regional twice-weekly newspaper); the organization of open days based on specific problems such as the dynamics of vegetation; the holding of evening meetings, two or three times per year, at which local people and research workers would discuss topics such as local history and changes in agriculture; measures for bringing together the experience and viewpoints of specialists and local people with knowledge in particular fields (amateur historians, agriculturalists, etc.); arranging for scientists to share their experience and knowledge, by giving lessons on the environment in local primary and secondary schools in such fields as botany, sociology, regional economy, geography, etc; creation, as appropriate, of working groups on specific subjects (regional development, economic impacts of tourism, . . .) where the tasks would be divided between research workers and local people. Useful results have already been produced by working groups on simulation models (Association MAB-Pays d'Enhaut, 1981) and on hypotheses of touristic development.

In ways like these, it is hoped to develop a very special form of cooperation, whereby the local population is not only a source and use of information coming from the research effort but also becomes a joint producer of research findings.

In conclusion, the idea of integrating local populations in the MAB-Pays d'Enhaut project is a challenge. It will be successful if the scientists are prepared to come out of their ivory towers and if the people concerned agree to take an active role in the research work. When these conditions are met, it will no longer be utopian to hope that research may one day provide a useful basis for regional or local decision-making, nor inconceivable that fundamental research will serve to improve the general knowledge of the people most immediately concerned. And, since the flux of knowledge is not a one way stream from research worker to local population, but also occurs in the opposite direction, an additional merit of this process will be in the use made of local knowledge, thus giving the research both originality and comprehensiveness.

# 3. The Viewpoint of the Local Population of the Pays d'Enhaut Region

*Albert Chapalay*

## THE CHANGING ECONOMY OF THE PAYS D'ENHAUT

THE Pays d'Enhaut, at an average altitude of 1,000 m in the pre-Alps to the east of the Vaud canton, consists of three political communes. Its 18,000 ha area is occupied by some 4,000 inhabitants, an average of more than 4 ha per inhabitant. Part of the land consists of rock and uncultivated areas, and it is only the lower part of the valley with its pastures that is prosperous.

The valley, which initially was purely agricultural, received its first tourists about 100 years ago and the construction of the Montreux-Bernese Oberland railway line in 1903 brought a new breath of life to the economy of the region. In 1930, with 5,654 inhabitants, the Pays d'Enhaut benefitted from an expansion in tourism and nobody could imagine that large changes would occur to imperil its economy. But in the 1960s changes in agricultural practices, mechanization, the increasing size of farms and the ageing of the tourist infrastructures created ideal conditions for stagnation. The industrial and urban regions opened new horizons where the problems of development, energy, unemployment would never be present. Young people were attracted to the towns in droves. There was an exodus from rural areas. In one generation the dreams of prosperity which had been predicted for even the most remote regions had faded.

## PROBLEMS OF DEVELOPMENT

The next important event, in the late 1960s, focussed around proposals from a group of developers, who wished to create a large tourist centre in the Château-d'Oex commune. These proposals gave rise to certain problems which led the authorities to ask a study group composed of people from both within and outside the Pays d'Enhaut to examine the likely effects of such a project. The first decision was easy: to extend the study to the whole of the Pays d'Enhaut and enlarge the tourism-based study to include an agro-sylvo-pastoral study, a study of the skiable area and a study of a secondary education centre.

During these studies a federal law concerning help to investments in mountainous regions (LIM) was adopted by the Swiss Parliament. To conform with the federal legislation the Pays d'Enhaut created an Association for the Development of Pays d'Enhaut (ADPE), and complementary studies were started. The impact of the arrival of research workers was strengthened by the appearance of terms of reference that were as worrying as they were promising: analysis of the overall situation, socio-economic studies, specific questions, comprehensive planning, regional potential, etc. After five years work, the inhabitants who participated in the creation and then the work of the ADPE finally began to see the launching of the regional project.

## THE EFFECT AND RESPONSIBILITY OF MAB

Just as the period of enquiries, studies and research seemed to be at an end, a new sign appeared above the horizon: MAB. The population was just becoming aware of all the studies carried out in the Pays d'Enhaut when new research workers appeared with new concerns and questions. "Who are they, What are they doing?" were two of the questions posed. Answers to these questions were not slow in coming: "They are doing research that is useless, but it is essential for them all to have some work to do".

For the population in general, the MAB study remains of marginal interest. But through the creation of the MAB-Pays d'Enhaut Association, and the use of simple but effective means for exchanging information, the population have certainly been able to provide useful data for the research effort.

The population of the Pays d'Enhaut in all the studies undertaken is conscious of the role that it can itself play in all the studies undertaken. But in the short and medium term, the key issue for the local people is obtaining results that are useful to them. The local people are aware of the socio-economic problems of their region. They wish to take a genuine part in the development of programes or in decisions that allow them to maintain and reinforce their local structures and ways of doing things. They are ready to accept their responsibilities, to make choices in the light of a market economy and to make decisions based on research and studies insofar as these are applicable to existing structures. But they are not prepared to degenerate into a population that has to be assisted by subventions and other means of support in order to survive. They will, with the help of the authorities, do all they can to ensure a planned, well balanced, form of development where urban and mountain dwellers are associated together, each fully aware of their rights and obligations.

The MAB Programme is not specifically regional but its fulfillment is. For the population of the Pays-d'Enhaut, the results of enquiries and research at the regional level are of primary importance. The research effort is due to end in 1984.

If, by that time, practical results and solutions to problems have not been achieved, then the inhabitants of the Pays d'Enhaut will have been nothing but guinea pigs. It is hoped that this will not be so. Rather, it is hoped that the sorts of MAB approach to conflicting land use in mountain areas — illustrated in one of the posters of the "Ecology in Action" Exhibit (Unesco, 1981) — will prove of direct use and benefit to local planners and populations, in the Pays d'Enhaut but also in other parts of Switzerland and in other countries.

Certainly, the contacts that have been estabished, and the consultations and discussions that are in course, hold every promise that collaboration between scientists and local populations can be brought to a successful and mutually satisfying conclusion.

## References

Association pour le Développement du Pays d'Enhaut (ADPE) (1977), *Programme de Développement Régional: Analyse de la Situation et des Potentialités,* ADPE, Lausanne.

Association pour le Développement du Pays d'Enhaut (ADPE) (1978), *Programme de Développement Régional: Objectifs et Mesures de Développement,* ADPE, Lausanne.

Association MAB-Pays d'Enhaut (1980), Révision des Statuts de l'Association MAB-Pays d'Enhaut (1980), *Information MAB-Pays d'Enhaut,* No. 3, Association MAB-Pays d'Enhaut, Château-d'Oex.

Association MAB-Pays d'Enhaut (1981), Modèle de Simulation Pays-d'Enhaut, P.L. 1, *Information MAB-Pays d'Enhaut,* Rapport No. 1, Association MAB-Pays d'Enhaut, Château-d'Oex.

Bridel, L. (1980), La Recherche Scientifique, Pour Qui? Rapport du Groupe de Travail No. 2, *Information MAB-Pays d'Enhaut,* No. 3, Association MAB-Pays d'Enhaut, Château-d'Oex.

Castri, F. di and G. Glaser (1979), Ecology and the Development of Mountains and Islands, *Nature and Resources,* 15(3), 8-16.

Darbellay, C. (1980), A la Recherche de l'Accord Entre l'Homme et son Environment: le Projet MAB-Pays d'Enhaut, *Information MAB-Pays d'Enhaut,* No. 2, Association MAB-Pays d'Enhaut, Château-dOex.

Darbellay, C. (1982), Ecological Problems in the Swiss Alps: the Pays d'Enhaut Project, *International Social Sciences Journal,* 34(3), 427-39.

Gunzert, U. (1978), *Gefährdeter Erholungsraum Alpen,* Aktuelle JRO Landkarte 5, JRO Kartografische Verlagsgesellschaft, Munich.

Unesco (1981), *Ecology in Action — An Exhibit, Thumbnail Sketches of 36 Posters,* Unesco, Paris.

CHAPTER 25

# Integration of Natural and Social Sciences in Environmental Research: a Case Study of the MAB Programme

*Anne Whyte*

Abstract

Barriers to integration between the natural and social sciences include institutional constraints, which are found within Unesco, within MAB and at national level. The MAB Programme has been dominated by the natural and biological sciences, and the fourteen themes of MAB emphasize ecosystem thinking. The role of the social sciences has often been misperceived. There are also conceptual and methodological differences between the natural and social sciences which make integration difficult to achieve. Within MAB, several approaches to integration have been developed. These include a human ecology framework, human use systems, perception studies and integrated models. Each approach, as presently used, is limited as a basis for integration, although modelling is viewed as having the most potential. On the basis of MAB experience over the past 10 years, several suggestions can be made for improving the integration of social and natural sciences in environmental research such as that within MAB. These relate to a management orientation, local participation, modelling, communication within projects, and the role of the project leader. Finally, the need for small scale projects is emphasized.

*Résumé*

*Des difficultés institutionnelles tant à l'intérieur de l'Unesco que du MAB, et au niveau national constituent des barrières à l'intégration des sciences sociales et naturelles. Les sciences naturelles et biologiques ont dominé dans le Programme MAB et les 14 thèmes internationaux mettent l'accent sur le point de vue écologique. Le rôle des sciences sociales a souvent été mal compris. Il existe également des différences conceptuelles et méthodologiques entre les sciences naturelles et sociales qui rendent difficile l'achèvement de leur intégration. Dans le cadre du MAB plusieurs approches en vue de l'intégration ont été élaborées. Elles incluent un cadre d'écologie humaine, des systèmes d'utilisation par l'homme ("human use systems"), des études sur la perception et des modèles intégrés. Chacune de ces approches, telles qu'utilisées à présent, n'a qu'une portée limitée en tant que base pour l'intégration bien qu'on reconnaisse à la modélisation la plus grande potentialité dans ce domaine. Sur la base de l'expérience MAB au long des dix dernières années, on peut faire plusieurs suggestions pour apporter une amélioration à l'intégration des sciences sociales et naturelles dans la recherche sur l'environnement telle qu'elle se fait dans le MAB. Elles se rapportent à l'orientation vers l'aménagement, à la participation locale, à la modélisation, à la communication à l'intérieur des projets et au rôle du chef de projet. L'accent est mis pour finir sur la nécessité de projets à petite échelle.*

## INTRODUCTION

> MAB documents give an impression that the sponsors of MAB did not intend to involve social sciences.
>
> Because of the freshness of approach required in handling the problem of culture vs species, conventional training in social sciences has very little chance.
>
> Social scientists can play a useful role by explaining to people the government's programme and by convincing them to accept it.
>
> *Report on seven seminars in Asia on the Social Sciences in the MAB Programme, (Unesco, 1979a)*

THESE three views of the role of social science in MAB illustrate some of the problems in bringing together the natural and social sciences: the dominance of the natural (and biological) sciences in MAB, the techniques and concepts that social science can bring to the programme, and the (mis)perceptions of the role that social scientists could, or should, play.

The topic of this paper is one on which every social or natural scientist is to some degree an "expert"[1]. At the same time, it is a goal that eludes the MAB

[1]This paper reflects the views of only one such "expert". It is based first on the author's own experience in interdisciplinary field projects both within MAB and outside, and on her experience as a member of an interdisciplinary research institution. It also includes the views of individuals within Unesco who kindly allowed her to interview them during 1981. Finally, reference is made to many MAB project documents provided by the Secretariat.

Programme, despite the skill, effort and goodwill of many people over the last 10 years. How has integration between the natural and social sciences remained a largely unfulfilled objective?

MAB was conceived from its very beginnings as a programme of research aimed at

> developing the basis within the natural and social sciences for the rational use and conservation of the resources of the biosphere, and for the improvement of the global relationship between man and the environment.
>
> *(First Session of the International Coordinating Council for the MAB Programme, Paris, November 1971. Unesco, 1972a)*

By 1977, the MAB Council realized that integration between the natural and social sciences was proving more difficult to achieve than had been initially assumed. The Council continued to stress that

> the human and social sciences were not an isolated element in the MAB Programme, but constituted an integral part of all MAB projects.
>
> *(Fifth Session of the International Coordinating Council for the MAB Programme, Vienna, September 1977. Unesco, 1978)*

This view was further emphasized by the nineteenth session of the Unesco General Conference (Nairobi, 1976) which recommended that the human and social sciences component in MAB be strengthened. At its fifth session, the International Coordinating Council of MAB agreed that "henceforth attention should be focussed not on the need for integration but on 'how' to achieve it" (Unesco, 1978). Specific suggestions for the "how" included: research focussed on concrete practical problems of resource management; coopting of planners and decision-makers to work with the natural and social scientists on the research team; using the human-use system as an integrating concept; developing interdisciplinary methodological guidelines such as those proposed in MAB Technical Notes 3 and 5; international and regional workshops to deal with the methodological and practical problems of interdisciplinary research; increased cooperation between the Social Sciences Sector of Unesco and the MAB Secretariat based in the Division of Ecological Sciences of the Natural Sciences Sector.

This contribution adopts a similar position: that the *need* for integration of the natural and social sciences in MAB is clear and agreed. The problem is *how* to realistically achieve it. The contribution first discusses some of the barriers to integration both within the MAB Programme and more generally. It then evaluates some of the integrating frameworks that have been associated with the MAB Programme. This discussion leads to suggestions for the future.

## BARRIERS TO INTEGRATION

### Institutional constraints

It is generally recognized that the disciplinary tradition that developed in education and scientific research, particularly since the mid-nineteenth century, has brought with it structural barriers to cross-disciplinary studies. Institutions reflect the training of the people in them and generally have a sectoral, or discipline-based structure. With a few notable exceptions, educational institutions are organized along nineteenth century disciplinary lines, and perpetuate a disciplinary view of the structure of knowledge for people trained at all levels.

The division between the natural and social sciences has been more significant, and therefore more difficult to bridge, than the boundary between botany and geology, or anthropology and sociology. Educational institutions divide academic disciplines into "Arts" and "Science" and separate both of them from training in medicine, engineering, law and management. In governmental, intergovernmental and private sector institutions, departments tend to be staffed with people from similar disciplinary backgrounds. Thus, the communication problems between disciplines become replicated as departmental divisions.

The strength of the disciplines is also maintained by career paths and rewards that encourge scientists to remain within the disciplines, rather than to move freely between them or to straddle across their boundaries. Despite recognition of the need for interdisciplinary research by universities, governments and research funding agencies, the size of the budgetary allocations and the complexity of the review process and administration, make interdisciplinary research more difficult to undertake than to stay within one discipline. The situation for interdisciplinary studies is improving, but many researchers still view it pessimistically.

**Within Unesco:** Strong institutional constraints to integration are also found within Unesco. The MAB Programme is located within the Natural Sciences Sector, Division of Ecological Sciences, despite the rationale for its very existence being the need to integrate the social and natural sciences. From its inception, MAB has therefore faced internal institutional barriers to reaching its objectives. Communications between sectors within Unesco is not easy; it has to be routed up to the top of one sector and then down through the organizational hierarchy of the other. Budgetary and organizational power is structured along sectoral lines. Personnel located within one sector have no power within another so that cross-sectoral activities are difficult to administer. Intersectoral committees exist, but have little or no budgetary or political influence.

To make MAB work within Unesco, would require that Unesco be a balanced matrix organization in which objectives (or projects) cut across sectors, and are able to call upon sectoral resources directly (Figure 25—1). In reality, Unesco has allowed the sectors to become dominant so that it operates as a traditional functional

| Interdisciplinary projects / Functional departments | Water supply | Rangeland management | Small scale cooperative ventures | Air pollution from industry | Floods in urban areas | Development of national risk profile |
|---|---|---|---|---|---|---|
| Public health and welfare | x | x | | x | x | x |
| Agriculture and forestry | | x | x | | | x |
| Animal husbandry | x | x | | | | x |
| Trade and industry | | x | x | x | x | x |
| Energy | | | x | x | | x |
| Labour | | | x | x | | x |
| Natural resources environment | x | x | x | x | x | x |
| Housing, urban affairs | | | | x | x | x |

**Figure 25—1.** Matrix organizational structure (Whyte and Burton, 1980)

organization. It has built-in rigidity which inhibits it from adapting its organizational structure to problems.

**Within MAB:** The MAB Programme inherited the mantle of the International Biological Programme (IBP). It also inherited, at national and international levels, the enthusiasm and biases of the scientists, many of whom were biologists, who had worked with IBP.

The fourteen project areas of MAB emphasize the ecosystem thinking that created them. Most are defined in terms of specific ecosystems or geographic entities (e.g. tropical forests, island ecosystems), including one on urban systems. Two are "refuges" for social scientists: Project Area 12 ("Interactions between environmental transformations and the adaptive, demographic and genetic structure of human populations") and Project Area 13 ("Perception of environmental quality").

Most MAB national committees did not have the resources to duplicate this structure of fourteen international themes or project areas. The USA National Committee did establish separate directorates and experienced problems of integration and communication between them (Wohlwill and Zube, 1980). Given the initial biological emphasis, the location of MAB in Unesco within the Division of Ecological Sciences is understandable. Indeed, the perceived need for increasing involvement of social sciences in MAB has itself been part of the intellectual evolution of the programme.

The history of MAB over the past 10 years has been one of changing attitudes towards the social sciences: from at first largely ignoring them; to seeing their input as a means of gaining interdisciplinary credibility; to frustration on the part of natural scientists at the competitiveness of social scientists who have too long felt neglected, misunderstood or mistreated.

Compared to the national MAB activities and to the scientific world in general, the MAB Secretariat emerge from this history, covered in interdisciplinary glory. They have continued to press for more involvement from the social sciences in MAB and have facilitated the activities of individual social scientists in the programme. Within their constraints, the Secretariat exemplify an adaptive response towards the interdisciplinary needs of MAB.

**At national level:** Some MAB national committees have also recognized the virtual absence of social scientists among their ranks and have evolved towards a more multidisciplinary composition. In 1974, when a questionnaire was sent from Unesco (Social Sciences Sector) to countries to find out the degree of social science input to MAB, they found that there was very little, if any, involvement. At the national level, MAB was dominated by natural and biological sciences. The situation is changing, but the *image* of domination remains.

A similar situation obtains at the level of MAB field projects. Of the 884 field projects listed in the July 1979 compilation of the MAB Information System (Unesco, 1979b) probably less than 5 per cent are interdisciplinary in that they integrate the work of natural and social sciences. Most are ecological studies which are geographically located in areas of resource potential and/or vulnerability to degradation and human impacts.

## Perceptual constraints

Within many organizations, the social sciences are placed apart, as a kind of disciplinary annex. They may be allowed to pursue their own objectives, but often they are also expected to play a special role in relation to the goals of other divisions — that of an advisor on the acceptability to the public of engineering or scientific projects; or a translator of technical information into educational materials.

> The Social Scientist is seen here as an animator, a campaigner, or an extension worker.
>
> *Nepal Seminar Report in The Social Sciences in the MAB Programme (Unesco, 1979a).*

There are historical as well as intellectual reasons why this situation has developed. They include the relative "youth" of the social sciences as disciplines; their emulation of scientific method and their search for scientific respectability through the discovery of "laws" of cause and effect — at a time when sciences like physics were moving away from mechanistic laws towards the notions of chance and probability.

> Social scientists are discipline-bound and they assume a rather technocratic view of

society. They have tried to emulate the style of natural scientists and engaged themselves in the search for something like natural laws.

*Rajni Kothari, Chairman of the Indian Council of Social Science Research (quoted in the Indian Seminar of The Social Sciences in the MAB Programme, Unesco, 1979a).*

The limited success of the social sciences at discovering scientific laws has placed them relatively low on the scientific totem pole. Another contributing factor has been the commitment felt by some social scientists to the societies or social processes which they have been studying. It has not been uncommon for social scientists, including some in MAB field projects, to simultaneously play the roles of investigator and participant; to become the spokesman for underprivileged groups of people or social and political causes.

The rights and wrongs of their position within a system of social ethics or a scientific code of conduct are not at issue here. But one outcome is clearly to erect barriers between the natural and social sciences. The social scientist committed to a social cause may see political action as a natural follow-up to his research. If an engineer finds a design fault, he will work to correct it. If a social scientist finds a grave injustice or "flaw" in society, he may feel a similar need to implement improvements. The difficulty he faces as a scientist is that we are more sensitive to any attempts to change society than we are to changes in technological systems. The nature of his subject matter and his method of approaching it has therefore led to a common perception of the social scientist as lacking objectivity, and (at worst) as a potentially disruptive social agitator and political activist; or (at best) as a social welfare worker.

## Conceptual diferences

The MAB objective of "rational use" of biosphere resources implies a search for normative models of resource management. The emphasis is on *rational outcomes*. This follows from a scientific perspective in which desired management outcomes, or system outputs (e.g. energy efficiency), are specified and the operations of the system can be measured against them. The social scientist is more likely to be concerned with *rational processes*; rational here meaning "thoughtful, considered, not *ir*rational". The models are thus descriptive, rather than normative: they seek to describe human resource use rather than to prescribe it. In social science terms, most decisions *are* rational, whereas outcomes may or may not be.

It is not yet clear that natural and social sciences are *able* to be integrated at a conceptual level. Some would argue that there are fundamental differences in concepts and in the language used to describe them that may prevent integration. Klausner (1972), Bennett (1976) and Wohlwill and Zube (1980) all sound cautionary notes against too-glib assumptions that the inclusion of natural (e.g. water quality) and social (e.g. tourist behaviour) data in the same frame of reference will produce an integrated model. In particular, Wohlwill and Zube (1980) wonder how one can interpret the results of such models as the "Obergürgl model" (Franz and Holling, 1974) which include (they argue) incommensurate variables.

Differences between natural and social sciences are also found in the precision with which concepts and values can be defined; in the level of generality or abstraction of theory; in the use of statistical procedures and mathematical concepts; in the time and space scales measured; and in the whole notion of inference between cause and effect. There is not a clear dividing line with all the natural sciences on one side, and all the social sciences on the other. Ecology shares much of its concepts and methods with the social sciences; some aspects of psychology, on the other hand, are concerned with highly specific measures of environmental cause (stimulus) and perceptual response.

### Differences within the social sciences

So far, the emphasis has been on the chasm, real and imagined, between the natural and social sciences. This is both an oversimplification and a distortion. Too often, the social sciences are spoken of as a "collective" (including in MAB reports) as in: "hydrology, geology, zoology and the social sciences". Bennett (1980) has argued that the term "social science" is a misnomer.

Within the many social sciences, are separate disciplines which have their own languages, paradigms of human behaviour, and beliefs in the major factors in social processes (Sherif and Sherif, 1969). Anthropology, geography, political science, social psychology and sociology are well established social sciences which each have recognized bodies of theory, methods and institutional bases. Disciplines such as economics and demography are regarded as social sciences by some, and not by others. Sociobiology, environmental studies, development studies, management studies are newer areas of specialization which have one foot in social science, and at least one more somewhere outside. Any amount of discussion about the gap between the natural and social sciences is multiplied many times over in the amount of internecine argument between the social sciences themselves. In general, it is a no-win situation because each disciplinary interpretation is correct, as far as it goes. However, some social science concepts and methods are more appropriate to MAB objectives than are others. There is a need for greater awareness within MAB of different disciplines' strengths and weaknesses to put the discussion on a more specific and useful level.

## APPROACHES TO INTEGRATION

Within the MAB Programme, several approaches to integration have been proposed and some innovative projects have been undertaken in the last 10 years. These include human ecology, human use systems, perception studies, modelling, and the involvement of decision-makers and local populations. None of these are without difficulties, but they represent some very real achievements in interdisciplinary research. The measure of success already achieved encourages one to apply some rather more critical assessment of their potential as integrating concepts for MAB.

## Human ecology

Human ecology has been proposed as an integrating framework for MAB projects (Boyden, 1979). One of its most important principles is that the

> state of health and well-being of an individual is largely a function of the quality of the environment which he experiences and of his behaviour pattern in that environment. This principle was clearly recognized by Hippocrates about 600 BC and has therefore been called "the Hippocratic postulate".
>
> (*Boyden, 1979*)

Figure 25—2, taken from MAB Technical Notes 12, illustrates the principle. The biological and psychological state of an individual (or population) is affected by the personal environment to which he/she is exposed. Within this personal environment, the individual will act in certain ways (the behaviour pattern). The personal environment is, in turn, influenced by the properties of the total environment. It differs from it through the action of filters, especially economic and cultural ones. The personal environment of the individual therefore consists of those aspects of the total environment which pass through the filters and directly affect him or her.

Perception of the personal environment acts as a second filter to modify the impact of the environment on human health and well-being. For example, there may be several sources of water in the total environment of a village; a stream, spring, pond and standpipe. The personal environment of any household may filter out the alternatives of standpipe (on economic grounds of cost) and spring (because it is on land owned by a rival faction). Perception of costs and benefits of the remaining two sources, stream and pond, may determine the water quality and quantity, and the work involved in obtaining it, that will affect the human health and behaviour of each individual in the household.

As described by Boyden (1979) in MAB Technical Notes 12, in human ecology "the basic rationale for the analysis of the system in question lies in an interest in its characteristics as they affect human well-being and the viability of life-supporting ecosystems". Furthermore, the natural scientist is warned that "The approach must be one that safeguards against the tendency in the study of systems for investigators to become excessively fascinated by the system or subsystem which they are studying, *per se*".

The principle that, in selecting environmental systems for study, due priority should be given to those aspects which appear to be most important for human health and well-being, can be readily accepted. In practice, it is more difficult to apply.

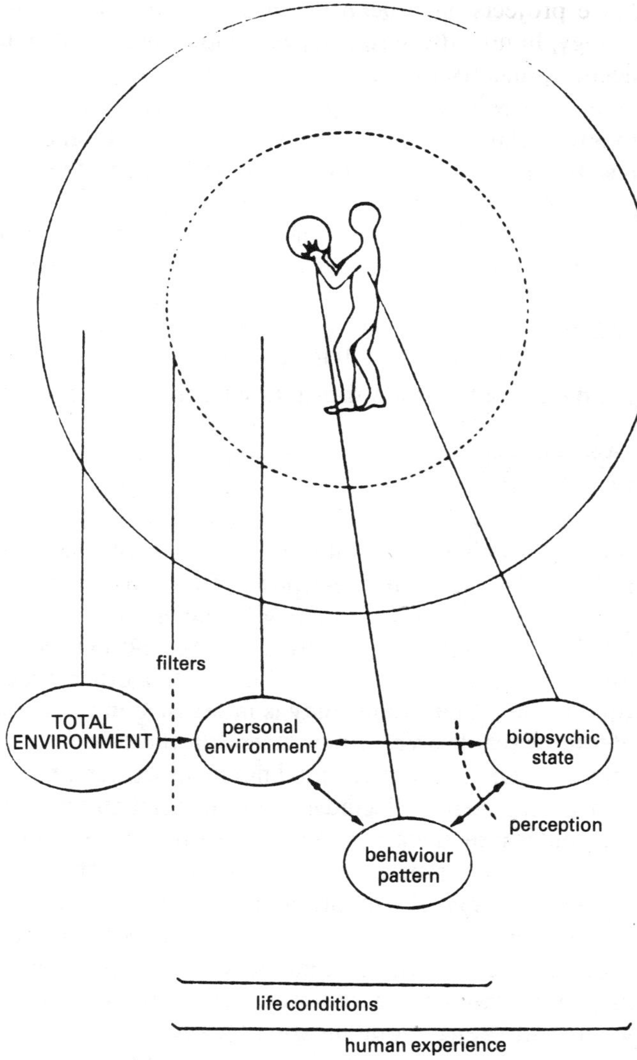


**Figure 25—2.** Conceptual framework for human ecology (Boyden, 1979)

For example, the population dynamics and spawning habits of an economically unimportant fish species in a relatively isolated lake in northern Canada may seem far removed from human health and economic well-being of urbanites in the megalopolis of the northeastern United States. However, a "purely" biological study of that fish species in the 1960s produced some key information that led to the recognition of the phenomenon now known as acid rain. Increasingly, evidence is accumulating of its economic effects on forest and fisheries productivity, and there are indications that acid rain also directly affects human health, by releasing heavy metals into drinking water supplies. The point here is that man-environment systems are very open; there can be long distance transfers of environmental factors, as well as complex interrelationships within an area.

When some political, or particularly, regulatory response is wanted to improve the situation, each link in the cause and effect chain from the total environment to human health and well-being needs to be documented with some degree of scientific acceptability. Otherwise, governmental action is unlikely.

This leads to the main difficulty in using a human ecology framework for integration in MAB field projects. While the framework is valuable as an educational and orientation heuristic device, it does not, as presently developed, provide a working model for *integrating* natural and social science data in a form that can be incorporated into management decisions. The human ecology framework is defined at a general level and does not provide sufficiently rigorous criteria for the design of research projects.

One aspect of this generality is the dual orientation (or two distinct levels of analysis) that the framework incorporates — the macro-level of the total environment and the micro-level of the individual human being (Figure 25—3). The Hong Kong study, for example, collected quantitative data at both levels (see Figure 25—4 and Millar, 1979). However, in the reports of the project, the macro-level models for energy, food and water flows in Hong Kong, are not integrated into the analysis of human health and well-being undertaken in the biosocial survey.

The Hong Kong study is also instructive in highlighting other difficulties faced by MAB field projects. The Hong Kong study is impressive in the amount of data collected, in the depth of scholarship achieved, and in the degree of cooperation between the scientists involved. However, it did not measure human health directly (reported indices of health were collected in the biosocial survey). Nor did the project have the resources to undertake environmental measurements like air quality and noise levels, or an adequate nutrition survey. Some of these weaknesses are pointed out by Millar (1979) in the report on the biosocial survey.

There is, here as elsewhere, a trade-off to be made between comprehensiveness and rigour. By including several environmental factors (housing, crowding, noise), socio-economic variables (socio-economic status, family structure, neighbourhood structure, employment) and behavioural measures (daily activities, drug consumption, recreation, mental and physical activities, social interaction), the biosocial survey in Hong Kong demonstrates the interrelationships of variables. It is, by its nature, less successful in providing a scientific basis for management decisions.

This is partly because it is usually difficult to ascribe any measurable human health effect to an environmental cause. Similar effects can result from different environmental causes; there are complex synergistic and antagonistic interactions between environmental factors in producing health effects; and a single environmental factor can affect the individual through several environmental pathways. Even the nature of the dose-response relationship can be uncertain. Similar problems are faced in analysing socio-economic impacts. Figure 25—5 illustrates the sequence of environmental and behavioural data that are involved in these kinds of studies. To provide such data for one environmental factor on human health is a

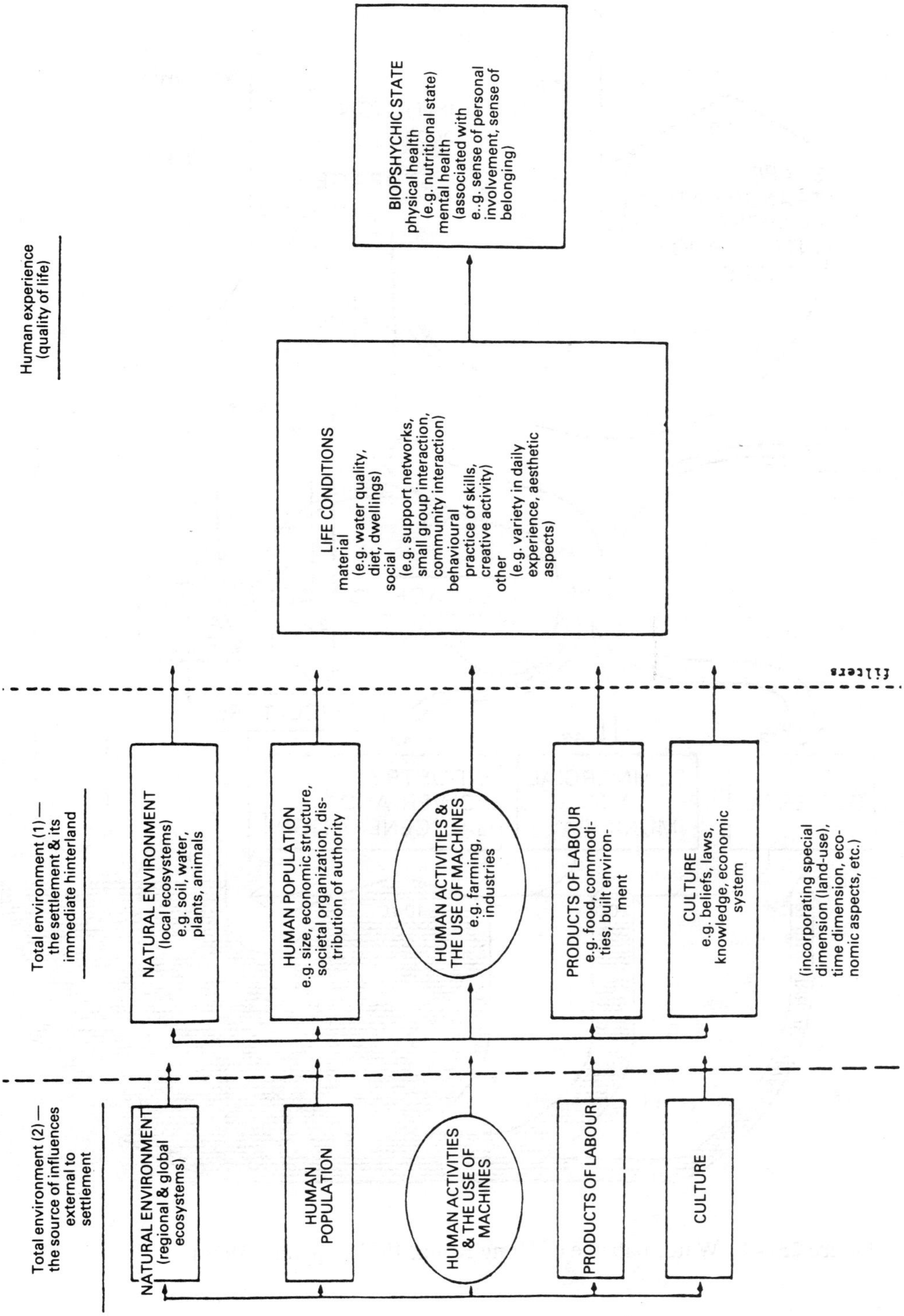


**Figure 25—3.** Conceptual model for human ecology study (Boyden, 1979)

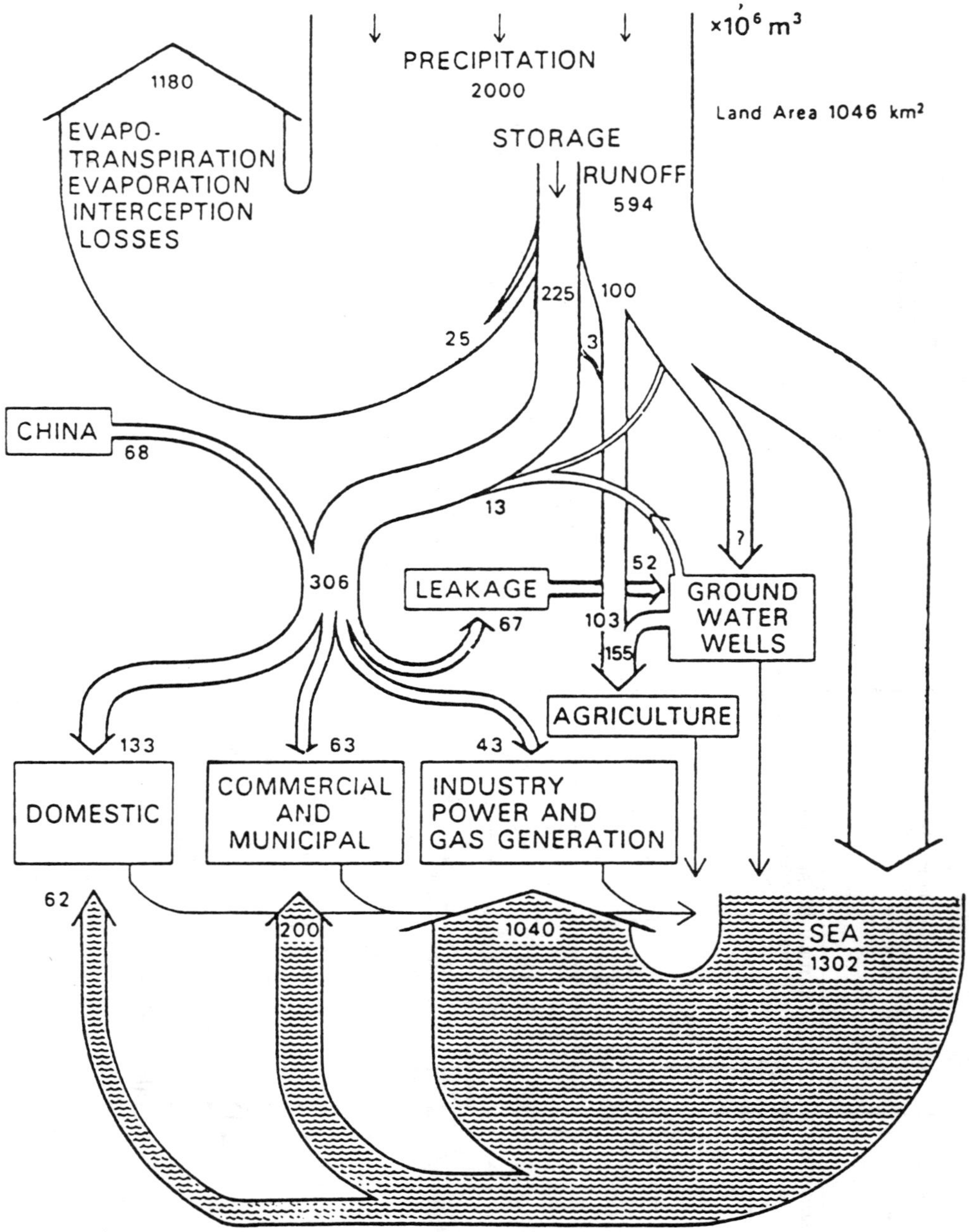


**Figure 25—4.** Water balance of Hong Kong, 1971 (Boyden, 1979)

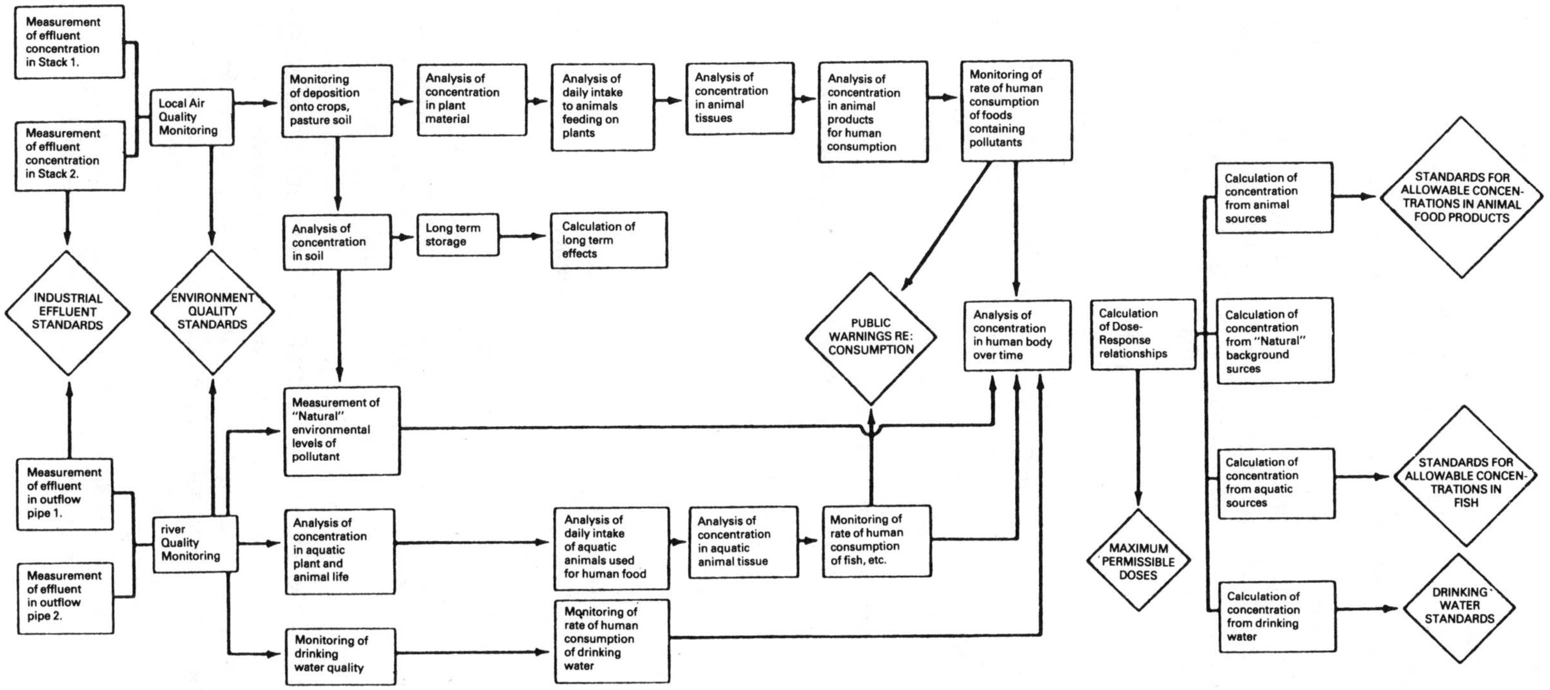


**Figure 25—5.** Generalized model of environmental pathways for an industrial pollutant, and points of entry for regulation (Whyte and Burton, 1980)

major research task. It is one of the activities being undertaken by ICSU/SCOPE within projects on ecotoxicology, risk assessment and biogeochemical cycles. It implies a different cut at the man-environment pie from human ecology — one that focusses on an individual risk system rather than on an individual person or population.

Although the human ecology framework for looking at the total burden on a defined population of environmental stresses (and their beneficial equivalents) is conceptually appealing, it therefore seems to founder on the rock of pragmatism. It does not take sufficiently into account the difficulties of determining cause and effect for specific factors, nor the need for data of this specificity for management decisions.

To compound this problem, government bureaucracies are not generally structured as matrix organizations in which departments would interact to form their own human ecological framework. They are composed of competing sectoral units which require that the results of human ecology be translated into: first, information that can be used and defended even in court by individual sectors; and second, arguments for rational management that are convincing in terms of economics or human health — these are generally the only two arguments that decision-makers find compelling.

At present, neither of these two disciplines — economics and medical sciences — are well represented in the MAB Programme. Human ecology presents us with a challenge, but in terms of the measurement techniques and modelling capabilities available to us, it may be ahead of its time.

## Human use systems

Human use systems were first described in the context of the MAB Programme in the task force on the contribution of the social sciences to the MAB Programme (Unesco, 1974).

> Human use systems may be defined as organizations through and by which resources are managed. They vary in size and composition from the household or tribe to the nation state or multi-national corporation. In their spatial expression, they are rarely, if ever, congruent with ecosystems. Indeed, they are often expressly organized to cut across several natural ecosystems in order to take advantage of the complementarities and contrasts of different zones.

This recognition that the appropriate unit of study may not be an ecosystem but another system focussed on the relationships between ecosystems, was an important step in the evolutionary thinking of MAB. In particular, the human use system appealed to ecologists as a social science concept that was related to their own concept of ecosystems, and was one which they felt they could work with.

Unfortunately, while the "human use system" contains the useful insight about the purposive incongruity between socio-economic and natural resource systems, it says nothing about the conceptual and language differences inherent in measurements of biological versus social variables. Indeed, by using the same term "systems" and by alluding to general systems theory, the discussion in MAB Report Series No. 17 (Unesco, 1974) probably conveys more of an aura of commonality than is warranted. Again, the discussion of human use systems has not been taken far enough, and conducted at a specific enough level within MAB, to take us very far forward, in conceptual terms.

## Perception studies

One of the fourteen project areas in MAB (Project Area 13) is devoted to the perception of environmental quality (Unesco, 1973). In addition to identifying six priority research areas, the expert panel charged with elaborating the content of MAB Project 13 emphasized that the perceptions of local populations need to be taken explicitly into consideration, along with those of experts or officials. The panel concluded that the success of MAB Project 13 would depend mainly on the degree to which perception studies could be incorporated into other MAB projects. To this end, MAB Technical Notes 5 (*Guidelines for Field Studies in Environmental Perception,* Whyte, 1977) reviewed field techniques developed in various social sciences and attempted to provide a framework for selecting which aspects of perception to study (Figure 25—6).

Following the nomenclature suggested by the expert panel on systems analysis and modelling in the MAB Programme (Unesco, 1972b), perception variables were ordered along two axes; the individual to collective resource manager, and distance from the point of management decision. Thus, as one moves from the right to the left of the flow chart in Figure 25—6, the variables more directly impinge upon the output variables, although they may not necessarily be more influential.

The framework proposed is no more than a simple heuristic device to help organize the research planning task. Again, it does not help to integrate natural and social science concepts and methods. It does no more than suggest ways in which social variables can be measured and thereby perhaps encourages scientists to include them in research studies.

Perception studies have also introduced into MAB the phenomenon of the perceptual filter, or operator, between environmental variables and expected behavioural responses. MAB projects such as the Hong Kong, eastern Fiji and eastern Caribbean projects have provided some of the best field data to show that people do not necessarily feel crowded when their living conditions rate high on a physical density scale, or perceive their environment as noisy when instruments measure high noise levels. Where perceptions are adequately measured in addition to environmental variables, the descriptive model of how people feel and behave can be compared to the environmentally based one. The comparisons can act as object lessons in human rationality and in researchers' misperceptions.

However, it is probable that perception studies have been more successful as an integrating factor *indirectly,* by facilitating the involvement of local populations and allowing some expressions of their views. In studying people's perceptions,

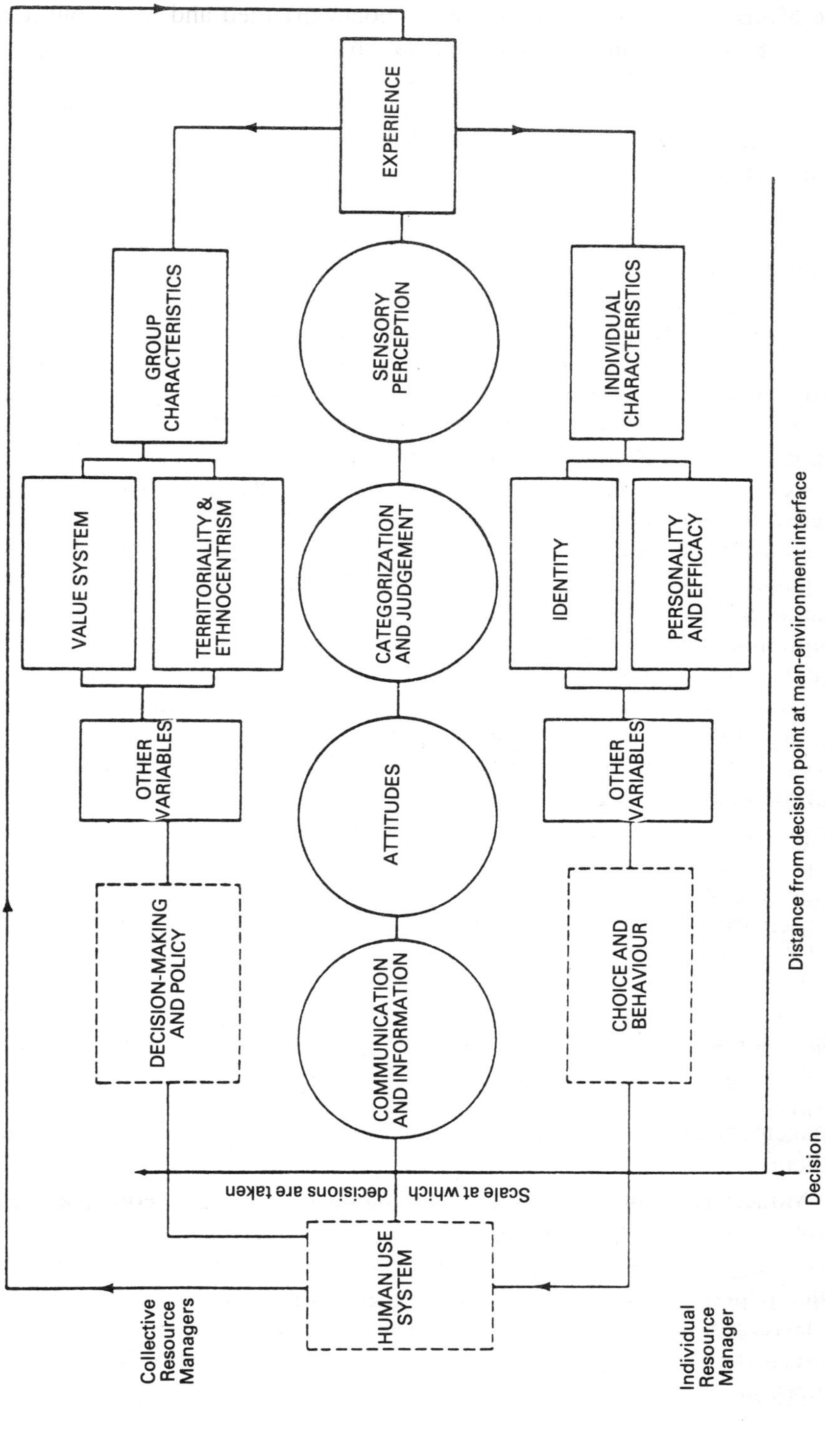


**Figure 25—6.** Simplified model of environmental perception (Whyte, 1977)

some MAB projects have become more policy-oriented and have collected data that are relevant to, and sought by, decision-makers.

## Integrated models

Modelling has been emphasized throughout the history of MAB as a crucial research objective and research tool. Some notable achievements within MAB are to be found in the Austrian MAB/IIASA Obergürgl project (Franz and Holling, 1974), the Integrated Project in Arid Lands (IPAL) in northern Kenya (Lamprey, 1978), and the sensitivity model for urban and regional planning (Vester and von Hesler, 1980). Some of these studies have used quantitative measures and sophisticated computer modelling facilities. Other MAB field projects have used qualitative models to articulate research hypotheses or have developed simplified, small-scale models to test concepts like carrying capacity, for example, the eastern Fiji project (Bayliss-Smith, 1980).

One less well known example of a simple quantitative model is that being developed by the Instituto de Ecología in Mexico as part of a project on livestock development in the region of Mapimí (Halffter, 1981; Montaña, 1980; Whyte and Burton, unpubl.). The model is initially designed to show the relationship between annual rainfall variability and range stocking levels (Figure 25—7). It assumes that the rangeland managers' decisions are based on their perceptions of the animal carrying capacity of the land over the coming dry season and builds in known economic constraints (credit facilities) and policy factors (land reform and biosphere reserve policies).

Using data supplied by local *ejiditarios,* the reproduction rates, share arrangements and animal sales (Figure 25—8) and the development of herds can be modelled (Figure 25—9), and the impacts of droughts, etc. can be predicted (Figure 25—10).

The Mapimí model illustrates several features that are discussed further in the next section; the highly specific data requirements; small-scale of the empirical investigation; need for comparable measures of natural resources and perceived resources; inclusion of external socio-economic influences and the involvement of local populations. It is also simple enough to be used as a means of discussing management alternatives with the people involved because it was largely generated by them.

Models can not only be used to interact with local people, as has been exemplified in the Obergürgl project. They can also act as tools for decision-makers to evaluate policy alternatives, as has been shown by the Lower Main sensitivity model. In particular, they can be manipulated to show different futures, or scenarios. Models can be used to examine thresholds or stress points at which the relationships between the variables may break down. They can be specifically built to include (conceptual problems notwithstanding) natural and social science components.

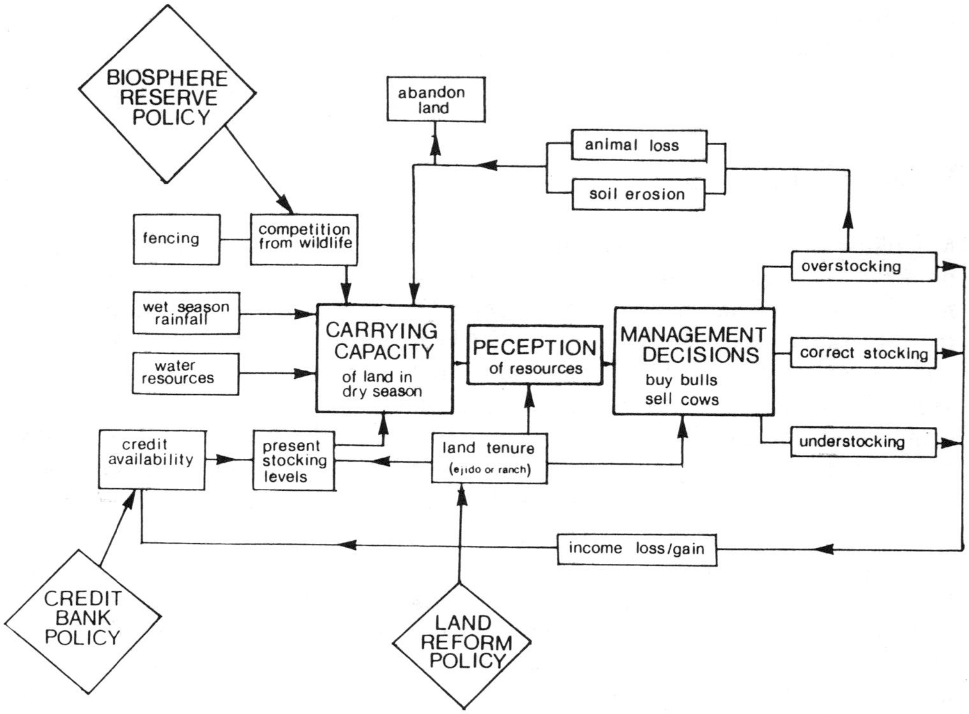


**Figure 25—7.** Model for annual range management decisions based on perception of seasonal carrying capacity

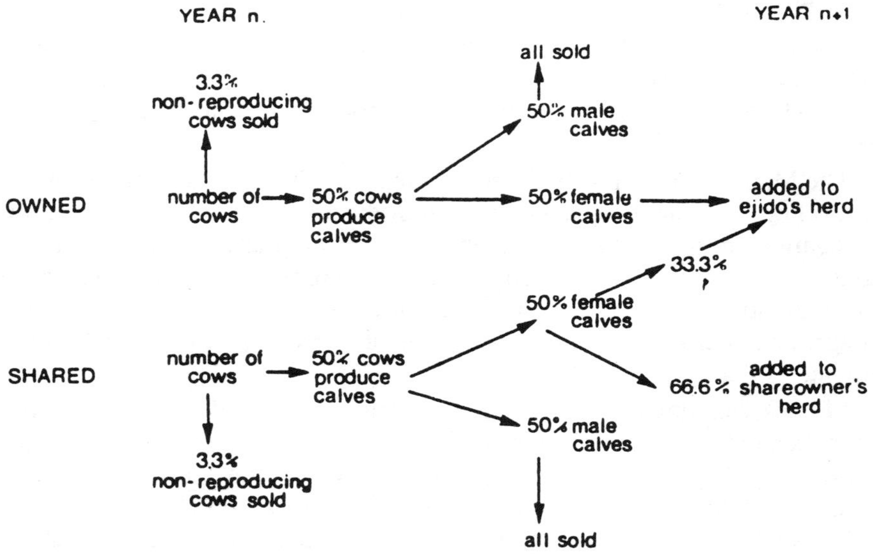


**Figure 25—8.** General model of livestock build-up based on data from Ejido La Flor in the Mapimí region of Mexico

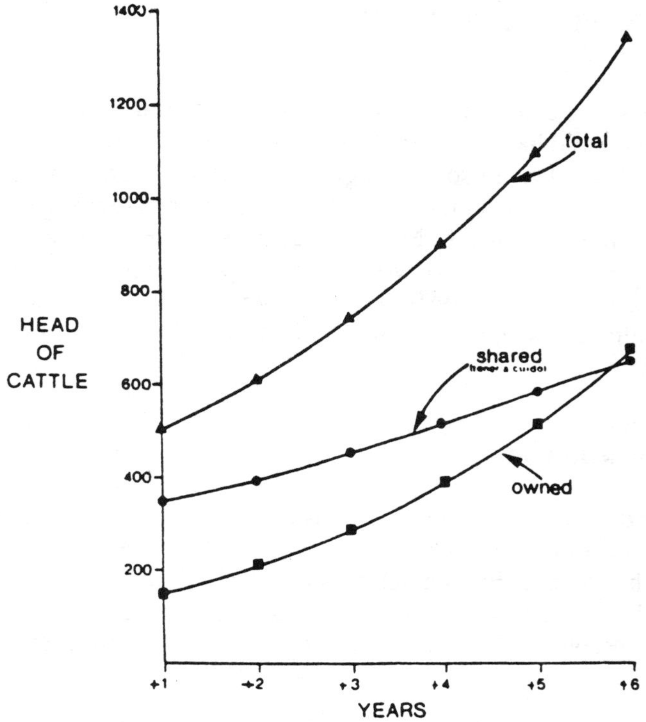


**Figure 25—9.** Build-up over time of cattle owned and shared at Mapimí (Mexico), ignoring impacts of climatic variability

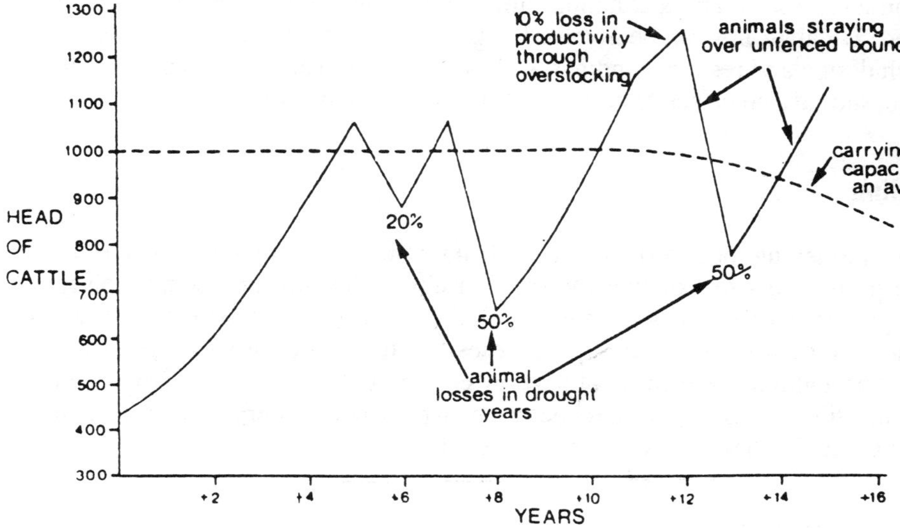


**Figure 25—10.** Model of herd build-up at Mapimí (Mexico) to show impact of drought and overstocking

## SUGGESTIONS FOR MAB PROJECTS

This short review of a complex subject has been able to touch on several important topics only briefly, and has not been able to do justice to those MAB field projects which have already achieved some integration of natural and social sciences. The review of MAB experience so far leads to several, interrelated suggestions for the future. They are not unknown to MAB researchers, MAB national committees or to the MAB Secretariat, who have written several papers on the evolution of MAB thinking (e.g. di Castri *et al.*, 1980) and have defined the terms interdisciplinarity and transdisciplinarity within the context of MAB (di Castri and Hadley, 1978) (Figure 25—11).

### Management orientation

Projects directed towards management decisions or policy alternatives have several benefits. Their information is likely to be used by decision-makers because it is responsive to their needs. For similar reasons, decision-makers are more likely to provide input data. The problem orientation also provides a common purpose and external test of relevancy for improving the integration of all disciplinary inputs.

### Local participation

The involvement of local people can similarly provide a better information base, a focus on local needs, and a common rationale for the natural and social scientists involved. Local groups can be involved as project advisors to provide feedback on research design and research findings as they emerge. Local people can and, where possible, should also be active participants in the research team.

### Modelling

A strong modelling focus provides an evolving framework for all natural and social science participants to examine the relationships, relevance and compatibility of their inputs to each other, and to the overall objectives of the project. Modelling exercises can be used to develop hypotheses and to identify research needs at the outset. Throughout the project's life, they can serve as the rationale for frequent communication of ongoing findings between all project members. The other advantages of modelling have already been discussed.

### Communication within projects

It is often argued that bridges between disciplines are built in the field and at the

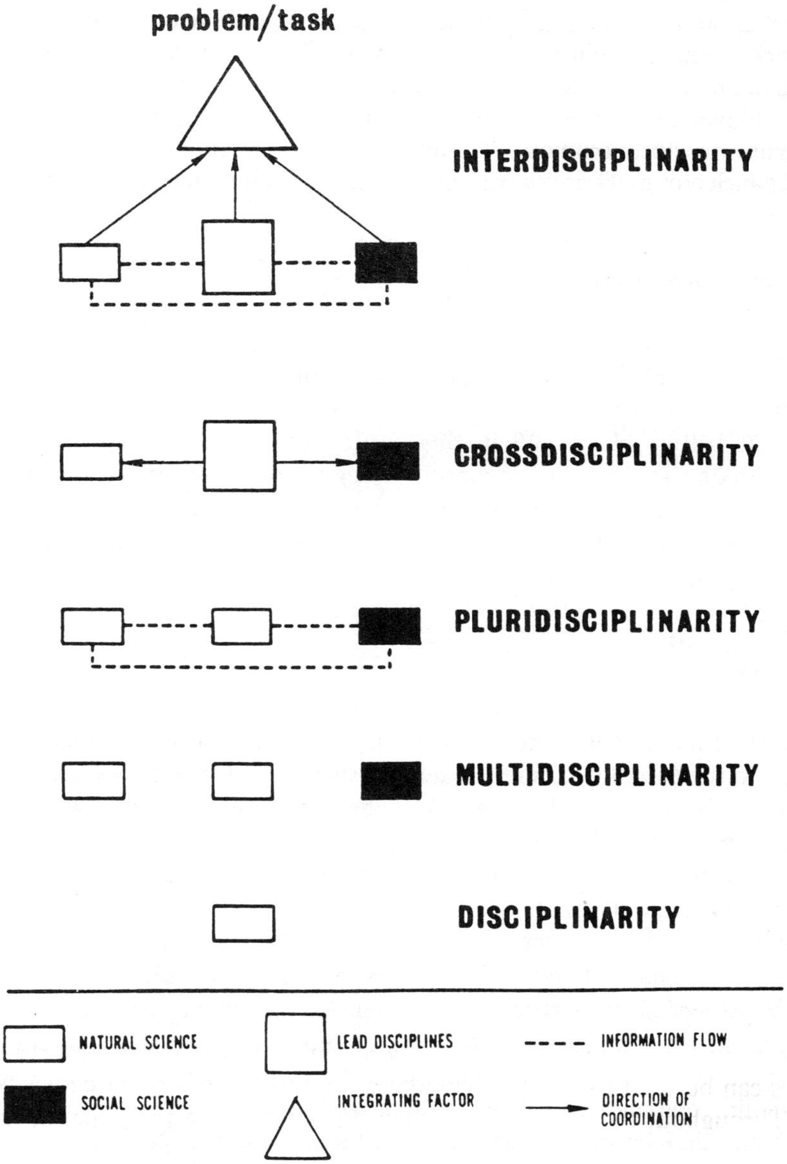


**Figure 25—11.** Definition of terms for projects involving several disciplines

work bench. This is true only if efforts are made to keep researchers talking to each other. Field projects can continue for several years without the researchers involved sharing a common definition of the problem.

In addition to the overall purpose, intermediate objectives need to be defined, agreed and reiterated among project members. This provides a transdisciplinary structure *internal* to the research project (Figure 25—12) which mirrors the overall transdisciplinary structure described by di Castri and Hadley (1978) which is based on integration at technological, planning and policy-making levels. Figure 25—13 illustrates such a transdisciplinary structure for the Mapimí project in Mexico.

## Role of the project leader

The project leader(s) play a crucial role in the integration process. While the broader their disciplinary background and skills the better, what seems to be more important is their attitude, personality and ability to provide the right project structure and atmosphere. When this occurs, the research team communicate directly with one another, and are more willing to adapt their individual objectives to the needs of others in the group, and to the overall objectives. The team adopts a mutually responsive mode which does not need to channel information only through the project leader(s).

## Small scale

The quality of the modelling component, the work with local people and decision-makers, and the (usually underestimated) costs of social science research all argue for projects which are small in scale. It is instructive that, together with Austrian MAB, the International Institute for Applied Systems Analysis (IIASA), one of the leading institutions in the world for systems modelling, selected a village of 300 inhabitants (Obergürgl) for analysis.

Other MAB field projects have been unduly optimistic in the scale and complexity of the natural and social systems that they have selected. This includes the MAB project on population-environment-development in the eastern Caribbean. Five islands were selected for study in order to examine social and natural system variability along several dimensions. From the outset, the data collection and modelling demands were greater than could ever be met with the resources. The problems then become somewhat akin to those of human ecology.

These suggestions all derive from a concern to put "Ecology into Practice". They do not tackle what some people would regard as the fundamental issue of the degree of conceptual and methodological difference between the natural and social sciences. That issue is not going to be resolved within the short term, and probably not within the life of MAB. Rather than wait for new conceptual solutions to develop, the MAB Programme can contribute towards our insights into the relative

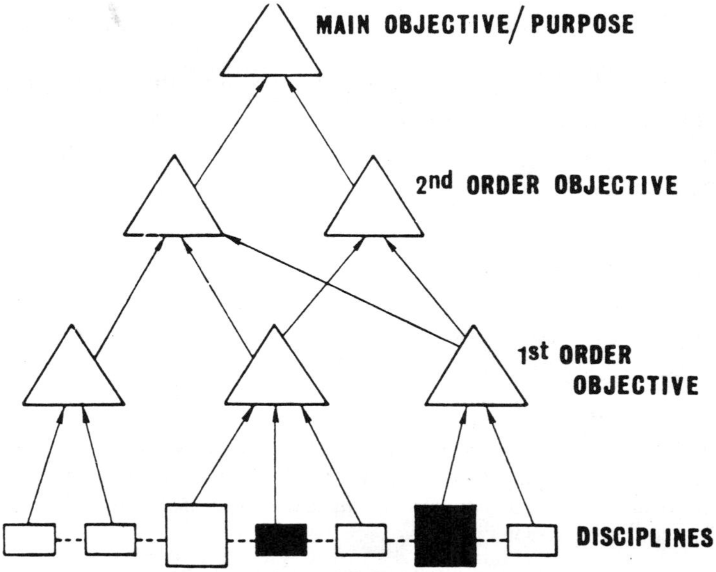


**Figure 25—12.** Transdisciplinary structure for MAB research projects

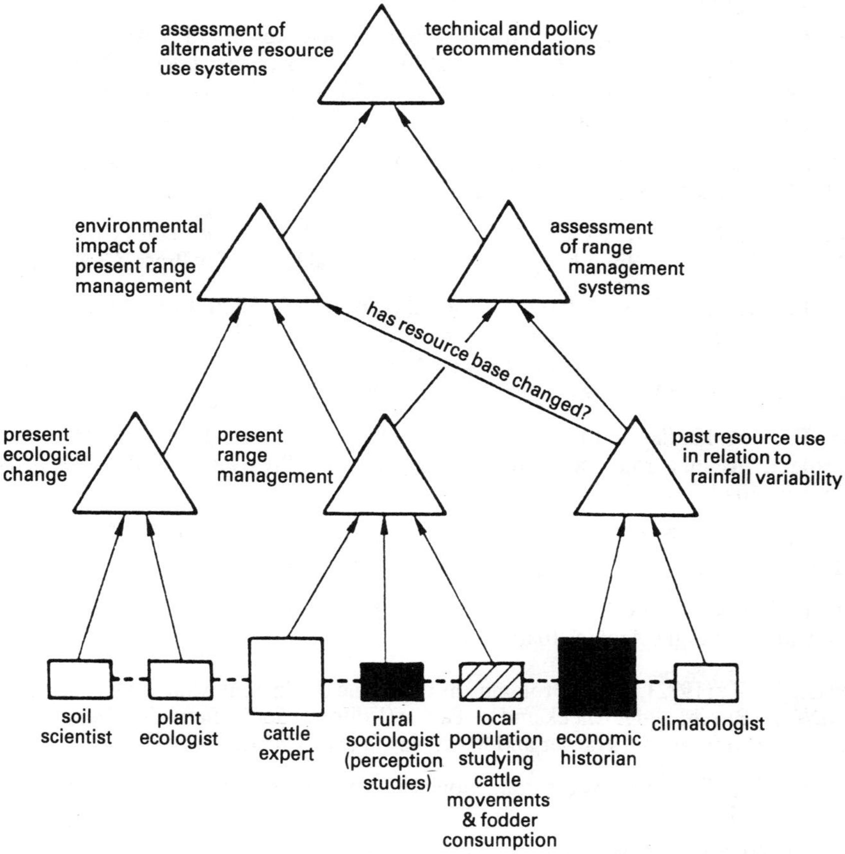


**Figure 25—13.** Transdisciplinary structure for range management project in Mapimí (Mexico)

natures of ecosystems and human use systems by providing the empirical basis for, and practical experience of, integration between the natural and social sciences. The challenge is for Unesco and MAB to achieve the transdisciplinary approach that they are advocating.

**References**

Bayliss-Smith, T. P. (1980), Population Pressure, Resources and Welfare: Towards a More Realistic Measure of Carrying Capacity, In H. C. Brookfield (Editor), *Population-Environment Relations in Tropical Islands: the Case of Eastern Fiji,* MAB Technical Notes 13, Unesco, Paris, 61-93.

Bennett, J. W. (1976), *The Ecological Transition: Cultural Anthropology and Human Adaptation,* Pergamon Press, New York.

Bennett, J. W. (1980), Social and Interdisciplinary Sciences in US MAB: Conceptual and Theoretical Aspects, In E. H. Zube (Editor), *Social Sciences, Interdisciplinary Research and the US Man and the Biosphere Program: Workshop Proceedings,* US-MAB, Department of State, Washington, D.C., 17-39.

Boyden, S. (1979), *An Integrative Ecological Approach to the Study of Human Settlements,* MAB Technical Notes 12, Unesco, Paris.

Boyden, S., S. Millar, K. Newcombe and B. O'Neill (1981), *The Ecology of a City and its People: the Case of Hong Kong,* Australian National University Press, Canberra.

Castri, F. di and M. Hadley (1978), Ecological Approaches to Land Resources Management in the Tropics: Some Case Studies from the Man and the Biosphere (MAB) Programme, In J. S. Singh and B. Gopal (Editors), *Glimpses of Ecology,* (Professor R. Misra Commemoration Volume), International Scientific Publications, Jaipur, 569-86.

Castri, F. di, M. Hadley and J. Damlamian (1980), The Ecology of an International Scientific Project; Insights from the Man and the Biosphere Programme, *Impact of Science on Society,* 30 (4), 247-60.

Franz, H. and C. S. Holling (1974), *Alpine Areas Workshop on the Obergürgl Model,* International Institute for Applied Systems Analysis, Laxenburg.

Halffter, G. (1981), The Mapimí Biosphere Reserve: Local Participation in Conservation and Development, *Ambio,* 10 (2-3), 93-6.

Klausner, S. Z. (1972), Some Problems in the Logic of Current Man-Environment Studies, In W. R. Burch, N. H. Cheek and L. Taylor (Editors), *Social Behavior, Natural Resources and the Environment,* Harper and Row, New York, 334-64.

Lamprey, H. (1978), The Integrated Project in Arid Lands (IPAL), *Nature and Resources,* 14 (4), 2-11.

Millar, S. (1979), *The Biosocial Survey in Hong Kong,* Australian National University, Canberra.

Montaña, C. (1980), *Projecto de Investigacion sobre el Desarrollo Ganadero en el Bolson de Mapimí Durango y en la Region del Pinacate (Sonova),* Mimeographed.

Sherif, M. and C. W. Sherif (Editors) (1969), *Interdisciplinary Relationships in the Social Sciences,* Aldine, Chicago.

Unesco (1972a), *International Coordinating Council of the Programme on Man and the Biosphere (MAB),* First Session, Final Report, MAB Report Series No. 1, Unesco, Paris.

Unesco (1972b), *Expert Panel on the Role of Systems Analysis and Modelling Approaches in the Programme on Man and the Biosphere,* MAB Report Series No. 2, Unesco, Paris.

Unesco (1973), *Expert Panel on MAB Project 13; Perception of Environmental Quality,* MAB Report Series No. 9, Unesco, Paris.

Unesco (1974), *Task Force on the Contribution of the Social Sciences to the MAB Programme,* MAB Report Series No. 17, Unesco, Paris.

Unesco (1976), *Human Population Problems in the Biosphere: Some Research Strategies and Designs,* MAB Technical Notes 3, Unesco, Paris.

Unesco (1978), *International Coordinating Council of the Programme on Man and the Biosphere (MAB),* Fifth Session, Final Report, MAB Report Series No. 46, Unesco, Paris.

Unesco (1979a), *The Social Sciences on Man and Biosphere (MAB) Programme. Report on Seven Seminars held in Asia,* Unesco, Bangkok.

Unesco (1979b), *MAB Information System,* Compilation 4, July/Juillet 1979, Unesco, Paris.

Vester, F. and A. von Hesler (1980), *Sensitivitäsmodell,* Regionale Planungsgemeinschaft Untermain, Frankfurt am Main.

Whyte, A. (1977), *Guidelines for Field Studies in Environmental Perception,* MAB Technical Notes 5, Unesco, Paris.

Whyte, A. and I. Burton (1980), *Environmental Risk Assessment,* SCOPE 15, Wiley, Chichester.

Wohlwill, J. F. and E. H. Zube (1980), The Role of the Social Sciences within US MAB, In E. H. Zube (Editor), *Social Sciences, Interdisciplinary Research and the US Man and the Biosphere Program: Workshop Proceedings,* US-MAB, Department of State, Washington, D.C., 2-16.

CHAPTER 26

# The MAB Approach: Problems, Clarifications and a Proposal

*Brian Spooner*

Abstract

Despite MAB's general success in promoting research on ecological problems from various points of view, the goal of integration still seems beyond reach. To the degree that integration has worked it has invariably involved the domination of one scientific discipline over others that could be persuaded to cooperate — especially of natural over social sciences — both in definition of research design and in formulation of questions. This problem derives partly from the relative scarcity of social scientists professionally interested in ecological problems, but also from the fact that social science comprehends a variety of approaches, based on different assumptions, all equally valid, leading to different but complementary results. A closer look at the variety and complementarity of a selection of different social science approaches suggests a promising model for the future development of the MAB Programme in the 1980s, which would take it closer to the goals it had previously sought through integration, and also towards the goal of improved communication and application of research findings. Not only the various social science approaches and different natural science approaches, but also the approaches of the other relevant actors — planners, politicians, extension workers, and, especially private (i.e. local) people — must be seen as complementary from the initial stage of the definition of the research problem through to the synthesis and application of the results. In this way, not only will the social sciences be effectively integrated into MAB but the more important goal of communication and application will also be assured.

*Résumé*

*Malgré le succès rencontré par le MAB dans la promotion de la recherche sur les problèmes écologiques vus de différents angles, l'intégration semble encore lointaine. Dans la mesure où l'intégration s'est faite, elle s'est toujours accomplie par la domination d'une discipline scientifique sur les autres disciplines appelées à coopérer — en particulier celle des sciences naturelles sur les sciences sociales — à la fois dans la définition de la recherche et dans la formulation des questions. Cela provient en partie du petit nombre relatif de professionnels des sciences sociales intéressés par les problèmes écologiques. Cela provient aussi du fait que les sciences sociales assemblent un éventail d'approches, basées sur différentes hypothèses, toutes valables, et aboutissant à des résultats différents mais complémentaires. Un regard plus attentif sur la diversité et la complémentarité d'une sélection de diverses approches des sciences sociales suggère un modèle plein de promesses pour le futur développement du Programme MAB dans les années 1980, modèle qui le rapprochera des buts recherchés par l'intégration et aussi pourvoira à l'amélioration de la communication et de l'application des données de la recherche. Non seulement les différentes approches des sciences sociales et celles des sciences naturelles, mais aussi les approches de tous les autres participants — planificateurs, politiciens, ceux chargés de l'application et surtout les individus (c'est à dire au niveau local) — doivent être perçues comme complémentaires depuis le stade initial de la définition du problème de recherche jusqu'à la synthèse et l'application des résultats. De cette façon on atteindra en plus d'une réelle intégration des sciences sociales dans le MAB le but très important de la communication et de l'application.*

## THE FAILURE OF INTEGRATION

THE publication of *Ecology in Practice* confirms that Unesco's Programme on Man and the Biosphere has been unique as an international programme and as a vehicle for the promotion of cooperative ecological research. This success can be assessed in the intangible effects of intellectual cooperation, and in the ratio of results to costs, as well as the results themselves in terms both of information and increased understanding. The present book gives contributors the opportunity to take stock and suggest ways and means whereby efforts might be modified or redirected to capitalize on the results so far and maximize the future potential of the Programme. Any criticisms made or implied in what follows are meant in this spirit.

As the title suggests the purpose of this contribution is to increase awareness of what has been attempted so far, in order to increase the potential for more significant achievement in the future. The focus is the "social science problem" in MAB, but it is pursued in a somewhat larger and different context than usual, and the conclusion restates the rationale of MAB in terms that may help to launch us into the new decennium.

A major emphasis at the coordinating and planning level of MAB has been integration. Here the suggestion is made that integration is not quite the right concept. In order to develop the argument it may be best, to begin with, to change the emphasis from integration to communication. But this is a heuristic change only, which will lead to a more promising formulation later.

Increased success of MAB requires more systematic communication, both of orientation and of results, with more different types of people, and this communication must be built into each concrete project. Communication so far has been an afterthought. After a project is over, or at least after it has been designed and its course is set, thought is taken (and then not always) for communication of the significance of the results beyond the immediately concerned sectors of the scientific community. The formulation of the research problem, or the planning of the research programme, is conducted according to perceptions of significance that have nothing to do with communication. But integration is more important at this level than later. The problem of a research effort in the MAB context should be defined as a problem of understanding *and* communication. The understanding is invalid unless it can be communicated to the various types of people to whom it is significant.

Ecological research reports both within MAB and outside tend inherently to be organized for communication with other ecological scientists at best, on occasion also with planners, but not with the other types of people who may be concerned with or affected by either the problem or the results of the research — of which the most important are local, private (i.e. not official or expert) people.

The integration of social sciences in the research process, even thorough integration at all levels from the formulation of the problem to synthesis and interpretation of the results, would not necessarily change this situation. The problem lies in the predominance of a particular approach, which happens generally to be a natural science approach, but it is not peculiar to the natural sciences. The social scientist focusses on different subject matter, but in dealing with that subject matter he may proceed from the same assumptions as the natural scientist. He may, for example, rely primarily on the concept of adaptation. Another social scientist, however, may eschew the concept of adaptation because his attention is taken by situations which it does not satisfactorily explain. A basic division in the social sciences is between those who argue as though history caused ideas and those who argue as though ideas cause history. Both have a handle on reality, but not the whole of reality. It is not certain if there is such a fundamental division among natural scientists — but that does not mean that natural scientists are closer to a monopoly of the truth!

Some people have hoped that MAB Project Area 13, the study of environmental perception, would take care of this problem. However, despite early ambitions the study of perception has been built into very few MAB field projects and even if it had been pursued more widely, no one has worked out how the results should be integrated with other project components. The perception movement in MAB did not get beyond methodology. However good the methodology, it did not help us to *understand* other points of view, let alone help us to decide what to do about them. It has a further drawback: it takes us from the position where we have only the scientist's interpretation of a situation to one where we have the scientist's and the people's viewpoints. Since we understand the scientist's viewpoint and we do not understand the people's viewpoint, the scientist's appears obviously right and the people's seems at best quaint, at worst pernicious.We must avoid this polarization. It has been suggested in recent years that human beings tend always to polarize, are even biologically programmed to polarize. As scientists we have to be on our guard against this tendency.

If the charter of MAB is taken seriously, the problem is already half solved. It can be read to imply that any MAB field project should involve both (ex officio) natural and social scientists, and planners and decision-makers, and (private) local people.

The MAB charter, however, has generally been operationalized by the natural scientist defining the research problem, and coopting the others — if at all — at various stages down the road. The defence of this process is that since MAB is an ecological research programme, the research problem should be defined by the ecologists. It is exactly here, with this apparently logical deduction, that the trouble starts. Other people do not share the ecologist's definition and either do not cooperate at all, or do not do so on an equal basis. They seek to answer the ecologist's questions (sometimes in order not to forfeit a job!) rather than asking their own, and their results tend not to be satisfying to them professionally.

This problem could be surmounted if we think of MAB not simply as ecological research, but as research on ecological problems. This difference is important. Ecologists are not the only people who see ecological problems. Social scientists, planners, politicians (often called decision-makers), extension workers, and private people — in different sectors of society in both developing and developed countries — see ecological problems and have approaches to them. The approaches of scientists are not always thoroughly rationalized; those of private people can be surprisingly rational. Ecologists may have a better understanding of natural processes — though as argued in certain other contributions to this book, they are not infallible — but an ecological problem as defined in MAB is a problem in the relationship between natural processes and human activities, and the other types of people have a better title to understanding human activities than ecologists do. What is more, ecologists are a minority and they cannot operationalize their research results without persuading the other types of people of their significance. They will stand a much better chance if they pay serious attention to other points of view, and do so from the initial planning stage of each project.

One of the problems in bringing about this type of (not integration, but) cooperation is in finding social scientists who are professionally interested in ecological problems. There is a long history to ecological interests in the social sciences, and they range in approach from arrant determinism to arguments that "hard" ecologists would not recognize as ecological at all. They have tended generally to be "pure" rather than "applied". MAB has in fact succeeded in generating one of the most important landmarks in the history of the relationship between the social and natural sciences in this context. The report of the 1974 MAB task force on the contribution of the social sciences to MAB (Unesco, 1974)[1] formulates the human use system as the social scientist's equivalent of the ecologist's ecosystem, and thus establishes that the social scientist defines his unit of study differently from the ecologist, and that the two units of study should not be expected to be coterminous. This formulation was a very important step forward, but unfortunately its implications have not been worked out, and MAB as a whole has not assimilated its significance.

This formulation has in fact not one set of implications but several. Another reason why MAB has had less success than was hoped in integrating the social sciences is that there is a great variety of approaches in the social sciences — which makes social science as a whole a very slippery creature for the non-social scientist to grasp (no doubt one reason why it is dubbed "soft science").

Variation in approach is independent of variation in discipline. The following brief characterizations illustrate four specific approaches commonly found in one or more disciplines of the social sciences.[2] These four approaches are based on two types of argument, one that discounts social and cultural variation and one that focusses on it. Each of these types breaks down again into two approaches. They are:

1. A technocentric approach, that pays little attention to cultural variation, but combines a conventional positivist orientation with a methodological emphasis on measuring and quantification of the various aspects of *human behaviour* and its economic and ecological context, and a theoretical focus on hypotheses of limited scope, so-called middle range hypotheses;

2. A naturalistic approach that seeks to reconstruct universals of *human nature* and to explain and predict behaviour by reference to them;

3. A sociocentric approach, which explains variation in behaviour primarily as the result of the coincidence or opposition of interests among social groups and individuals in different statuses, as they interact in the social process, especially in relationships of relative *power*;

4. An ideocentric or semiological approach that focusses on interaction at the level of ideas — especially values and symbols — and attempts an explanation of cultural variation in terms of the *meaning* of behaviour in its social and cultural context.

The technocentric approach is distinguished by the assumption that all people in all geographical, cultural and historical situations can be treated as rational, in

[1]By the way, the task force appears to have been dominated by geographers rather than social scientists, possibly because geography has a foot on both sides of the divide.

[2]See Acknowledgements.

the sense that so long as we try hard enough and bring about the requisite economic and political conditions they will see the logic of the implications of developments in science and engineering as we see them, and respond accordingly by adopting them, in order to maximize their socio-economic options in a manner that we would consider rational. Technology, thus narrowly defined, is assumed to be essentially transferable.

Approaches led by a focus on one or other of the remaining three categories — human nature, social process, thought — do not preclude belief in the transfer of technology, but they make greater and more conscious allowance for differences in its occurrence. They tend to interpret technology more broadly to include first technique, then the necessary social and economic provisions, and finally the ideas that motivate the practitioners. As we proceed from the consideration of one approach to the next, the definition of technology becomes broader. It starts as little more than tools and machinery, and emerges at the end as the total life of a particular population seen through the window of its production-related activities.

The approach from human nature, while still discounting the significance of variation in human thought and behaviour, implies an underlying equality of subject and object — us and them — and allows for quirks of personality and emotional needs that a straightforward faith in rationality would not explain, but that at least constrains us to treat all people the same.

If we allow our enquiries to be led by the concept of *social* process, the research task becomes more difficult because we have to deal with variations in social status, but the rewards are accordingly greater because we can also deal with the differences in response to the rich and the poor, the influential and the unfranchized, and we are still saved from the difficulty of dealing with the ultimate complexity of *cultural* variation and the differences in values between different societies, especially between developing countries and the Western countries where most modern technology and resource management guidelines originate.

The fourth approach listed above, which might also be called the value-led approach, is by far the most difficult to develop methodologically since it deals with intangibles, but its advantages are that if successful it is the most humanistic and humanitarian and allows the greatest and most sympathetic control for human variation. It considers perceptions and feelings.

## SELECTED APPROACHES

### The technocentric approach

The following paragraphs introduce some examples that are representative of each of these general approaches and discuss some of their advantages and limitations, as a basis for arguing their complementarity for purposes of research on ecological problems.

The technocentric approach includes all the various orientations that start from the assumption that change is exogenous and is effected by means of the transfer of technology, especially (for example) land management regimes. The vast majority of ecological projects exemplify this approach. This approach focusses on externals that can be touched and observed, and uses methods that are basically statistical. It imples that everyone is basically alike and acts and reacts as rationally as we would like to think European and American populations do, according to the predictions of mechanical models that take little or no account of unquantifiable variables. The major advantage of this approach is that it can be methodologically sophisticated and supremely "operationalizable;" its major disadvantage is that since it concentrates on behaviour that lends itself to rational explanation it tends to lead to tautologous interpretations and is very limited in its ability to explain human behaviour.

It is important to note that the thinking behind this type of enquiry is basically functionalist. It breaks the situation down into components, which are defined either arbitrarily or in terms of scientific disciplines, subdisciplines and subsidiary distinctions derived from Western research experience, such as production, distribution, consumption, energy, seasonality, nutrition, risk, innovativeness, cooperativeness, etc., rather than in terms of the internal logic of the particular situation. These components are assumed to interact with each other in a restricted way, such that any one component serves the purposes of the others, change in one is balanced by changes in others and only exogenous factors can bring out real change. This conceptual framework has been elaborated in different ways in social science, in ecological studies and in general systems theory.

The dangers implied in functionalism will always be present, so long as the social sciences model themselves on the natural sciences. It is, however, possible for one science to learn from another without falling prey to either reductionist or determinist fallacies, so long as this is done not by direct borrowing (which often degenerates into intellectual subjection) but by conscious analogy.

In the context of the study of arid lands a profitable example of such an analogy is suggested by the work of the biologist Noy-Meir (1973, 1974). Noy-Meir models arid ecosystems in terms of "pulses" and "reserves". The controlling factor in these ecosystems is, as it happens, always exogenous — unpredictability in the spatial and temporal availability of water. The input that drives the system comes in stochastic pulses of precipitation. In order to survive the periods between pulses, flora must opt for a reserve strategy and either restrict their growth to the brief and unpredictable wet periods (in the case of annuals or ephemerals), maintaining dormancy in between, or with the aid of sophisticated root systems and specialized physiology find and carefully husband a subsurface source (as perennial shrubs). Fauna must either do the same (as many invertebrates do), or alternatively adapt to the spatial aspect of the system by maximizing their mobility (as is the case with large fauna).

Human systems cannot be explained simply as direct adaptations to these conditions. However, a similar model of human systems in arid lands is instructive.

Populations in arid lands are never socially or culturally entirely self-sufficient. The socio-cultural equivalent of precipitation, as the exogenous input into the system, is investment. When dryland production systems are modelled in terms of the exogenous pulses of investment, our understanding of their internal relationships and of the relationship within them between human activities and natural processes is considerably enhanced — without any implication that human activity is determined by natural causes or follows natural laws.

This type of model illustrates the point of seeing human social and cultural factors on a level with natural biological and physical factors for analytical purposes. Any universe for ecological research comprehends both natural resources and social and cultural resources. Although the social and cultural without the natural are doomed, the natural without the social and cultural are in the long run of no interest to us! It is important to realize that when we do not give appropriate emphasis to the social and cultural resources, we are in effect interpreting the ecology in terms of the values that go with Western science rather than in terms of local values. We need a theory to interrelate natural, social and cultural resources, which is not dependent on functionalist assumptions. For a while we may have to be satisfied with separate disciplines and separate theories. We should not be fooled into thinking that modelling or simulation is the answer, since these are methodological not theoretical approaches.

The most significant deficiency in the technocentric approach is its inability to take into account and analyse the attitudes of the people under study, and its emphasis on rationality is in fact a very particular view of rationality. It emphasizes rational behaviour rather than thought, and implicitly defines as rational any behaviour that can be predicted by the technocentric approach; what cannot be predicted is therefore irrational and inadequacies in performance are automatically blamed on human agents.

The first prerequisite of any study of ecological problems is good measurement and correlation from a number of disciplinary angles but these quantitative methods lead nowhere unless they are pursued in the framework of a theory that relates them also to human perception, decision-making and institutionalization.

## The naturalistic approach

Is it possible to make the transition from generalization about the behaviour we are familiar with in Western society to universals of human nature generally? If so, the advantages of such an approach would be in the opportunity it would afford at least to begin to get inside the people who are the object of study and investigate them not simply after the manner of the natural sciences (so much social science has in fact been emulation of the natural sciences in relation to different subject matter), but as sentient, thinking beings. It is perhaps paradoxical that this approach, while being distinguished from the technocentric, is glossed as naturalistic, but this heading is chosen because in order to detect universals of human nature it is necessary first to identify what can only be called natural instincts or properties.

There are other reasons why this approach is introduced at this stage. In searching for universals in human behaviour by appealing to the biological and "psychic" unity of the human species that is assumed to lie behind the variation with which we have to deal, we are able to construct a basis from which to explore that variation in two different ways in the following two sections. It is important at the outset to note that this approach has no special methods of its own and depends heavily on assumptions that cannot be directly verified, though in very general terms they underlie all social science to some degree.

Perhaps the most basic question to be asked about human nature has to do with cooperativeness, and the same question is probably the most important in relation to ecology. Is man naturally good, but corrupted by society (as Rousseau argued), or is he naturally uncooperative and in need of forms of social or administrative organization to enable him to achieve anything (as Hobbes would have it). This Hobbesian view of human nature lies behind some interpretations of the prevailing ethos of peasant life as "amoral familism" (Banfield, 1958) or the "image of limited good" (Foster, 1965), according to which, irrespective of cultural variation, each family is supposed to see itself engaged in a life-long struggle with its neighbours for *its* portion of the total finite amount of prosperity and well-being. The single most important contribution to the study of universals of this type is probably that of Freud who explored the relationships which he assumed develop in any person's life as a result of infant dependency and sexuality and the psychological conflicts that develop as the infant matures and transfers those relationships into the social universe of adult life. Anthropologists have shown how much of culture can be understood as elaboration upon the experience of physiological processes and needs such as birth, infant dependency and socialization, hunger, thirst, elimination, sex, rest, play, warmth and the developmental cycle (e.g. van Gennep, 1960). Any analysis of variation if taken far enough can lead to an underlying actual or presumed psychological or biological universal — which suggests that however alike we believe we all are we might expect to find significant differences at the level of experience. We should therefore *expect* to find behaviour that does not — at least from a positivist point of view — appear rational!

The naturalistic approach treats concepts like rationality at the level of thought rather than behaviour, and makes use of principles such as those of game theory to develop models of social interaction which in their cross-cultural form are best known under the name of transactionalism and are especially associated with Barth (1966, 1981). Barth's models are empirical, and although they have offered significant insights into the dynamics of some of the situations where they have been applied, they have come under considerable criticism precisely (though not solely) because of the view of human nature that they imply as essentially competitive. An important advantage of this type of approach, however, is that it forces us to see people as active participants in events and somewhat less as reactors to forces greater than themselves: they are seen as "planners and strategists, rational at times, impetuous at others, but above all as men and women engaged in the process of trying to make the world livable" (Foldes, 1968).

Despite the interest of these inductive forays into the study of human nature, the most promising postulations so far have come from deductive approaches. Ethology has generated hypotheses about human aggressiveness while sociobiology has concentrated on the explanation and prediction of altruism on the basis of hypotheses from evolutionary biology and genetic processes, but neither of these directions has yet produced very convincing findings. Rather more promising is the structuralism developed by Lévi-Strauss into a theoretical basis for the explanation of human thought and behaviour in general as the result of a pan-human proclivity to order thought and experience according to a basic reconstructible programme.

Lévi-Strauss' structural theory was developed with the aid of an analogy from linguistics of the same type as the analogical use of Noy Meir's work suggested above, and was further corroborated by a similar development in linguistics which posits a universal structure underlying all languages (Chomsky, 1968). Both of these structuralisms imply a single cause at the level of the neurophysiological structure of the brain — which brain research is not yet able to verify. But even without that verification they are both valuable to the extent that they enable us to see similarities in thought processes behind the considerable variation in practice.

Besides these essentially biological approaches it is possible to take a somewhat different tack, which might be called logistical, in the search for universals of human nature that might help us to understand and predict human behaviour in relation to the environment. Apart from "the biological and psychological qualities of our species" human nature can be used to mean that which all men "inevitably have as human beings, regardless of their particular biological or social heritage" (Kluckhohn *apud* Redfield, 1957). In this sense human nature includes a number of traits which might be put down to the easiest solutions to problems of mediating between properties of the human body and properties of nature — the environmental as distinct from the genetic component of evolution. This idea has been taken a step further by a hypothesis that *Homo sapiens* is psychologically capable of handling only a certain number of face-to-face relationships (Forge, 1972) — an idea that has been made use of in an attempt to explain similarities in community size among nomads in Southwest Asia (Tapper, 1979). There is in fact a more sophisticated hypothesis available that would explain this consistency in terms once again of the neurophysiological structure of the brain which apparently tends to limit the number of taxa in any non-scientific, everyday taxonomy, such as the kinship terminology that such populations often use as a primary classifying mechanism in their day-to-day social relations, at six binary oppositions, or $2^6$, which is equivalent to 64 taxa (Wallace, 1961). Since in a normal socio-demographic situation an individual fits a number of persons — but not too many — into each kinship slot, 64 or fewer kinship terms could produce an organizing framework for a society of up to or even somewhat over 500. Larger societies would have to develop additional taxonomies, or schemes of classifications such as descent, generation, class, and (in the largest or most complex societies) institutional affiliation.

Although this work may tell us something about community viability at one level, it does not help us to understand aspects of human behaviour more immediately related to resource management, such as variation in cooperativeness. For this we can turn back again to work by economists and sociologists on the theory of groups which focusses on the conditions under which individuals can be expected to contribute or not to contribute to "collective action" (Olson, 1971). But, despite the definite relevance and interest of this literature, it depends again on assumptions of rationality of the type which is specific to Western culture and does not take account of the dynamics of group action among peasant and tribal societies — though it could serve as the basis for hypotheses for research.

A final tack in the use of the human nature approach is more social. It looks at the degree of psychological variation that might be expected within any human group, and psychological interdependence or variation over time that might be expected in any cultural context. Little work has yet been done in this direction but a few pioneering studies are promising, especially on entrepreneurship (Barth, 1963), on opportunism and conservatism (Sandford, 1982) and on styles of management (Bennett, 1980).

## The sociocentric approach

The sociocentric approach focusses on human variation — not variation in general, but variation in behaviour, and then not from culture to culture, but within the range that might reasonably be expected in many cultures, depending for example on factors such as absolute and relative numbers of people, age and sex structure of social groups, and comparative affluence. It emphasizes those aspects of human variation that have to do with the accumulation and exercise of power, cooperatively and competitively, and although it implies that human nature is basically acquisitive it focusses not on individuals but on the formation and interaction of social groups and explains the other aspects of human life as generated by these social processes. It is collectivist and functionalist.

All ecological problems eventually reduce to the question: which human beings have rights over which others in relation to which resources (Passmore, 1974)? According to this approach technological innovation would be expected to occur when it will enable an individual or a group to increase their power either through increasing their income or enhancing their status. A research design based on this approach will aim to explain how these rights are generated, maintained and changed.

The justification of this approach is precisely that it deals with the more immediate, tangible and translatable phenomena of human variation. Social factors are more immediate than values or symbols because they are the basis of the distribution of power in any cultural context. Whatever the system of values — Hindu, Muslim, Marxist, capitalist or tribal — values are manipulated socially by individuals making use of differences in sex, age, and group size and alliance in relation to resources. It is these manipulations that are played out in the events and processes of history.

## The ideocentric approach

The most promising antidote — or complement — to functionalism has come from a focus on ideas, values and symbols. The ideocentric approach is concerned with meaning and studies other phenomena as the context of meaning. It is pursued in individualistic and collectivist forms.

The best known example of this approach is the study of perception carried out mainly by geographers (cf. Whyte, 1977). This type of study is largely descriptive and designed to inform environmental and technological studies. Another type, less well known outside anthropology, and more comprehensive in its aims is an individualistic example known as ethnoscience, which is an essentially ethnographic attempt to understand non-Western and non-literate knowledge and technology, as heterocultural equivalents of science as we know it. Ethnoscience has been a logical development of ethnographic method but has been developed, since the 1950s, mainly by American ethnographers with a linguistic rather than a social or technological bias. It has paid particular attention to folk taxonomies and has produced highly detailed and valuable descriptions of the natural environment and other areas of knowledge, as perceived by particular individuals in their local cultural context. The best known example is the work of Conklin (1980), based on his research among the Ifugao tropical wet rice cultivators in the Philippines which has produced an Ethnographic Atlas of Ifugao.

There is no doubt about the value and validity of this meticulous and sympathetic description. In this context, ethnoscience is interesting because it facilitates ecological description according to vernacular perception. But since it is not feasible to produce such descriptions of every system that might turn out to be of significance in relation to larger human problems, it is not always clear how we might put them to use. Lévi-Strauss has exploited this type of work (cf. Lévi-Strauss, 1966) for his own purpose, which is to reconstruct and explore the underlying structures of human thought. And he is able to show how Western science and non-Western systems of knowledge show analogous concerns. For example, he quotes Edgar Anderson on the concern with purity in plant breeding among Guatemalan maize cultivating Indians and among the Naga in Assam (Lévi-Strauss, 1966).

The most practical example so far of an ideocentric approach comes from a study of horticulture, carried out between 1914 and 1918 in the Trobriand Islands off the East coast of New Guinea (Malinowski, 1935). In this work, Malinowski studies the totality of the society and culture of the Trobriands through their use of natural resources. The work has been described as going "to the furthest limit of institutional study by means of correlating one set of activities, in this case agricultural institutions, with the whole" (Richards, 1957).

A more explicit concern with meaning has been developed quite distinctively over the last twenty-five years by another anthropologist working in insular South East Asia, Clifford Geertz. Unlike Malinowski, Geertz's primary concern is consciously and consistently with meaning. To operationalize this concern he has taken the idea of "thick description" from the philosopher Gilbert Ruyle and uses it to help him understand everything that is going on in the heads of the actors in relation to each particular activity or experience.

As a subsidiary part of this programme he takes our concept of common sense, and suggests that everyone everywhere has a comparable concept but that it is in fact culturally specific. We can recognize it as a basic component of thinking in non-scientific interaction everywhere. Although it is somewhat slippery it constitutes what we in fact mean much of the time by rationality, and it can be grasped — he suggests — in terms of "naturalness", "practicalness", "thinness" (trenchantness), "immethodicalness" and "accessibleness" (Geertz, 1975). He thinks that the comparative investigation of common sense "should lead . . . to a movement (one actually already well begun [in anthropology]) away from functionalist accounts of the devices on which societies rest toward interpretative ones on the kinds of lives societies support". A closely related approach may be found in "ethnomethodology" (see, for instance, Leiter, 1980), which sets out to explain non-specialist problem-solving behaviour, and has found a fertile field of study in jury consultations.

Apart from a little ethnoscience and the attention paid to ecology in some of Geertz's writings, there is very little ecological work based on an ideocentric approach. Malinowski's work remains the best published example of an environmental application of this approach. This situation in itself is an excellent illustration of the underlying theme in Western science that discourages ideocentric approaches to scientific and technological studies. A growing reaction to this theme is represented by people who have grouped themselves under the quasi-disciplinary banner of semiology or semiotics. In relation to agricultural research and development the position of the semiological version of the ideocentric approach has been stated in an article in the United Nations Environment Programme's journal *Mazingira* (Baquer *et al.*, 1979).[3]

Briefly, this semiological approach to agriculture begins from two principles which distinguish it from conventional Western science. These are the significance of the full context of any object of study, and the dialectical relationship between subject and object. The immediate implications of these principles in the study of agriculture are that all science is ethnoscience; that the field is the farmer's equivalent of the scientist's laboratory; that the farmer is like the scientist a man of knowledge, and the scientist like the farmer also a man of faith; that there is a theoretical basis to the farmer's farming as much as to the scientist's experimentation. The difference between the farmer and the scientist is a difference in the meaning to them of the phenomena they experience, and that difference derives from their social and historical context. The task of agricultural research is to bridge that difference and give an account of it that will allow others — farmers, scientists, planners, politicians — to join the dialogue and improve the quality of the political process.

[3]The theoretical basis of this position is worked out more fully by one of the authors in a book (Uberoi, 1978).

## PROPOSAL FOR COMPLEMENTARITY

Each of these approaches has a valid contribution to make to research on ecological problems. It is important to note that the first two have much in common with natural science. The latter two depend exclusively on social theory. Of these the third is more practical than the fourth, because social processes are more tangible than thought processes. But it is restricted in that it cannot deal with feelings, except by assuming that they are always determined by social variables.

The important point to derive from this brief caricature of social science approaches is that there is no one approach that is capable of giving full weight to each of the major relevant variables in the analysis of everyday life — let alone in the relationship between everyday life and natural processes. Each approach unavoidably and arbitrarily subordinates thought to behaviour or vice versa, and assumes that the one is caused by the other. Any one approach can produce a fascinating story — which might be the best story for the purpose — and it may be true, but it will not be the whole truth. The only way to approach more closely to the whole truth is to produce a dialogue between a number of approaches and see them not as competing but as complementary interpretations.

In order to be more successful in its second decade MAB should therefore focus not on integration but on complementarity, and research projects should be designed with the explicit final aim of explanation *to* a range of particular other types of people. An essential part of the methodology of each project should be a dialogue between different approaches to the central ecological problem. These approaches should include those of the natural and social scientists, planners, politicians, educators and non-officials — the involved people. Each will define the general problem in their own specific terms. These definitions must be accepted as complementary. The basic assumption must be that any approach is by definition selective. No approach can comprehend everything. Any approach is likely to have something to learn from others.

MAB is uniquely suited to put this plan into practice because of its peculiar structure that encourages flow of ideas in both directions between autonomous national committees and the central secretariat. If this concept of complementarity is accepted MAB will be able to bring together broader variety of expertise and thinking for the solution of ecological problems in the coming decade than it has so far. And by so doing, it will not only increase the range of basic research under its aegis but will provide more fertile ground for new ideas. It will be on its record in the production of new ideas that MAB is most likely to be assessed at the end of the next ten years.

**Acknowledgements**

These characterizations of the four main approaches described in this contribution (technocentric, naturalistic, sociocentric, and ideocentric or semiological) are adapted from a paper by the same author entitled Rethinking Agricultural Research, which was written for the United Nations University's Programme in the Use and Management of Natural Resources (UNUNR), based on an informal series of discussions funded by UNUNR at various times during 1980-1981. The author acknowledges with gratitude the stimulation of the following persons in these discussions: L. P. Bharara (Central Arid Zone Research Institute), V. S. Doherty (International Crop Research Institute for the Semi-Arid Tropics), K. L. Khanna (International Crop Research Institute for the Semi-Arid Tropics), S. P. Malhotra (Central Arid Zone Research Institute), K. Mamkoottam (Delhi School of Economics), H. S. Mann (Central Arid Zone Research Institute), V. K. Ohri (Delhi School of Economics), M. L. Purohit (Central Arid Zone Research Institute), M. Savyasaci (Delhi School of Economics), J. P. S. Uberoi (Delhi School of Economics). His appreciation of the potential and of the complementarity of various approaches has been increased enormously by their ideas and arguments and their research experience. They are not, however, responsible for his formulations in this contribution.

## References

Banfield, E. C. (1958), *The Moral Basis of a Backward Society*, The Free Press, New York.

Baquer, A., A. Nandi, J. P. S. Uberoi, H. Y. Mohan Ram and N. Reynolds (1979), A Lesson for Science, *Mazingira*, 10, 69-74.

Barlett, P. F. (Editor) (1980), *Agricultural Decision-Making. Anthropological Contributions to Rural Development*, Academic Press, New York.

Barth, F. (Editor) 1963, *The Role of the Enterpreneur in Social Change in Northern Norway,* Universitetsforlaget, Bergen.

Barth, F. (1966), *Models of Social Organization*, Occasional Paper No. 23, Royal Anthropological Institute, London.

Barth, F. (1981), *Process and Form in Social Life. Selected Essays of Frederik Barth,* Volume I, Routledge and Kegan Paul, London.

Bennett, J. W. (1980), Management Style: a Concept and a Method for the Analysis of Family-Operated Agricultural Enterprise, In P. F. Barlett (Editor), *Agricultural Decision Making. Anthropological Contributions to Rural Development,* Academic Press, New York, 203-37.

Chomsky, N. (1968), *Language and Mind*, Harcourt, Brace and World, New York.

Conklin, H. C. (1980), *Ethnographic Atlas of Ifugao: A Study of Environment, Culture and Society in Northern Luzon*, Yale University Press, New Haven.

Foldes, L. P. (1968), A Note on Individualistic Explanations, In I. Lakatos and A. Musgrave (Editors), *Problems in the Philosophy of Science*, North-Holland Publishing Company, Amsterdam, 322-37.

Forge, A. (1972), Normative Factors in the Settlement Size of Neolithic Cultivations, New Guinea, In P. J. Ucko, R. Tringham and G. W. Dimbleby (Editors), *Man, Settlement and Urbanism*, Duckworth, London, 363-76.

Foster, G. M. (1965), Peasant Society and the Image of the Limited Good, *American Anthropologist,* 67(2), 293-315.

Geertz, C. (1975), Common Sense as a Cultural System, *The Antioch Review*, 33, 5-26.

Gennep, A. van (1960), *The Rites of Passage*, Routledge and Kegan Paul, London.

Leiter, K. (1980), *A Primer on Ethnomethodology*, Oxford University Press, New York.

Lévi-Strauss, C. (1966), *The Savage Mind*, (English translation of 1962 original in French), Weidenfeld and Nicolson, London.

Malinowski, B. (1935), *Coral Gardens and Their Magic,* Vol. I — *Soil Tilling and Agricultural Rites in the Trobriand Islands*, Vol. II — *The Language of Magic and Gardening*, Indiana University Press, Bloomington (Vol. I), George Allen and Unwin, London (Vol. II).

Noy-Meir, I. (1973), Desert Ecosystems: Environment and Producers, *Annual Review of Ecology and Systematics*, 4, 25-51.

Noy-Meir, I. (1974), Desert Ecosystems: Higher Trophic Levels, *Annual Review of Ecology and Systematics*, 5, 195-214.

Olson, M. (1971), *The Logic of Collective Action*, Harvard University Press, Cambridge, Mass.

Passmore, J. (1974), *Man's Responsibility for Nature*, Scribner's, New York.

Redfield, R. (1957), The Universally Human and the Culturally Variable, *Journal of General Education,* 10, 150-60.

Richards, A. I. (1957), The Concept of Culture in Malinowski's Work, In R. Firth (Editor), *Man and Culture, An Evaluation of the Work of Bronislav Malinowski*, Routledge and Kegan Paul, London, 15-31.

Sandford, S. (1982), Pastoral Strategies and Desertification, Opportunism and Conservatism in Dry Areas, In B. Spooner and H. S. Mann (Editors), *Desertification and Development,* Academic Press, London, pp. 61-80.

Tapper, R. (1979), *Pastures and Politics. Economics, Conflict and Ritual Among Shahsevan Nomads of Northwestern Iran*, Academic Press, London.

Uberoi, J. P. S. (1978), *Science and Culture*, Oxford University Press, Delhi.

Unesco (1974), *Task Force on the Contribution of the Social Sciences to the MAB Programme*, MAB Report Series No. 17, Unesco, Paris.

Wallace, A. F. C. (1961), On Being Just Complicated Enough, *Proceedings of the National Academy of Sciences*, 47, 458-64.

Whyte, A. (1977), *Guidelines for Field Studies in Environmental Perception*, MAB Technical Notes 5, Unesco, Paris.

CHAPTER 27

# System Approaches to Research Design and the Communication of Results: "Don't Confuse Me with Facts"

*John N. R. Jeffers*

Abstract

The problems of communication of research results are seldom discussed, and it is too readily assumed that the traditional ways of publishing scientific information are adequate for this important task. This contribution uses a systems approach to the identification of the roles of the many participants in the MAB Programme and to the exploration of techniques for the communication of research results. The implications of new technology, for the developing countries especially, are emphasized.

*Résumé*

*Les problèmes de la communication des résultats de la recherche sont rarement discutés et il est trop facilement admis que les voies traditionnelles de publication de l'information scientifique suffisent pour cette tâche importante. Cette contribution utilise l'approche de l'analyse des systèmes afin d'identifier les rôles des nombreux participants du Programme MAB et pour explorer les techniques de communication des résultats de la recherche. Les répercussions de la nouvelle technologie, en particulier pour les pays en développement, sont mises en relief.*

## INTRODUCTION

THE Man and the Biosphere Programme (MAB) is almost certainly the most ambitious programme of research on man's interaction with his environment that has ever been attempted. More than 5,000 scientists from 78 countries have contributed some 1,000 projects, many of which involve the collaboration of scientists from two or more countries. The range of these projects is awe-inspiring, from studies of natural or semi-natural systems in tropical and temperate biomes to research on the problems of alpine villages and of urban centres like Rome and Hong Kong. The results of this vast programme of research are beginning to be published in scientific journals and executive reports, and even in the popular press. How will these results be used to fulfil MAB's long-term objectives of: producing new information of intrinsic scientific value; providing research results which can be applied in land management; building up the research capacity of developing countries; and breaking down institutional barriers to the conduct and application of interdisciplinary research?

Two common misconceptions can be identified as being at the root of the difficulties of communication between research scientists and the wider community which might make use of the results of research. The first misconception is that scientific research is primarily concerned with the solution of practical problems: in reality, scientific research is a self-motivating activity, with its own carefully-structured reward system. The second misconception is that decision-makers, administrators, and politicians are eager to receive objective advice about the ecological consequences of their policies: in reality, their response is closer to that of "Don't confuse me with facts — just confirm my preconceptions!"

## WHO ARE THE ACTORS?

One of the interesting developments of decision-making with multiple objectives is the analysis of actors, objectives and alternatives to demonstrate the wide span of elements which could, or should, have entered into the decision processes. This analysis begins by identifying the main actors or organizations involved in the problem. Sets of objectives and sets of alternatives are defined for each of the actors. These steps comprise the structuring phase of the analysis, and, in the ensuing modelling phase, multi-attribute evaluation models are developed for each actor in order to evaluate their respective alternatives in the light of their own objectives. Finally, the evaluations of the different actors may be integrated into an overall model that attempts to find a solution to the problem (Edwards, 1977; Keeney and Raiffa, 1976).

Some of the principal participants in the MAB Programme are indicated in

Figure 27—1. They include the research scientists, on whom the Programme depends for the research in the fundamental and applied sciences, and in sociology, psychology and economics. Some of these scientists are in the universities and technical colleges which combine research with teaching and training. Others are in regional, national, and international research institutes for which the primary function is that of scientific research. Also included are the managers of the natural resources of the world, and especially in agriculture, horticulture, forestry, energy and minerals. In most countries, the driving force behind the exploitation and management of natural resources is industry, embracing the whole range of industrial activities of manufacturing, engineering, transport and the provision of services of various kinds, including health, safety and welfare. Our use of natural resources, and our interaction with the biosphere, is also greatly affected by the ways in which public affairs are managed by the administrators in central and local government. It is these administrators who have to implement policies for the use of natural resources and for the control of pollution, and who provide advice for politicians, the elected representatives who have the task of creating legislation for the management of public affairs. Administrators exist at national, regional and local levels, and, as countries increasingly recognize the need to work together to solve the major problems of the biosphere, politicians are also creating international legislation. Last, but by no means least, is the participation in the MAB Programme of all those concerned with education at primary, secondary and university levels, and also the education of the general public so as to create a body of lay opinion which is increasingly better informed about the steps that need to be taken to resolve environmental issues.

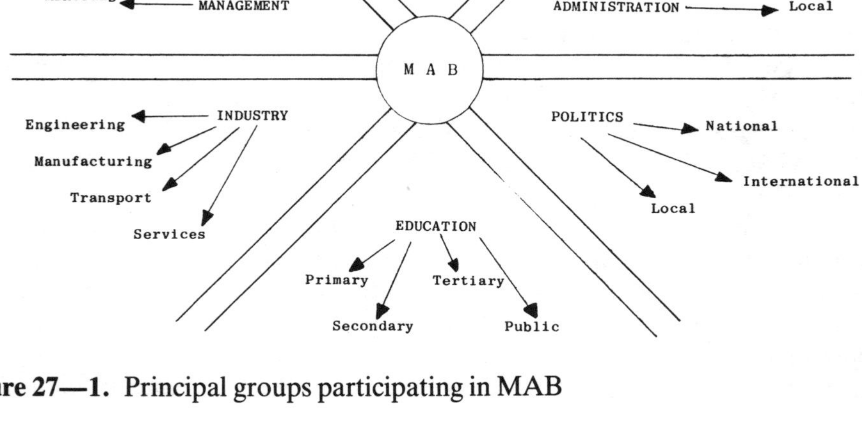


**Figure 27—1.** Principal groups participating in MAB

Figure 27—2 gives a slightly different orientation to the participation in MAB by the six groups, and distinguishes, in particular, some subclasses of individuals who may be regarded as primary participants in the Programme. Thus, within the group of research participants, the research scientists working in the field or the laboratory are separated from the research directors, who, while they may retain some active interest in actual research, have as their main task the provision of the necessary resources and facilities for research, and the recruitment, training and supervision of research scientists. Similarly, in resource management, it is desirable to distinguish between the decision-makers and their scientific advisors. In industry, the principal actors are the industrial managers (again the decision-makers) and the workers in the industries concerned. Within the ranks of the administrators, very different perceptions and objectives will usually exist between the senior administrators, who have already achieved positions in which they are able to influence political and resource management decisions, and their more junior colleagues, who have yet to reach such positions. Within politics, for every successful politician who has been elected to a responsible position in government, there are several candidates, often with widely conflicting opinions, who are, or will be, seeking election at the earliest possible opportunity. In education, Figure 27—2 distinguishes between teachers and university lecturers who are mainly concerned with formal education, and the journalists and popular science writers who educate the general public informally.

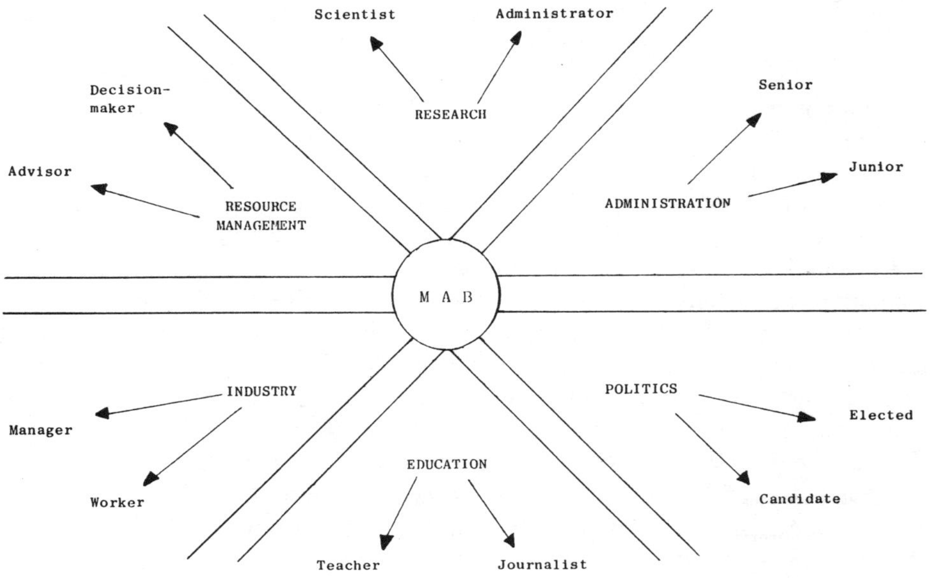


**Figure 27—2.** Principal actors in MAB Programme

In order to keep this contribution within reasonable limits, several important groups of actors have been omitted. Landowners, for example, have a major effect on land use policies by the ways in which they accept or reject advice from resource managers and administrators. Particularly where land plays an important role in the inheritance of property, land tenure and land ownership may be one of the principal modifying influences in the management of resources. The tenants of land, either managing land for the landowner, or managing farms or forests on their own behalf, should therefore be included in any assessment of the roles of actors in the MAB Programme. Finally, although the general public has been included, by implication, in the educational role of MAB, those who support the activities of national and international organizations, including Unesco itself, through the taxes that are collected by national governments should perhaps be mentioned explicitly. Without their, perhaps unwitting, support, none of the MAB Programme would be possible.

## WHAT ARE THEIR OBJECTIVES?

Having identified the principal actors in the MAB Programme, we now need to identify the objectives of these actors, and the end-products of their activities. For the research scientists, the principal objective is to do research. The whole of his, or her, training has been devoted to creating an attitude of mind which can only be satisfied by an active participation in research. All other activities are secondary, and, by definition, unsatisfactory, although some, like administration, management (and even sleeping and eating), are necessary adjuncts to the research. The objective of research is essentially to create more research; otherwise the research scientist would find himself in the position of no longer being able to do the one activity for which he was trained, and for which he has the passion, flair, and all-consuming interest. In order to succeed as a research scientist, however, he also has to make regular contributions to the publication of scientific papers in journals and conference proceedings. These papers follow a very special set of rules and conventions, but we should not make the mistake of imagining that they are the principal means of communication between scientists. It is only by writing such papers, and by having them accepted in journals refereed by other scientists in his own field of science, that anyone can become a member of the "scientific club". Merely possessing a degree in science is not enough (it is not even necessary)! Publication in journals refereed and read by one's peers is both necessary and sufficient.

The actual communication between research scientists is chiefly informal, by correspondence, and by meetings at each other's laboratories and at conferences. Alerted by the existence of their published material, research scientists communicate with each other in informal groups, and the price of remaining in one or more of these groups is that of continuing to publish relevant papers in the appropriate journals. One of the persistent difficulties of people who have trained as scientists, but never done sufficient research to publish their results, is that they are confined to what they read in the scientific literature, which is perhaps only one tenth of the information which flows between active research scientists through the informal channels of communication. Scientific advisers in the resource management agencies and in administration are, therefore, virtually excluded from the sources of information that they would like to have, even though they have a scientific training and qualification.

Of course, from time to time, research scientists are requested to write "Executive Summaries" of their research, most usually to satisfy the requirements of the committees that may be expected to give them the necessary funds for their research, or to present the results of the research to the agency that provided the funds. Most scientists write such reports with a degree of cynicism, knowing that very little of such summaries will be read beyond the first page, and expecting that even less will be understood. This statement may sound somewhat arrogant, but, as affairs are now arranged, there is very little merit to be gained from one's scientific peers from the expression of complex relationships in a language which is readily understood by non-scientists — and such relationships are anyway better expressed by mathematics. A relatively small number of research scientists do turn ultimately to the writing of books on science for non-scientists, but almost inevitably slip out of the "club" because they no longer contribute to the refereed literature.

One additional end-product of the research scientist's activity is that of advice, and a surprising amount of scientific information is passed on informally in this way. Obtaining such advice, however, necessitates knowing whom to contact. Sometimes, such knowledge comes from the reading of scientific papers or abstracts; more usually, it comes through a complex web of contacts similar to the traditionally British "old-boy network". However, as the solutions to most present-day problems are required immediately, there is little or no prospect of basing these solutions on new research. The informed opinions of scientists who have worked on similar problems in the past are therefore often the best help that can be hoped for, provided that access to the relevant and most informed scientists can be obtained. Knowing the key research administrators in a particular field is often the quickest way into this network. In addition to maintaining their own research projects, if they can, research administrators have to obtain the necessary funds, and resources to enable their colleagues to do their work, and consequently usually have an excellent overview of what is being done and who is doing it.

Scientific research is, therefore, a self-motivated activity, with its own carefully structured end-products, designed to ensure that research scientists have access to the informal communication which takes place between those scientists accepted by their peers. Although research scientists will often respond to requests for advice, they will usually avoid, as far as possible, any activity that limits or constrains their ability to do research, including interpreting their results for non-scientists.

A point that confuses the non-scientist is that the decision-makers and their advisers in the resource managing agencies are also trained as scientists, in the sense that they usually have a first or post-graduate degree in one or more of the sciences. However, once they join the resource management agency, they usually cease to do any active research, even if they have done some before. They then drop out of the "club", and are confined to the small and relatively unimportant scientific communication which takes place through published papers. As a result, access to current research is constrained, unless the decision-makers or their advisors can make use of the "enquiry network". Many agencies do succeed in obtaining advice quickly and efficiently in this way, but there is a constant danger that they will continue to consult a "known" expert long after that expert should have been replaced by one with more relevant and more recent expertise. The end-products of the resource managers are day-to-day management, and the formation of policies which guide the day-to-day decisions, and which decide the practical outcome of competition between the agencies for common resources, like land, water, manpower, etc. Ideally, the resource management agencies should also provide the necessary funding for research which will help to solve the problems which are likely to confront them in 5, 10, or 15 years' time. It is too late to attempt to fund the research to solve the problems of today.

Industry and commerce are in a similar position to the resource management agencies, except that their end-products are the goods and services characteristic of the particular concerns. Industrial and commercial economics also usually demands that each company shall make sufficient profit to satisfy its investors and to plough back for future investment. In addition, the environmental effects of industry and commerce are increasingly being considered in the profit and loss balance sheet which must be drawn up for the company, and both managers and workers are likely to have a lively interest in the levels of all these end-products.

Administration, or the management of public affairs in central and local government, has a rather similar set of objectives, although fewer administrators have any scientific training, and hence much access to the informal "enquiry network". In addition to day-to-day decisions in the management of affairs, administrators will be particularly concerned with the development of policies to govern the conduct of future management, especially where there is likely to be conflict, and hence disagreement. Indeed, administrators may have to advise politicians who have been elected, or who are seeking election, on the basis of their support for particular policies. Much of the success, or failure, of administration lies in selective support of a limited number of policies which do not foreclose future options (Walters, 1975).

In popular mythology, resource managers, industrial managers, and administrators, together with their advisors, are inevitably open-minded and approach every problem without pre-conceptions or conditioning. In fact, any successful manager or administrator has a whole "set" of experiences and responses which have stood him in good stead in the past. Almost certainly, he will already have strong views about the policy which is "correct" or about the decision which should be taken. He would hardly be where he is today without some positive ideas, based on past experience.

In general, therefore, he is not overjoyed to be told by some "expert", no matter how scientific or reputable, that the decision to which he is already committed, in his head if not on paper, is unsound or unwise. Advice from scientific research can usually be rejected quite easily on the grounds that "the research does not take into account all of the relevant factors", and most of us have learnt from experience that it is wiser to couch scientific advice in rather general terms if it is not to be rejected as being too specific! There is, therefore, a strong resistance to the use of the results of scientific research unless those results confirm the pre-conceptions of those who will make the decisions, when, of course, the results will be eagerly accepted and quoted. This resistance, taken with the fact that there is relatively little advantage to the research scientist, at least in his terms, in presenting the results of his research in a format in which they are likely to be understood by the decision-maker or his advisers, is sufficient to inhibit the transfer of scientific information from one to the other.

Junior administrators, and the scientifically-trained advisers of resource managers, have an additional problem. Their chance of reaching a senior position depends mainly on their not making any serious mistakes in the advice that they give to their senior colleagues. Even when right, causing a senior colleague to abandon a cherished policy by showing that the results would be disastrous is not guaranteed to result in rapid promotion!

The principal end-product for the politician or the political candidate, is, of course, election to the required official position. Once in office, the principal problem is to stay in office, in competition with those who would also like to be elected to that office. In support of this principal end-product, it is necessary to produce legislation and political manifestos, but, again, successful politicians are likely to have strong views about the policies which should be adopted, or the decisions which should be made. Politicians in office will usually be guarded by their administrators, and, again, scientific advice is most likely to be welcome when it confirms the pre-conceptions of both.

Of the actors shown in Figure 27—2, those concerned with education are almost neutral. Like the resource managers, administrators, and industrial managers, teachers will not have access to the important informal channels of communication between research scientists unless they are themselves doing active research. In most universities, many lecturers are also engaged in research, and are able, therefore, to pass on to their students the latest information about research methods and discoveries. Informal education of the general public, and much of the training in science by science graduates who have never done any research, is almost inevitably 5 to 10 years out-of-date, and, in some subjects, even more out-dated.

The above account of the objectives of the principal actors in the MAB Programme is more descriptive than analytic. We need a model which emphasizes the interrelationships between components in the system, which, interacting with the structure of the system, produce interdependencies. Kelly (1955) proposed the technique of the repertory grid as a way of examining and bringing into awareness the conceptual system built and held by an individual. His formal presentation of his theory of personal constructs was in the form of a fundamental postulate and 11 corollaries. Shaw (1980) has developed the use of Kelly's repertory grid technique, and some of the ideas suggested by this development can be used to show the interrelationships between the objectives of our actors.

In the repertory grid, the universe of discourse is represented by a particular, though not necessarily specific, problem or need. In our case, the problem is to show the interrelationships between the objectives of the groups of persons involved in the MAB Programme. From the area mapped out by the universe of discourse, a set of "observations" or "elements" is chosen. Table 27—1 gives the list of elements chosen in this case, essentially a list of "actors". Note that the list has been extended to include some of the actors suggested as "missing" from Figure 27—2. The technique now requires a series of constructs, each of which is a bipolar dimension which to some degree is an attribute or property of each element. For example, we may regard the objectives of the actors as being distributed along a dimension which is characterized by having "research" as one pole and "management" as the other. If, therefore, we characterize each actor on a 5-point scale along this axis, we would give the value "1" to the research scientist, and the value of "5" to the resource managers, and each of the other elements would be given an integer value from 1 to 5. Shaw (1980) has developed an effective method on interactive computers to elicit constructs and to assign values from these constructs to the selected elements. Her method greatly enhances the power of the repertory grid to probe the constructs and elements and the relationships within them, in a way that is not easy to describe to anyone who has not experienced it. Table 27—2 gives a set of constructs which has been elicited by Shaw's methods. In order to encourage active participation in this evaluation of the objectives of the actors, the actual values assigned to each actor for each construct are not given. Interested readers may like to assign their own values, using a 5-point scale, to the table of Appendix 1, and add any further constructs that occur to them.

**Table 27—1.** List of elements or "actors" in the MAB Programme

| |
|---|
| Research scientist |
| Research administrator |
| Resource manager |
| Resource adviser |
| National administrator |
| Regional administrator |
| Local administrator |
| Politician |
| Political candidate |
| Industrial manager |
| Industrial worker |
| Teacher |
| Journalist |
| Landowner |
| Taxpayer |

Analysis of grids falls into two major categories: methods of exhibiting pattern and structure in the grid responses, and methods for psychometric scaling in general. The traditional methods of grid-analysis have been the $D^2$ (non-metric) method of factor analysis (Bonarius, 1965; Kelly, 1955: Osgood *et al.*, 1957), other methods of factor analysis, both metric and non-metric (Coombs, 1964; Cronback, 1955), principal component analysis (Slater, 1964), and multi-dimensional scaling (Coombs, 1964; Kruskal, 1964; Shepard, 1962; Torgenson, 1958). More recently, cluster analysis has also been used to identify patterns in the grid responses (Rosenberg, 1976). Shaw's (1980) method of the focussed grid has been used here, however, because of its greater facility for interactive interpretation of the grid.

**Table 27—2.** Constructs for objectives of MAB participants

| | | |
|---|---|---|
| Research | — | — Management |
| Resources | — | — Human welfare |
| Appointed | — | — Elected |
| Education | — | — Application |
| Active | — | — Passive |
| National | — | — Local |
| Self-interest | — | — Altruism |
| Informed | — | — Uninformed |

Figure 27—3 shows the interrelationships between the constructs derived from one such analysis. The constructs fall broadly into three groups. The first group consists of the two axes "Research-management" and "Education-application", reflecting the close relationship between research and education and between management and application of theory. The second and larger group links the "Resources—human welfare", "Informed and uninformed", and, "Active—passive" constructs, and, more loosely, the "Appointed—elected" construct. The third group consists of the two axes of "National—local", and "Self-interest—altruism".

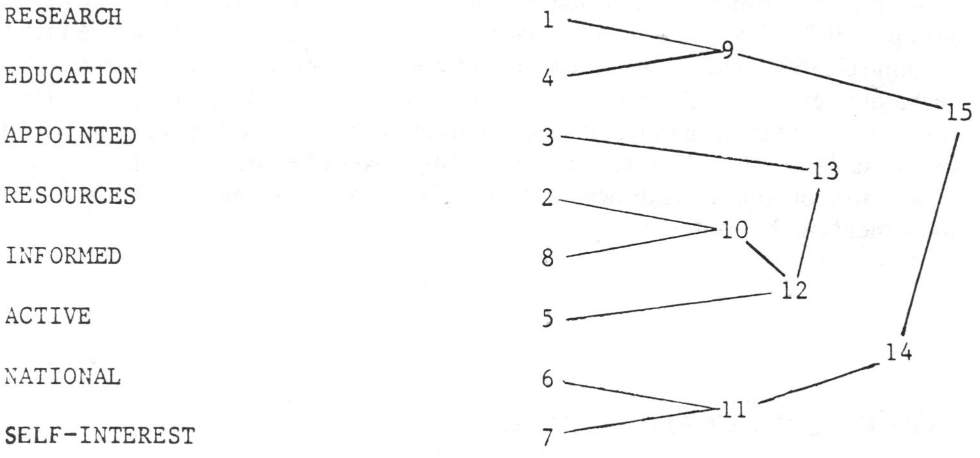


**Figure 27—3.** Classification of constructs related to participants in the MAB Programme

**Table 27—3.** Elements of methods of communicating R and D

| |
|---|
| Books |
| Journals |
| Research reports |
| Conference proceedings |
| Microfiche |
| Abstracts |
| On-demand reports |
| Data-base |
| Algorithms |
| Word processors |
| Viewdata |
| Videodisk |
| Simulation |
| Lectures |
| Workshops |

Figure 27—4 showing the matching of the elements, or actors is perhaps more interesting. In this classification, the separate community of the scientists is quite clearly revealed, the closest matching of objectives being between the research scientists and the research administrators, joined first by the resource adviser and then by the resource manager. This structure matches quite well with the strength of communication between the actors. Elsewhere in the classification, workers, candidates, and taxpayers form another close group, joined to an equally close group of local, regional and national administrators. Both of these groups are then joined by the politicians to form the macro-group of the non-scientific public. Teachers and journalists are closely matched and linked only remotely to the public group, with landowners and industrial managers as "outliers" to the whole group of non-scientific actors. Too much should not be claimed for the repertory grid technique, except for the extent to which it reveals the constructs used by at least one observer to relate a range of participants in the MAB Programme. What would be particularly interesting would be the comparison of constructs obtained independently from observers drawn from each of the different categories of actors used as elements in this example.

## HOW DO THEY COMMUNICATE?

Using a similar approach to investigate the methods by which the actors can, or

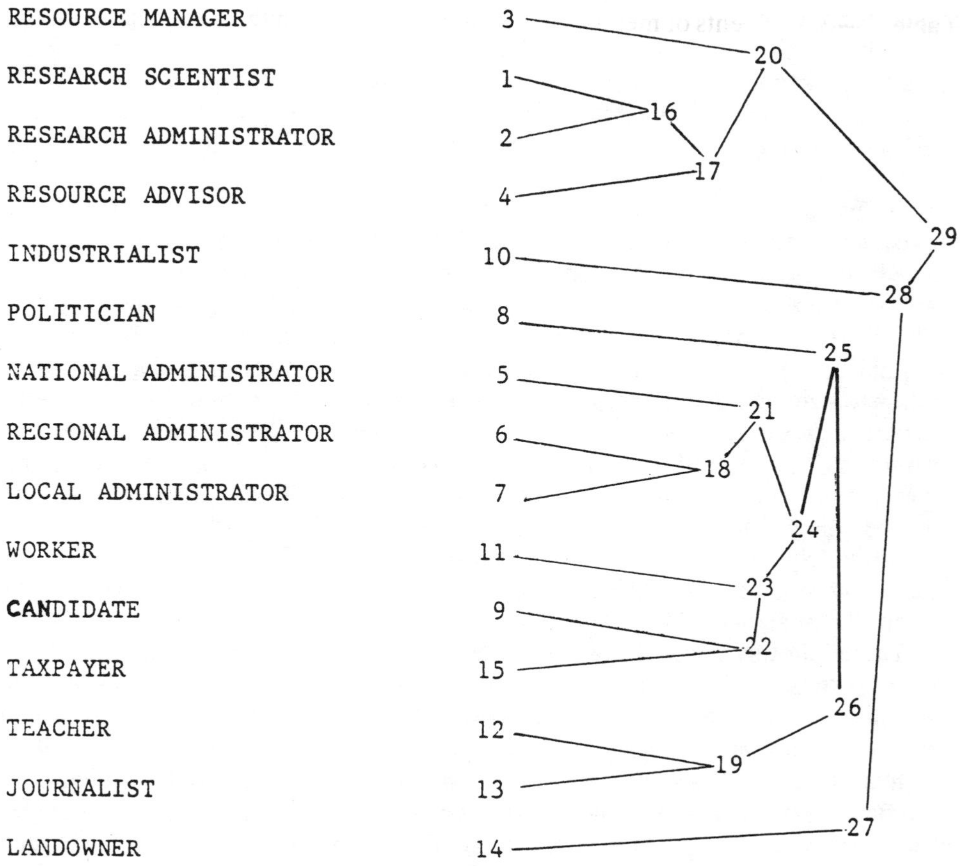


**Figure 27—4.** Classification of participants in the MAB Programme

could, communicate, Table 27—3 lists the elements of a repertory grid designed to reveal the interrelationships between methods of communicating research and development. The most conventional methods of communicating the results of scientific research and development are those of published books, papers in scientific journals, research reports, and conference proceedings. The papers in scientific journals are subjected to a careful scrutiny by editors and referees—indeed, as has been stressed above, this refereeing is part of the peer review which is necessary to establish the reputation of the research scientist. As a result, there is usually a long delay between the submission of a paper and its publication—sometimes as much as 2 or 3 years. Research reports are also submitted to critical review by the director or senior staff of the research organization producing the report, but the delays in publication will usually be less. Such reports also have less value to the research scientist in furthering his career, and hence his ability to do more research. The policy of the research organization may also place constraints on what can be said in the report. The publication of the proceedings of conferences is often intended to short-circuit the delays in formal publication by scientific journals, but the proceedings will usually still be refereed and edited. Finally, many scientists do write books, principally collecting together research results obtained over several years. The economics of book publishing, however, are such that highly specialized topics are less likely to be published than general text books because they have only a limited sale.

As pressure on space in libraries mounts, as well as the cost of traditional publishing, some results of scientific research are now appearing on microfilm. A limited amount of material appears in this form alone, but microphotographic readers are not popular with anyone except librarians and the logical development of this more condensed form of information presentation has been slow. In contrast, the publication of scientific abstracts has become extremely popular, enabling readers to scan a large number of papers very rapidly through their titles and short abstracts of their contents. It has even been suggested that, in future, only the abstracts should be published, and that anyone interested in reading the complete paper should then write to the author for a copy. This form of publication would not meet the requirements of the peer review demanded by scientists, however, and is unlikely to be adopted. The same reason has inhibited the production of on-demand reports, although short reports of this kind may be produced as a result of requests for advice.

There follows a range of computer-based methods of communication. The best known of these is the familiar data-base, an organized collection of information which may be consulted by authorized enquirers, either directly or through a remote terminal. Several large data-bases of this kind already exist, and provide valuable information to requests which can be structured in ways relevant to the structure of the data-base itself. An even more direct way of using the computer as a medium of information exchange is through computer programs or algorithms. One of the strange paradoxes to emerge from the development of computers is that the need to write completely unambiguous and precise sets of instructions to enable the computer to do anything useful at all has also enabled the human beings who use computers to communicate more easily between themselves through the medium of these programs. Many scientists now transfer their ideas to other scientists through algorithms — systematic mathematical procedures enabling a problem to be solved in a finite number of steps expressed in one or other of the internationally accepted computer languages. Heuristic problems for which no algorithms exist are often solved by simulation, and, again, many scientists transfer ideas by means of complete computer programs which provide simulations of complex processes or problems.

Closely related to computers are the rapidly expanding types of word-processors. These are essentially computers, but they are provided with special programming facilities to make them particularly adapted to the handling of text, tables and figures. Although these machines cannot be programmed so flexibly as the general purpose computers, they are especially useful for the tasks of editing, updating, and revising written information. Linked together, word-processors also act as a complete communication system, enabling the scientist, administrator, or resource manager to consult, edit or update material which has been prepared by other participants many miles away. Much experimentation is currently being done with the linking of computers and word-processors by telephone lines so that material can be displayed on television screens, or on special display devices. These viewdata systems already offer access — sometimes interactive access — to considerable amounts of information and may, ultimately, revolutionize the transfer of information from one location to another if the current problems can be solved of indexing the information so that it can be retrieved quickly and economically. It is not even necessary to link the receivers of information by telephone lines if the information does not require to be updated very frequently. Videotape, or videodisks, already offer considerable potential for transferring written, visual, and audible information to remote receivers.

Finally, where the participants can be brought together, in reasonably small numbers, the communication of information can be done face-to-face. When one person gives a talk and the rest listen, the use of lecturers can be effective, providing the transfer of up-to-date information is in terms which are understandable to the audience. During the last two decades, the more interactive "workshops" have become popular, enabling many people to communicate information about their problems and the solutions to these problems.

As before, Table 27—4 gives a set of constructs which has been elicited for the elements of Table 27—3. Again, in order to encourage active participation in this evaluation, the actual values assigned to each element for each construct are not given. Interested readers may like to assign their own values, using a 5-point scale, to the table of Appendix 2, adding any further constructs that may occur to them.

Figure 27—5 shows the linking of the constructs derived from an analysis of the element scores. Six of the constructs are closely matched, beginning with the close

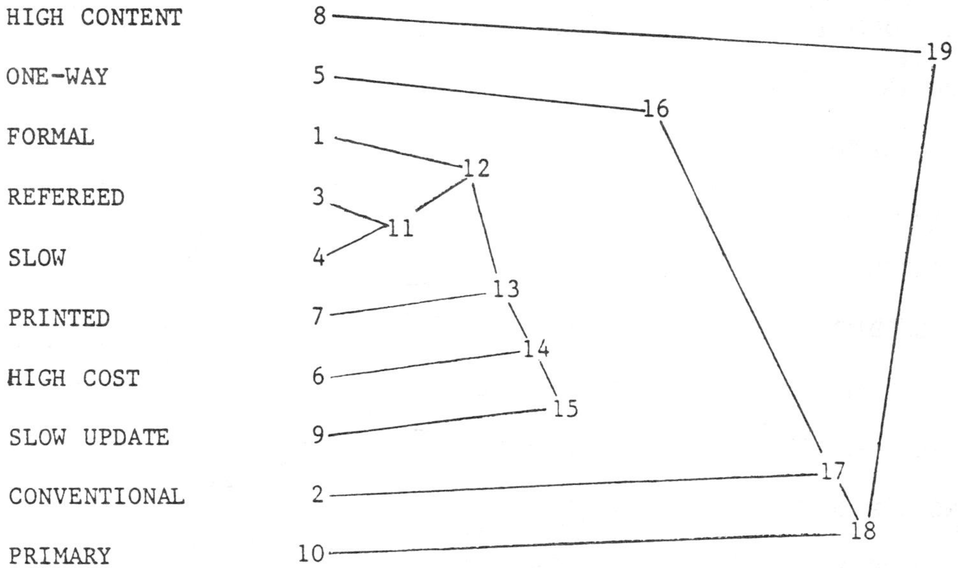


**Figure 27—5.** Classification of constructs related to methods of communicating R & D

association between the formal methods of publishing refereed material which are slow, and then linking to the fact that these methods result in printed material, produced at high cost, and slow to update. Most of this communication represents communication in one direction only, usually from the research scientist to the resource manager or administrator. The other three constructs are not closely matched with any other constructs, namely whether the method is conventional or unconventional, primary or secondary, or has a high or low information content.

Figure 27—6 shows the linking of the elements derived from an analysis of the construct scores, and suggests three principal groupings of the methods of communication. The conventional methods of communication, by papers in journals, conference proceedings, abstracts and formally published books, are matched together. An even more closely matched group consists of those methods which are dependent in some way on computers, namely data-bases, algorithms, simulation, viewdata, and videotape/videodisk. Interestingly, microfiche, which usually contains the same kind of material that appears in journals and conference proceedings, but

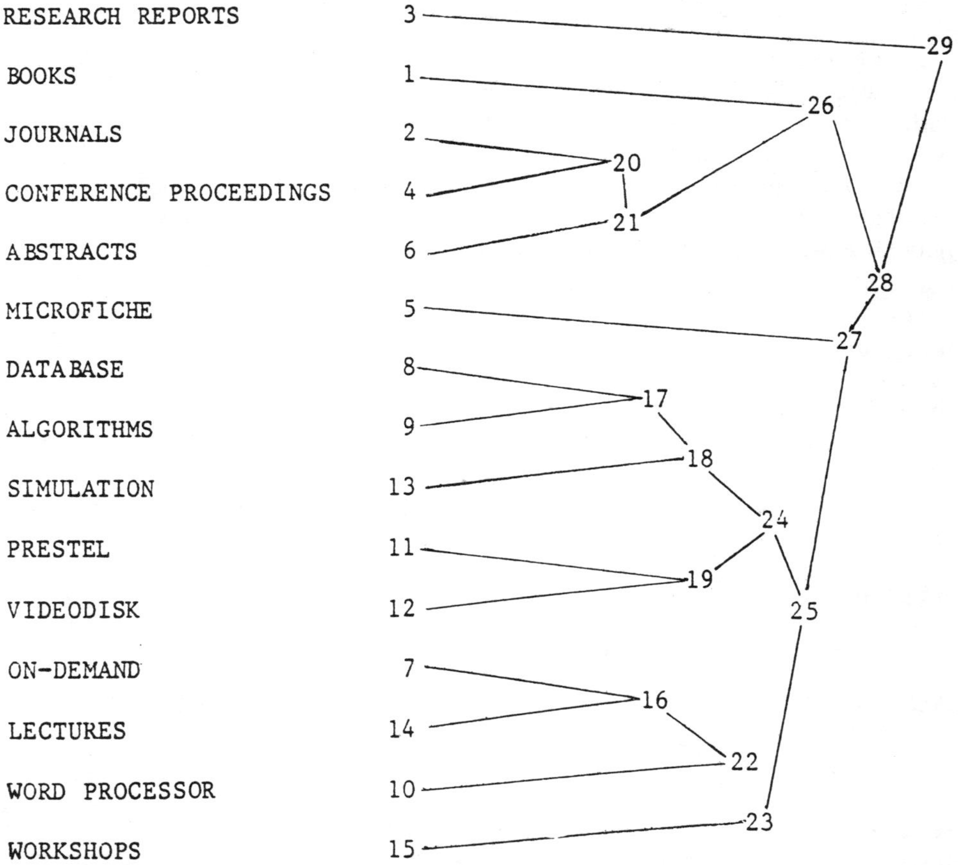


**Figure 27—6.** Classification of methods of communicating R & D

which requires special readers, falls between these two groups. The third main group consists of the less conventional, less formal methods of communication, that is lectures, workshops and on-demand reports to which has been added — surprisingly perhaps — word-processors.

**Table 27 —4.** Constructs for methods of communicating R and D

| | | |
|---|---|---|
| Formal | — | — Informal |
| Conventional | — | — Unconventional |
| Refereed | — | — Unrefereed |
| Slow | — | — Fast |
| One-way | — | — Interactive |
| High cost | — | — Low cost |
| Printed | — | — Ephemeral |
| High content | — | — Low content |
| Slow update | — | — Fast update |
| Primary | — | — Secondary |

Research reports have the least connection with any other form of communication. Incidentally, the technique of the repertory grid should itself perhaps have been included as a method of communication. Any method which identifies the mental constructs that an individual uses to classify a set of elements, and then compares the constructs of different individuals might be regarded as being of particular importance, especially if the individuals represent different categories of participants. It would be of critical importance to know if the various groups of participants have markedly different constructs for the same set of elements. Even if they used the same constructs, it would be desirable to know that they used the same names for the poles of those constructs. The repertory grid technique may be the only way in which these important facets of the communication of research results can be investigated effectively.

## PARADIGMS

Are there any examples of successful communication between the participants of a programme like MAB? Clearly, dependence upon the publication of results by research scientists is likely to be only accidentally successful. The forms of publication used by scientists are intended mainly for the quite different purpose of peer review and to ensure that the scientist publishing the results can continue to do research. Transfer of information through this route is slow, and reaches any potential user only indirectly, even where that user has been trained to use abstracting journals, etc. In any case, most of the really important information is being transferred informally from scientist to scientist, and is not readily available to resource managers or administrators unless they can tap into the informal network. A more serious disadvantage, however, is that very little information travels in the other direction, that is from resource manager or administrator to the research scientist.

Various attempts have been made, in several different countries, to develop an effective link between research scientists and the users of research. Some of these attempts have focussed on the idea of contract research, with resource managers or administrators defining the objectives of the research they want done, and the research scientist being offered a contract to do the necessary research to meet the objectives. The idea has a seductive feasibility, but has failed consistently, most recently in Britain following the implementation of the "Rothschild principle" in 1973. Rothschild advocated the appointment of Chief Scientists in government departments, supported by scientists with appropriate expertise, to identify the research needed for practical decision-making in the departments and to commission the necessary research from universities and research institutes. With rare exceptions, however, these scientists quickly lost touch with the community of research scientists from which they were drawn — as they were no longer contributing to the published research, they were no longer members of the "club". The result, in general, has been a slow decline in useful research, with an extensive and counter-productive discussion of the research that might be done, this discussion often costing as much as, or more than, the amount actually invested in research. Experience elsewhere has been similar, resulting in expensive and ineffectual attempts to control research, or a complete lack of control which enables any research, relevant to practical problems or not, to be supported. Little or no improvement in communication has resulted from the attempts to "commercialize" the creative spirit necessary for good research.

One of the most successful interactions between research scientists and the users of research occurred in Britain during the Second World War in the linking of scientists with a wide range of expertise to military personnel concerned with tactical and strategic problems in the conduct of the war. The scientists worked in loosely organized teams and discussed current problems regularly and informally with the military officers, notably at meetings on Sunday mornings, with suitable alcoholic refreshment, known as "Sunday Soviets"! Both users and research scientists therefore had access to an informal level of communication capable of passing information both ways. Confidence was established by regular meetings, and by the sequential nature of the frequent interactions. However, it would be difficult, if not impossible, to duplicate these conditions in peace-time outside of very exceptional circumstances. Over long periods of time, the normal commitments of family life and organizational structures would almost certainly break down the informality of the interaction.

Holling (1978) has suggested, and successfully implemented, an approach to the solution of clearly identified environmental problems, known as adaptive environmental assessment and management. In a later paper, Holling (1981) summarizes what has been learnt from the use of this approach in a number of practical applications. The approach involves four groups of actors:

1. *The "Project Team"* representing the client with a genuine concern to evaluate alternative policies for natural resources and the environment.
2. *The "Workshop Staff"* consisting of a group of four to six analysts who jointly have backgrounds in several different resource disciplines, are familiar with a range of modelling and evaluation techniques, and have the aptitudes and experience to lead workshop discussions.
3. *The "Core Planning Group"* consisting of the leader of the "Project Team", one or two of his senior staff and the workshop staff. This group has the primary responsibility of planning the sequence of activities, of identifying the key problems to be solved, and of identifying the participants in the various workshop sessions from among the research scientists, resource managers and administrators.
4. *The "Participants"* are the scientists, managers and administrators, usually from several institutions, who have major roles to play in technical or decision aspects of the problem. They constitute the wider audience which helps to define the problem and assess the priorities, as well as monitoring the results of the research.

The sequence of activities begins with a meeting of the core planning group to provide some preliminary ideas about the problem, and to identify key participants and the information needed for the first workshop. This workshop, over a period of five days, establishes the policy framework by discussion with the participants, and sets the priorities for research and for management and administration. Typically, the workshop is followed by a period of two or three months during which independent work is done by the research scientists in an attempt to refine and solve the problems that have been identified. This period is followed by a second workshop of the same participants to produce an improved analysis, model and policies. Alternate periods of independent work and workshop discussion follow, controlled by the core group and by the identification of problems and solutions during the workshop sessions.

The whole scheme of adaptive environmental assessment and management works by ensuring that resource managers and administrators participate in the informal communication which normally takes place only between research scientists. The presentation of the results of research is fast and interactive, and leaves the scientist free to publish his results formally in the way that guarantees him the necessary peer review. Careful choice of topics for research can help to avoid premature commitment by resource managers and administrators to policies which they might later wish to abandon, but which they will probably continue to support because of the human characteristic of being unwilling to admit, and pay for, our mistakes (Walters, 1975). However, it is clear that only major problems are likely to be tackled by such a procedure. The concept of the normal workshop disturbs the normal models of communication and the routines of institutional life. The lack of any guarantee of support for cherished policies or preconceptions will ensure that many of the more pressing environmental problems of the world will not be approached in this way.

## ROLE OF NEW TECHNOLOGY

It follows that resource managers and administrators, if not politicians, industrialists and the general public, could best achieve the kind of interactive communication with research scientists that is desirable for the informal exchange of ideas while they are still fresh in the minds of the scientists and before resource managers and administrators have committed themselves to policies to be advocated. Very large problems may be tackled in the future through Holling's adaptive environmental management and assessment, and, even here, the computer and the word-processor will have a valuable part to play in helping to establish the dialogue between the scientist and the user of scientific results, and in continuing this dialogue during the periods of intensive work by scientists on problems which have been identified by the resource managers and administrators. However, there are many problems which are unlikely to be identified for solution by so formal a method. Informal interaction, via the computer and word-processor, may therefore help to identify those areas which would ultimately be tackled in this way.

The repertory grid for methods of communicating the results of research and development identified one group of computer-related techniques, including databases, algorithms, simulation, viewdata and videodisk. One of the common misconceptions about computers today is that they are principally used for computing. In fact, the most important use of the computer today is communication, and the rapid development of very cheap mini- and micro-computers has resulted in a reduction of the cost of communication to almost nothing. Information processing is now virtually free (Witten, 1980). Scientists who have learnt to use computers for themselves, rather than through packages and management systems, now find that their principal transfer of information to and from other scientists is through the data which they have stored in the computer, and through the programmes they have written to analyse or simulate these data. Once data have been made machine-readable, they can, in theory at least, be transferred to any other machine. Similarly, once a concept has been captured in an algorithm, or simulated, using any programming language, it can be reproduced in any other computer language. Because the program has to be complete and unambiguous if the computer is to work at all, the program also provides for an unambiguous transfer of information between human minds.

One of the most powerful weapons in the search for a more informal and more productive interchange of information between research scientists and the rest of the world may, indeed, be through the use of simulators. After all, we do not teach airline pilots to fly by giving them a real airliner in the first instance. They learn to fly on a simulator, which is essentially a computer programmed to represent the cockpit of the aircraft and to respond to the controls in the same way that the aircraft itself responds. Only when our novice pilot has demonstrated that he understands the principles by which the plane can be flown, is he allowed to take up the real airliner. The simulator also has some additional advantages, in that it is possible to arrange for various minor or major incidents to occur, like one of the engines catching fire, or the undercarriage failing to engage, which it would perhaps be unwise to attempt on the real airliner. The pilot can then be taught to handle these emergencies in the correct manner, and, if he takes an incorrect action, the results are not disastrous. The trainee pilot learns on the simulator rather than on the real aircraft. Presumably, one of the reasons for training in this way is that the safety of human beings is obviously involved in exposing passengers to an untrained pilot.

Astonishingly, however, we allow resource managers to train themselves on the real environment. Admittedly, the mistakes that they make are unlikely to be recognized during the period of their management or administration, but the results may be equally disastrous in the long term. Since it is possible now to prepare simulations of a wide range of environmental situations, we are entitled to ask why those who seek to manage the earth's resources should not be first invited to manage a simulator to show that the policies and methods of management which they intend to adopt will not do irreversible harm to the environment, and to the people who are dependent upon that environment. With an accelerated time-scale, it is possible for a resource manager or administrator to gain valuable experience about the complex interactions between various components of the environment as they respond to policies and management practices. At the very least, such training should bring greater awareness of the manifold interdependencies of plants, animals, and the physical and chemical components of the environment. It should also bring greater awareness of the ultimate results to human populations and stimulate a search for less destructive policies.

Lest this should be thought a somewhat far-fetched notion, it should be pointed out that, in many parts of the world, the young people with little formal education are already playing games on computers with every sign of enjoyment and understanding. Admittedly, many of these games are concerned with fantasy, such as space invaders and galactic warfare, but the games are interactive and teach coordination, anticipation, and the marshalling of resources. It requires very little imagination to extend these kinds of games to more practical use. There would be nothing strange to someone who has played space invaders to play a similar game concerned with the management of pests or diseases. Such a player would quickly learn, by repeated trials, which strategies were unlikely to lead to a successful conclusion to the game, and, through interactive learning, would come to identify the successful strategy even where there were considerable uncertainties about both the spread and virulence of the disease. Existing games already incorporate the kinds of uncertainties which are inherent in ecological and environmental problems, and it is this repeated practice in learning to cope with uncertainty which is essential for the training of anyone who has to handle problems of the environment. So much of the conventional training of scientists and administrators is concerned with the management of clearly defined problems, where uncertainty is eliminated or ignored for the sake of a tidy solution. It is the injection of uncertainty into the role-playing situation which encompasses every real task in the world that provides the principal advantage of the kinds of simulations which could now easily be written for a great many practical applications.

A little experience with such simulations would quickly lead to a demand for efficient procedures for handling the kinds of problems which emerge. We might regard these procedures as algorithms, and, again, the computer is an ideal tool for passing on proposed algorithms from one user to the next. The matching of algorithms with simulations would enable the algorithms themselves to be tested under a wide variety of different circumstances, and for inefficient algorithms to be eliminated before they have an opportunity to do too much damage in the real world. Indeed, the emphasis on the idea of an algorithm which is capable of handling a complex situation, and adapting to the information which is available about that situation, would do much to remedy the approach to problems which is so characteristic of today's environmental management, that is that of "driving by the seat of one's pants". At a time when it is probably safer in fog to allow an aircraft to be landed by the ground-landing system of the automatic pilot, so it may be safer to allow some at least of our environmental problems to be solved by a computer-based algorithm rather than the attempts of human beings to muddle through the situation. At least such algorithms should be tried and tested against the simulations that we are now able to build of environmental problems. The ways in which computers work focus attention upon algorithms and sets of instructions, many of which can be modified according to the information derived sequentially from the problem itself. The concept of maintaining options, and keeping these options open at successive stages in the solution of a problem, is one which can be extended. So, too, can the use of heuristic algorithms which learn from experience. We already have examples of chess-playing computer programs which improve as they play more and more opponents. These programs "learn" from their experience and adapt their strategies to what has been learnt in playing human opponents. It is not inconceivable, therefore, that heuristic algorithms for the solution of environmental problems could also learn from the applications which have been made of the programs, either to simulations or to real-life systems. The transfer of such algorithms or programs from one computer to another is now a relatively trivial task.

Communication between computers and word-processors is also already a simple task. Such machines can be linked by telephone lines, by satellites, and, in the near future, by fibre optics. It is already possible to link machines in different continents, and to transfer information at speeds which would have been incomprehensible only a few short years ago. Much of the transfer of information, however, need not require speeds of this kind. It is perfectly possible to send cassettes or "floppy disks" by post or by messenger, and this speed of transmission may be quite sufficient, particularly when we bear in mind that one simulation capable of being held on a single cassette or on a single disk might provide enough information to satisfy a user for several weeks, or even months. The transfer need not be all in one direction. Resource managers and administrators, with the aid of suitable programs, could define the problems and the uncertainties inherent in those problems for the attention of scientists, either through games, or through techniques such as the repertory grid. Diagnostic and search algorithms have already shown themselves to be an extremely powerful way of eliciting information even from people who have previously never had any contact with computers. We have convincing evidence that people will tell computers things that they would not tell to other human beings, even though they know that human beings are subsequently going to look at the information which they are providing to the computer. Techniques which stimulate creativity such as brain storming, delphi forecasting, and normative association are more easily managed by anonymous machines like computers, than in groups where people are afraid of making fools of themselves. In this way, the computer can be used to enhance the creativity of the human mind. It is precisely by stretching the human mind around the problem that the research scientist can be made aware of options which would otherwise not be considered. Similarly, the research scientist can present ideas which would otherwise be regarded as being so far-fetched as to be virtually lunatic.

The importance of all this is especially great for developing countries. It is yet another paradox that these ideas are more likely to find acceptance in developing countries than in the developed countries, in which there is already a structure for management which inhibits the development of new ideas. Unless problems are solved within an accepted framework, the solutions are unlikely to be accepted and implemented. In the developing countries, in contrast, the same infrastructure does not exist, so that it may be possible to use less conventional approaches to the same problems which occur in the more developed countries. Since the computers now cost so little, it will almost certainly be much cheaper to develop solutions through their use than to try to develop the same kind of infrastructure which is now hindering the adoption of modern scientific techniques for the management of the environment in the more developed countries. These techniques, therefore, have a particular importance to the developing world.

It is vital that we do not ruin the resources of the developing world by making the same type of mistakes that we have made in past history in the more developed parts of the world. It is not enough to transfer our disastrous experience to the new resources of the world. We need, now, to use the knowledge we already have in a more productive way so as to anticipate problems and to prevent those who seek solutions to those problems from imposing their preconceptions on to the environment and the human populations dependent upon that environment.

## IMPLICATIONS FOR MAB

What has all this to do with MAB? First, it is clear that the MAB Programme was envisaged as using the conventional transfers of information between research scientists and the decision-makers. At the outset, some attempt was made to establish a systems approach to the problems of MAB, and the first expert panel on systems analysis and modelling emphasized the kinds of approaches which would have aided or enhanced this transfer of information (Unesco, 1972). Regrettably, however, very few of the recommendations of the expert panel have been adopted in the practical implementation of the MAB Programme. The scientific community has preferred to rely on the more conventional methods of transferring information, and we are now beginning to see, some 10 years after the start of the MAB Programme, the slow publication of the first results. The results from MAB may therefore be no more than a collection of scientific papers in obscure and dusty journals, and MAB will have done very little to establish the interchange of information between the research scientist and those who should be using the information obtained from past research, or, more important still, the interchange of information which will lead to the solution of tomorrow's problems. Analysis of existing methods indicates that we cannot rely on the conventional publishing of research to effect this interchange. A new approach is needed.

It is suggested, therefore, that MAB should enter a new phase, devoted not to the support of research projects *per se,* but to the establishing of an effective interface between research scientists and resource managers and administrators, with a second interface to the general public through a more popular presentation of the MAB philosophy. In establishing a more effective communication between research and management, we should not be afraid of embracing the new technology of information processing and the use of algorithms and simulation as a basis for informal and interactive communication. In particular, we should encourage the developing countries to leap-frog the stultifying effects of developed-world institutions and bureaucracies. We have the technology and we have the necessary mathematical and logical theory to overcome most, if not all, of the barriers to communication. What we need now is the will and the energy to combat the present reaction of those who manage the environment which is too often characterized by the statement — "Don't confuse me with facts — just confirm my preconceptions!"

## APPENDIX 1: Blank grid for objectives of MAB participants

| | *Methods of communication* | | | | | | | | | | | | | | | | |
|---|---|---|---|---|---|---|---|---|---|---|---|---|---|---|---|---|---|
| *Construct* | Books | Journals | Research report | Conf. proceedings | Microfiche | Abstracts | On demand report | Data-base | Algorithm | Word-processor | Viewdata | Videodisk | Simulation | Lecture | Workshop | *Poles* | |
| 1 | | | | | | | | | | | | | | | | 1. Research | 5. Management |
| 2 | | | | | | | | | | | | | | | | 1. Resources | 5. Human welfare |
| 3 | | | | | | | | | | | | | | | | 1. Appointed | 5. Elected |
| 4 | | | | | | | | | | | | | | | | 1. Education | 5. Application |
| 5 | | | | | | | | | | | | | | | | 1. Active | 5. Passive |
| 6 | | | | | | | | | | | | | | | | 1. National | 5. Local |
| 7 | | | | | | | | | | | | | | | | 1. Self-interest | 5. Altruism |
| 8 | | | | | | | | | | | | | | | | 1. Informed | 5. Uninformed |
| 9 | | | | | | | | | | | | | | | | | |
| 10 | | | | | | | | | | | | | | | | | |
| 11 | | | | | | | | | | | | | | | | | |
| 12 | | | | | | | | | | | | | | | | | |
| 13 | | | | | | | | | | | | | | | | | |
| 14 | | | | | | | | | | | | | | | | | |
| 15 | | | | | | | | | | | | | | | | | |

## APPENDIX 2: Blank grid for communication of research and development

| Construct | Research scientist | Research administrator | Resource manager | Resource advisor | National administrator | Regional administrator | Local administrator | Politician | Political candidate | Industrial manager | Industrial worker | Teacher | Journalist | Landowner | Taxpayer | *Poles* | |
|---|---|---|---|---|---|---|---|---|---|---|---|---|---|---|---|---|---|
| 1 | | | | | | | | | | | | | | | | 1. Formal | 5. Informal |
| 2 | | | | | | | | | | | | | | | | 1. Conventional | 5. Unconventional |
| 3 | | | | | | | | | | | | | | | | 1. Refereed | 5. Unrefereed |
| 4 | | | | | | | | | | | | | | | | 1. Slow | 5. Fast |
| 5 | | | | | | | | | | | | | | | | 1. One-way | 5. Interactive |
| 6 | | | | | | | | | | | | | | | | 1. High cost | 5. Low cost |
| 7 | | | | | | | | | | | | | | | | 1. Printed | 5. Ephemeral |
| 8 | | | | | | | | | | | | | | | | 1. High content | 5. Low content |
| 9 | | | | | | | | | | | | | | | | 1. Slow update | 5. Fast update |
| 10 | | | | | | | | | | | | | | | | 1. Primary | 5. Secondary |
| 11 | | | | | | | | | | | | | | | | | |
| 12 | | | | | | | | | | | | | | | | | |
| 13 | | | | | | | | | | | | | | | | | |
| 14 | | | | | | | | | | | | | | | | | |
| 15 | | | | | | | | | | | | | | | | | |

*Participants* (column group heading over the participant columns)

## References


Bonarius, J. C. J. (1965), Research in the Personal Construct Theory of George A. Kelly: Role Construct Repertory Test and Basic Theory, In B. Maher (Editor), *Progress in Experimental Personality Research,* Volume 2, Academic Press, New York and London.

Coombs, C. H. (1964), *A Theory of Data,* Wiley, New York.

Cronback, L. J. (1955), Processes Affecting Scores, *Psychol. Bull.,* 52, 177—93.

Edwards, W. (1977), How to Use Multi-Attribute Utility Measurement for Social Decision Making, *IEEE Trans. Systems, Man, Cybernet.,* 7, 326-40.

Holling, C. S. (Editor) (1978), *Adaptive Environmental Assessment and Management,* John Wiley, Chichester.

Holling, C. S. (1981a), *Science for Public Policy: Highlights of Adaptive Environmental Assessment and Management,* Publication R-23, Institute of Resource Ecology, University of British Columbia.

Holling, C. S. (1981b), *Resilience in the Unforgiving Society,* Publication R-24, Institute of Resource Ecology, University of British Columbia.

Keeney, R. L. and H. Raiffa (1976), *Decision Making with Multiple Objectives: Preference and Value Trade-offs,* Wiley, New York.

Kelly, G. A. (1955), *The Psychology of Personal Constructs,* Norton, New York.

Kruskal, J. B. (1964), Multidimensional Scaling by Optimizing Goodness of Fit to a Non-Metric Hypothesis, *Psychometrika,* 29, 1-27, 115-29.

Osgood, C. E., G. J. Suci and P. H. Tannenbaum (1957), *The Measurement of Meaning,* University of Illinois Press, Urbana.

Rosenberg, S. (1976), New Approaches to the Analysis of Personal Constructs in Person Perception, In *Nebraska Symposium on Motivation,* University of Nebraska Press.

Shaw, M. L. G. (1980), *On Becoming a Personal Scientist,* Academic Press, London and New York.

Shepard, R. N. (1962), The Analysis of Proximities: Multidimensional Scaling with an Unknown Distance Function, *Psychometrika,* 27, 125-39, 219-46.

Slater, P. (1964), *The Principal Components of a Repertory Grid,* Vincent Andrew, London.

Torgerson, W. S. (1958), *Theory and Methods of Scaling,* Wiley, New York.

Unesco (1972), *Expert Panel on the Role of Systems Analysis and Modelling Approaches in the Man and Biosphere (MAB) Programme,* MAB Report Series No. 2, Unesco, Paris.

Walters, C. J. (1975), *Foreclosure of Options in Sequential Resource Development Decisions,* IIASA Research Report RR-75-12, International Institute for Applied Systems Analysis, Laxenburg.

Witten, I. H. (1980), *Communicating with Microcomputers,* Academic Press, London and New York.

CHAPTER 28

# Approaches to Evaluation: Insights from Analysis of a Large-Scale Ecosystem Project

*Bengt Lundholm*

Abstract

The development of the Swedish Coniferous Forest Project (Swecon) is described from the initial planning to the development of a truly integrated programme. The initial problems of organization are outlined and the utility of an international advisory council demonstrated. The role of the project in providing opportunities for education in ecology and as a framework for research by university and other scientists is stressed. Attention is drawn to the importance of the selection of team leaders and of team work. The contribution ends with an evaluation of the project's strengths and weaknesses and with a suggestion for the inclusion of better information handling and of socio-economic aspects in future projects.

*Résumé*

*On décrit le développement du Projet Suédois sur la Forêt de Conifères ('Swecon') depuis le stade initial de sa planification jusqu'à l'intégration complète du programme. On met l'accent sur les problèmes initiaux d'organisation et l'utilité d'un groupe international de consultation. Le rôle du projet comme point focal pour l'éducation en écologie et comme cadre de recherche pour l'université et les scientifiques en général est souligné. On attire l'attention sur l'importance du choix des directeurs d'équipe et du travail en équipe. La contribution se termine par une évaluation des points forts et faibles du projet et en suggérant d'inclure dans les projets futurs une meilleure diffusion de l'information et une meilleure prise en compte des aspects socio-économiques.*

At a number of MAB meetings, at both national and international levels, the author has argued strongly for a strict objective evaluation of MAB activities and projects. One principal result of this pleading is that he hereby makes a subjective evaluation of a Swedish project, the Swedish Coniferous Forest Project, or "Swecon". It had been hoped to have two independent evaluations as a basis for this assessment. In May, 1981, the Swedish Natural Science Research Council arranged for an international group of scientists to evaluate the scientific results of the project. But their report is not available at the time of drafting the present contribution. Thus the first point about evaluation is: if an international group makes the evaluation, give its members time to complete the report *before* they leave your country, otherwise it may be a very long time before the report is received. The second report expected is a critical review of the structure and function of the project, but it again is still under preparation. Both these reports will be in English, and will be available from the Swedish MAB committee.

Without these two reports, the author relies on his own experience within the project. He has been involved from the planning stage to the very end, first as a member of the group which initiated the project and later as a member of the project committee. During this time many mistakes have been made, but also a certain number of successes have been achieved. In presenting a personal view on these matters, it is hoped that our experiences will be of value to others concerned with planning and managing a large integrated project.

## PROJECT BACKGROUND AND PLANNING

In 1970, a commission reported to the Swedish Natural Science Research Council on the topic of terrestrial ecology. At that time, ecology was regarded as a key word in relation to environmental problems. Politicians, administrators and planners were asking the ecologists a lot of questions which could not be answered, as our knowledge base was insufficient. The commission stated that some of the knowledge lacking could only be generated by joint efforts from scientists representing very different disciplines. As a result of this report, the planning of the Coniferous Forest Project started. The duration of the planning period was two years and the cost considerable (between 10 and 20 per cent of the total project cost).

### Budget and structure

The project started in 1972 and terminated in the fiscal year 1980-81. It has had a total budget of 39 million Skr, or US$9,000,000. The first year it had a support of 2 million Skr or $500,000. In 1976-77, the support was 6.3 million Skr and in its last year it was again down to 2 million Skr. Thus because of inflation, the budget at the end was considerably less than at the beginning.

The main goal of the project was to study the structure and function of a forest ecosystem by means of systems analysis and modelling. The project was organized in six problem areas, and within each of these problem areas different processes were to be studied. The six problem areas were: energy exchange and microclimate; water turnover; dynamics of aboveground plant biomass; dynamics of assimilates and of nitrogen, phosphorus and potassium in plant biomass; dynamics of below-ground plant biomass; and soil processes.

## Management and coordination

The scientists within the project have been responsible for the detailed planning and running of the project, through a special planning group. The overall responsibility, especially in relation to financial matters, has been the task of the project committee, with representatives of the sponsoring agencies, organizations with special interests in relation to the project and one representative for the scientists. This committee had a great need for objective advice in relation to scientific matters. In a small country like Sweden almost all scientists were either engaged within the project or had, for different reasons, negative opinions about the project. It was therefore necessary to invite experts from abroad to serve on an advisory council for the project.

The initial task for the planning group was to state the general objectives of the project. In general terms, the project was described as a study of the ecology of the forest landscape, and "Ecology of the Forest Landscape" was also the Swedish name of the project. Here a mistake was made. We underestimated the complexity of the landscape. It has, in effect, been possible only to study *a part* of the forest landscape. In order to obtain good conditions for field work, an area with sandy soils had to be selected for the research site; these are comparatively rare in Sweden. The most common forest soils were too difficult for the purpose, even if most of the Swedish forests grow on moraine soils. The project should have been an integrated one, but initially was organized by small groups consisting of either zoologists or botanists or microbiologists and so on. Based on this plan we received two million crowns. Thus it was possible to call in the new advisory council. This group, drawing on experiences gained within the US International Biological Programme, advised that our basis for planning was completely wrong. The project should instead be organized around groups responsible for specific problems, and these groups should include botanists, microbiologists, hydrologists or other specialists, according to the disciplines needed for tackling the specific problem at hand. And the efforts of specialists from different disciplines had to be integrated.

The project committee followed the advice of the advisory council and decided to reorganize the project: half a year was lost. This example demonstrates the role and the importance of having advice from the international community. Another point was also learned. Call in the advisory group at the outset of planning and some mistakes may be avoided and money saved. However, when planning the project it is impossible to tell if money will be obtained for the project, and during this planning time the money may not suffice to pay for an advisory group.

From one point of view the most worrying thing about large projects such as Swecon is how to obtain sufficient money and the decisions necessary to cover the long timetable of the project. The process of scientific decision-making varies from country to country. In Sweden, political actions are needed in order to start these large projects, and in this respect the reports from the advisory council were of utmost importance in financial discussions.

Another difficulty which large integrated projects face when they are considered for funding is the division of the project into small components, with each piece being evaluated and judged by specialists. Inevitably, one or two of the pieces are not in the so-called research forefront, and are considered to be lacking in substance compared with other projects which compete for money. A particular piece might appear very trivial, but be essential for the project as a whole: if it is taken away the whole project might collapse. Therefore it is obligatory to stress the importance of evaluating these integrated projects as complete units and not as a number of small separate projects. The difficulty is, however, that at least in Sweden there are very few people capable of evaluating a project as an integrated whole. Here an advisory council may be of assistance.

It is of utmost importance to create a good research climate within the project. This depends partly on a proper interaction between the planning group and the project committee. The most important person in the project is the project leader and coordinator. Failure or success depends to a large extent on the leader. Swecon has been very lucky in having an efficient leader, trusted by both the scientists and the project committee. Though heading 50 scientists for nine years requires a strong nervous system.

Another point, overlooked in the planning, is the need for the participants to be able to work in teams. When scientists were selected for the project only their scientific merits were considered, not their ability to work together. As scientists are usually not trained in teamwork, problems can result if this question is not fully considered.

## EVALUATION OF PROJECT RESULTS

During the ten years of its existence the project has been the target of very strong criticism in Sweden. However, as the results have appeared this criticism has decreased with time.

## Practical significance

In the beginning it was said that the project was too theoretical, and that the results would be of no practical value. This was a serious accusation as one of the arguments, when we were pleading for money to the political and administrative bodies, was that the project results should be of practical importance to society. This criticism has, however, now abated, as the results from the project form the base for tackling Sweden's most urgent environmental problem — that of acid rain. The project has not only produced useful information in relation to this problem, but also scientists capable of undertaking research on acid rain.

The project has also contributed knowledge and competent personnel to such problem areas as the use of products from forests as an additional fuel resource and the establishment of short rotation forests for energy purposes. The results from the project have also been of importance for evaluating the consequences of large-scale clear-cutting and forestry practices that involve removal of the whole tree. Without question, the project has now proved to be of very large practical importance.

## Use of research resources

Another point of criticism has been that the project has taken economic resources from other fields of science, and that more and also better scientific results could have been achieved if the money had been divided up and used for small projects with higher scientific excellence. This might be true, but it is very difficult to prove or to disprove. In any case, the type of results generated by Swecon could possibly not have been achieved in any other way.

## Quantitative measures of project outputs

As mentioned earlier, the results of the independent scientific evaluation of the whole project are not available at the time of writing. However, some parts of the project have been evaluated, and some indications of research effort are available.

Very often research activities are measured by the number of reports and papers accepted in international journals. Even though this is a very blunt tool with which to measure research performance, some figures can be put forward. Up to March 1981 work on the project had led to the production of 200 papers and 130 reports (101 internal and 29 technical), with the *opus magnum* being a book, *Structure and Function of Northern Coniferous Forests* with more than 600 pages (Persson, 1980). Many scientific papers have thus been published and many more will follow. In addition, the project has produced 19 dissertations for the doctorate of philosophy degree, and three of the scientists working in the project have been appointed professors. There is no doubt that the project has been of great importance for the training of ecologists in Sweden. This might be, in fact, the most important result of the project.

## Scientific evaluation

This is not the place to present the scientific significance of the research results obtained within Swecon. Mention might, however, be made of one noteworthy finding — that 54 per cent of the annual photosynthesis is used for belowground growth while only 7 to 8 per cent is for stem growth. And it is stem growth that is the principal aim of forestry.

In the field of modelling, a series of models with very different time scales have been developed within the project. A plant physiological model deals with energy, water in plants, and water outside plants, with a time frame of 24 hours. In five models (on energy, water, tree growth, ground vegetation and nutrients), the time scale is between one and ten years. Finally there is a stand development model, where the time unit is 100 years. At the outset some project participants had ideas about developing one huge model covering the total forest ecosystem, but this turned out to be impossible.

Mention might also be made of an evaluation of Swedish activities in the field of ecological microbiology, sponsored in September 1980 by the Swedish Natural Science Research Council. The evaluation was carried out by an international group of scientists from Finland, Federal Republic of Germany, Netherlands, United Kingdom and United States of America. This is a quote from their report:

> A major strength, especially in terrestrial microbial ecology in Sweden, is the integrated multidisciplinary approach that has been developed principally through Swecon. Success of these large team efforts is equally dependent on the leadership of initiating scientists and the expertise and dedicated involvement of the supported scientists. It is our opinion that several parts of the microbial ecology portion of Swecon are among the most successful of these large projects anywhere in the world. It is also clear that the benefit of this project goes beyond the basic findings in that it has established a network and style for multidisciplinary research. It has also trained a new generation of investigators with an appreciation of the total system which should help insure that questions of major importance are not neglected.
>
> The special investment of funds in microbial ecology has brought Sweden an international reputation in some parts of this field and has produced a generation of young researchers which should further advance the Swedish position towards a center of excellence in this field. We have met several outstanding young people in this review who could become international leaders in this field.

This statement was encouraging, especially as at the start of the project the component on ecological microbiology was regarded as the weakest point.

## The importance of communication

At the beginning of the project, although the first grant of 2 million Skr. had been received and scientists recruited, we had no instruments because delivery had been delayed. One advisor, Dr R. H. Waring, however, advised that this would encourage discussion by enforced idleness rather than isolated working in the field. Internal communication within an integrated project is most important for its success. Nine years later Dr Waring visited us again and wrote a report on the project. One of his points concerns studies of the role of insect pests in different stands. He concludes with the following recommendation: "The studies most fruitful to pursue in the immediate future would be those, where health of the host can be experimentally manipulated and where populations of insects can be attracted or diseases inoculated." He continues: "In summary, the greatest opportunity for improved forest management may lie in discovering the ecosystem role that native insect pests and diseases play." That is a challenge and also calls for new integrated projects: should these materialize, the Swedish project will have fulfilled a useful function.

Finally a weakness in the project should be pointed out; one which probably will not be considered in the usual type of evaluation. The internal information flow, mentioned earlier has functioned well, as has that of scientific information in general. But there has been a great deal of misunderstanding in groups outside the project which could have been avoided if a proper information flow had existed. When the first project plan appeared there was a special information group, but it was the first thing to be cut. In hindsight this was an error. In the author's opinion information activities should be built into the project from the beginning. Even more important are actions to take care of the information flow after the project has been terminated. The results must reach those people who need them for practical purposes. These matters have not been properly handled in Swecon, but it is hoped to include proper information planning in future projects. It must be realized, however, that in general ecologists and scientists are not competent in the transfer of certain types of information. There must be specially trained people to do this and they must be involved in the research projects from the first planning stages.

## CONCLUDING REMARKS

Swecon has been a MAB project. It was based on the ecosystem concepts and experiences emerging from the International Biological Programme. It is an example of interdisciplinary integration within the fields of natural sciences. At a very early stage it had been desired to include socio-economic aspects also, but that proved to be too much to grasp in a single project. The MAB framework has been important to the project because it has made it possible to establish contacts and to exchange information with scientists and administrators from other countries. One of these conferences held in Sweden was entitled "Man and the Boreal Forest" (Tamm, 1976). This is within the MAB concept and in Sweden we are now trying to structure research to integrate ecological and socio-economic systems, a very difficult task. The MAB Programme has been useful to Swecon and we hope that Swecon also has given the MAB Programme something of value.

## References

Persson, T. (Editor), 1980, *Structure and Function of Northern Coniferous Forests — An Ecosystem Study,* Ecological Bulletins 32, Swedish Natural Science Research Council, Stockholm.

Tamm, C. O. (Editor), 1976, *Man and the Boreal Forest,* Ecological Bulletins 21, Swedish Natural Science Research Council, Stockholm.

CHAPTER 29

# International Governmental and Non-Governmental Scientific Cooperation

*J. C. I. Dooge and F. W. G. Baker*

Abstract

This contribution reviews the development of international cooperation over the last two hundred years, and particularly cooperation between the International Council of Scientific Unions, a non-governmental organization, and several inter-governmental organizations, especially in the United Nations over the last fifty years. The benefits from combining the best of both types of organization is examined and the important synergistic effects of contributions from the communities of governmental and non-governmental scientists are emphasized.

*Résumé*

*Cette contribution passe en revue l'évolution de la coopération internationale au cours des deux cents dernières années, et en particulier la coopération entre le Conseil International des Unions Scientifiques, organisation non-gouvernementale, et plusieurs organisations intergouvernementales, plus spécialement dans le cadre des Nations Unies pendant les 50 dernières années. On fait l'examen des avantages obtenus en faisant fond sur la meilleure alliance de ces deux types d'organisation et l'accent est mis sur les importants effets synergétiques des apports des communautés de scientifiques gouvernementaux et non-gouvernementaux.*

## EARLY INITIATIVES

THE first initiatives in cooperation between scientists of different nations began in the universities or other centres of learning, around a teacher whose renown was sufficient to extend beyond the frontiers of his native land. The beginnings of cooperation between nations, as such, can be placed in the middle of the eighteenth century when as a result of initiatives by internationally-minded astronomers observations of the parallaxes of the Moon, Mars and Venus were made in 1751 at the Cape of Good Hope, Berlin, Greenwich, Bologna, Paris and Stockholm.

One of the earliest experiments in cooperation to observe natural phenomena simultaneously at several points of the globe is that of the meteorological network initiated in 1781 by J. J. Hemmer of the Palatinate Academy at Mannheim. The network so established extended from Gotthaab in Greenland to Pyshminsh in Russia. Observations were made three times daily at a maximum number of 39 stations. The observations were published in standard form in *Ephemerides Societatis Meteorological Palatinae.* Although the Academy ceased to function as a result of the occupation of Mannheim during the Revolutionary War many of the stations continued their observations and the instructions of the Academy for the observations and publication of the results had an important influence on the first decisions of the International Meteorological Organization (IMO) when it was created in 1873.

Interest in the weather, which is probably about as old as thought, was at the origin of the first programme which brought together a mixture of governments, when a North-American M. F. Maury was responsible for initiating a programme of meteorological observations made voluntarily aboard ships of opportunity. In order to stimulate the collection and transmission to him of these observations Maury offered in exchange copies of his Wind and Current Charts, which helped the captains of sailing vessels make their voyages more rapidly because the publication improved their knowledge of winds and currents.

Thirty years later in 1882-83 the First International Polar Year provided an opportunity for an interesting mixture of governmental, non-governmental and private enterprise. Eleven nations established a network of circumpolar stations to make simultaneous observations over a period of 13 months in a wide range of disciplines, including biology. The Year was organized by an International Polar Commission established by the IMO. The sponsors ranged widely. The Austro-Hungarian expedition was largely financed by a Count, organized by the Royal Academy of Sciences in Vienna and transported to the island Jan Mayen in a boat of the Austro-Hungarian navy. The Swedish expedition was organized by the Academy of Sciences in Stockholm, transported in a vessel of the Swedish navy and floated on sales of brandy — the major part of the expenses were paid by a distiller called Smith.

The First International Polar Year is memorable not only for this mixture of patronage but also for the series of international multidisciplinary programmes it initiated (Baker, 1982). These include the Second Polar Year, the International Geophysical Year, the Global Atmospheric Research Programme, etc., some of which are mentioned later in this contribution.

Cooperation between governmental and non-governmental organizations really began to flower, however, with the creation of international organizations of the two types. Within the framework of ICSU, a non-governmental organization of scientists, cooperation with an international governmental organization began with the International Meteorological Organization in the 1920s, and when the Second World War began in 1939 ICSU (which had established a Committee on Science and Social Relations to survey the most important results obtained and the directions of progress in science with reference to their interconnections; the development of the scientific picture of the world in general; and the practical application of scientific results in the life of the community) was cooperating with the International Organization for Intellectual Cooperation of the League of Nations how best the findings of this Committee might receive support internationally. (It is interesting to note that some forty years later ICSU is still discussing with Unesco, a lineal descendant of the International Organization for Intellectual Cooperation, these same questions). This Committee on Science and Social Relations began its work again in 1945 and in 1946 the Fourth General Assembly of ICSU adopted a recommendation urging that "greater efforts be made by the United Nations to develop the potential benefits of nuclear energy and to avoid its misuse" (Stratton, 1946).

At the Fifth General Assembly in 1950 the committee submitted a report in which it drew attention to the dangers arising from unbalanced applications of the power of science and mentioned, inter alia, "the extermination of species of animals or plants; the depletion of mineral resources; erosion consequent upon developing an unsuitable structure of the soil; the contamination of soil, waters and the air by waste products; etc." and applied to Unesco and the United Nations for a concerted effort directed to investigations of these problems on an international scale.

Unfortunately these attempts by ICSU failed to make the United Nations Organization take greater notice of these important problems facing humanity: problems that seem more pressing today than they were 30 years ago. This failure did not however diminish ICSU's attempts to work with intergovernmental organizations in other fields of endeavour.

## SOME EXAMPLES

### The International Geophysical Year

Shortly after the Fifth General Assembly a proposal was made in ICSU for the organization of a Third International Polar Year to be held in 1957-58, seventy-five years after the First and twenty-five after the Second. ICSU naturally sought the help and advice of the World Meteorological Organization for the development of the meteorological programme and that of Unesco with regard to several other parts of the programme, in particular in oceanography.

The third year, which became, the International Geophysical Year, was a great success with more than sixty nations participating in a programme that cost several thousands of millions of dollars and within the framework of which the first artificial satellite of the earth was launched. Before the end of the Year it was evident that the arrangements for international cooperation made in some of the disciplines, such as oceanic, space and Antarctic research, had reached such a level that some long-term arrangements would be needed after 1958 in order to maintain the impetus created. ICSU therefore decided (Baker, 1980) to create three new Committees which later became the Scientific Committee on Oceanic Research (SCOR), the Committee on Space Research (COSPAR), and the Scientific Committee for Antarctic Research (SCAR). Each of these committees has in the last two decades created a unique working relationship with an intergovernmental body.

In the case of SCOR this developed first with Unesco, later with the Intergovernmental Oceanographic Commission (IOC) and the Advisory Committee on Marine Resources Research (ACMRR) of the FAO. When SCOR began to plan a five-year international oceanographic cooperative research programme, it was decided to choose the Indian Ocean because, first it was one of the least investigated oceans and is one of the controlling factors in the development and dissipation of the monsoons, and second over a quarter of the population of the world lives in countries bordering it — a population in which a protein-deficient diet is common and which could therefore benefit from a study which would provide a scientific basis for a wise expansion of fishing. A basic plan for the study was drawn up by a working group early in 1960. As this plan envisaged several tens of research vessels operating in the Indian Ocean over a period of five years it was decided to seek the cooperation of Unesco, which had itself dissolved its Advisory Committee on Oceanography shortly after the creation of SCOR, and later initiated the creation of the Intergovernmental Oceanographic Commission.

The effective way in which SCOR interacted with Unesco and IOC has led to the Committee becoming deeply involved both with the IOC and the Division of Marine Sciences of Unesco. An involvement that has been not only scientific but also personal: two of the Secretaries of IOC have become Presidents of SCOR. This close cooperation is evident today in many fields of oceanography and SCOR and IOC have recently created a joint Committee on Climate Change and the Oceans (CCCO): the secretary for the committee has been provided by ICSU and is housed in the IOC Secretariat.

In the field of space research COSPAR has maintained since its inception close cooperation with the United Nations following UN General Assembly resolution 1802 (XVIII) of 1962 which asked ICSU to play an advisory role on space issues. This task devolved on COSPAR, and when the UN Committee on the Peaceful Uses of Outer Space was formed, COSPAR was invited to attend the annual sessions of the committee. It has prepared a background paper on "Current and

Future State of Space Science" at the request of the Secretary General of the UN Conference on the Exploration and Peaceful Uses of Outer Space.

This first IGY Conference on the Antarctic was held in Paris in July 1955. In his opening remarks the chairman emphasized the technical character of the conference and stressed that questions of politics were excluded. This led to a motion, which was adopted unanimously, confirming that the overall aims of the conference were entirely scientific: this has formed the basis for international activities in Antarctica since then and has been maintained by SCAR since its creation in 1958. When the Antarctic Treaty was drawn up in 1959 it included in Article 2 the statement that: "Freedom of scientific investigation in Antarctica and cooperation towards that end as applied during the International Geophysical Year shall continue, subject to the provisions of the present Treaty."

At the first Antarctic Treaty Consultative Meeting the representatives recommended to their governments that "they should facilitate the continuation of the exchange of information regarding plans for scientific programmes, as now carried on through the SCAR and through other member unions and committees of ICSU and by such other means as may ensure the availability of this information." The SCAR continues to act as scientific adviser to the Antarctic Treaty and has been active in proposals for the conservation of Antarctic fauna and flora, and is in the process of promoting studies on the problems of mineral exploration and exploitation in the Antarctic and with SCOR on the Biological Investigation of Marine Antarctic Systems and Stocks (BIOMASS).

## The International Biological Programme

Another more surprising legacy of the International Geophysical Year was the International Biological Programme, called by a former Assistant Director General for Science of Unesco, the father or mother of MAB. The IBP was the result of a proposal in 1959 to have an international biological year, a proposal that resulted in a decade of biological studies carried out in cooperation with FAO, Unesco and WHO and, after the 1972 Stockholm Conference, UNEP. One of the outcomes of this cooperation was a technical conference on Plant Exploration and Introduction organized jointly by the IBP and FAO in 1967. This conference and the book that followed "Genetic Resources in Plants — their Exploration and Conservation" (Frankel and Bennett, 1970) began a slow process that is still continuing of spreading awareness and concern about the problem of genetic resources. Similarly the IBP work on nitrogen fixation, although less structured, resulted in a new impetus in such studies and affected the activities of both FAO and Unesco.

## Joint projects

In closing our review of fifty plus years of cooperation between ICSU, a non-governmental, and intergovernmental organizations, we should like to mention briefly four other undertakings which illustrate the effectiveness of cooperation between these two different types of organization. These are:

1. The Unesco-ICSU Study of the Feasibility of a World Scientific Information System (UNISIST);
2. The joint Unesco-IUGS International Geological Correlation Progamme (IGCP);
3. The joint ICSU-WMO Global Atmospheric Research Programme (GARP);
4. SCOPE-UNEP cooperation on the Global Environmental Monitoring System.

**UNISIST:** The practical development of UNISIST began with a recommendation to establish a committee to study the feasibility of creating a world science information system at the ICSU General Assembly in Bombay in 1966. The joint Unesco-ICSU Committee was formed as a result of discussions between representatives of ICSU and Unesco and this committee submitted its report indicating that such a system was feasible to the first intergovernmental conference on UNISIST in 1972. Following this a special division was created in Unesco which is now concerned with what is called the General Programme of Information.

**IGCP:** The International Geological Correlation Programme is a cooperative effort between the International Union of Geological Sciences (IUGS) and Unesco which encourages international research on geological problems related to the identification and assessment of natural resources and the improvement of man's environment. IGCP began in 1973. The Board of the IGCP, which is appointed jointly by IUGS and Unesco, decides on the research priorities for the Programme.

**GARP:** The Global Atmospheric Research Programme (GARP) was a response to United Nations General Assembly Resolution 1972 (XVI) of 20 December 1961 which requested the World Meteorological Organization, consulting as appropriate with Unesco and other governmental and non-governmental organizations such as ICSU, to submit a report on the worldwide benefits to be derived from international cooperation in weather research and analysis, particularly in view of the possibilities opened up by the advances as a result of the launching of meteorological satellites and of developments in the field of electronic computers and in the simulation of atmospheric motions using mathematical models. In June 1964, ICSU established an ICSU-IUGG Committee on Atmospheric Sciences (CAS) which at its second meeting in 1966 proposed to WMO, Member Nations, ICSU and IUGG and their committees, commissions and working groups that 1972 be designated as a twelve-month period for an intensive, international observational study and analysis of the global circulation in the troposphere and lower stratosphere and suggested that the proposed research programme be called the Global Atmospheric Research Programme.

The WMO agreed to cooperate and in 1967 a GARP Study Conference was organized near Stockholm jointly by the Committee on Atmospheric Sciences and

COSPAR with the cosponsorship of WMO. It was decided that it would be necessary to have a longer period of study with several experiments. The recommendations were accepted by WMO and ICSU and in October 1967 a formal agreement on the GARP was signed by the Presidents of WMO and ICSU. Thus entered into history the first formal agreement between an intergovernmental and a non-governmental organization for an extended series of experiments on regional and global scales, which terminated with the First GARP Global Experiment, the field phase of which took place from 1 December 1978 to 30 November 1979 (WMO-ICSU, 1980). The programme of GARP has been developed under the guidance of the Joint Organizing Committee composed of twelve distinguished scientists selected jointly by ICSU and WMO.

**GEMS:** In 1970, ICSU's Scientific Committee on Problems of the Environment (SCOPE) began a study which led to the preparation of a report that was submitted to the UN Conference on the Human Environment (SCOPE, 1971). After the creation of the United Nations Environment Programme this report was revised in cooperation with UNEP and published by SCOPE as *Global Environmental Monitoring* in 1973. These two reports have provided the basis for the present GEMS.

**Others:** The examples we have given are of ICSU's cooperation with intergovernmental organizations, which we know best. There are of course others: one which has played an important role in MAB, and in whose creation Unesco played an important part is the International Union for Conservation of Nature and Natural Resources (IUCN) which cooperates closely with Unesco and also with the UN Environment Programme. The Council for International Organizations of Medical Sciences (CIOMS) in whose creation Unesco was instrumental cooperates closely with the World Health Organization and with Unesco, although less than previously. Similarly there are many NGO's whose field of expertise corresponds to that of an intergovernmental organization and where cooperation is close and effective.

## ADVANTAGES AND DISADVANTAGES

We have given a review of some instances of international activities over more than a hundred years to illustrate different types of cooperation. What are the advantages and disadvantages of such types of cooperation? Perhaps we can start with a quotation from a Unesco publication entitled *Geological Correlation* which after giving details about some types of cooperation goes on:

> Against this background of non-governmental and intergovernmental cooperation, IGCP is truly unique. It combines the advantages of objective, nongovernmental

scientific participation and guidance though IUGS with the benefits of intergovernmental support through Unesco. It draws on the results of all these relevant programmes, and in turn contributes research results that can strengthen these programmes. But perhaps its most distinctive characteristic — and its main source of strength — is in the enthusiasm and support it has received from geologists around the world, without which the Programme could not succeed. The significance of this support is evident if one considers the fact that most of the personnel and financial commitments on IGCP have been arranged through the efforts of the participating scientists themselves, with relatively, less formal support from governments. Thus the achievements of IGCP can be regarded not only as a record of the Programme's contributions toward human needs but also as the result of personal commitment by a large number of the world's geologists towards IGCP objectives.

One of the disadvantages is the disparity in size and funding between most intergovernmental and non-governmental organizations. This tends to make the non-governmental organization appear to be the small, poor partner: a viewpoint that is not always borne out by reality.

The combination of the two types of organization provides two approaches to the resolution of problems: the individual scientists working in the non-governmental organization are able to provide a problem-oriented approach with an analysis of the studies being undertaken that is independent and has a minimum of political bias, while the intergovernmental organizations can provide political and financial support for programmes and can make available the time of scientists working in the national research councils and national institutes. Of course the differentiation between the two is small or lacking in some countries, where it is not always possible for the non-governmental organizations to obtain the help of a scientist unless he first receives governmental approval. Fortunately scientists, although aware of the importance of the political milieu, have a tendency to be international in outlook and conscious of the importance of international cooperation for the benefit of humanity. This is one of ICSU's prime objectives and we hope that working with Unesco and with MAB we shall be able to bring new benefits to those most in need.

## References

Baker, F. W. G. (1980), *The International Council of Scientific Unions: A Brief Review,* ICSU, Paris.

Baker, F. W. G. (1982), The International Polar Years: 1882/83, 1932/33 and 1957/58, *Nature and Resources,* 18 (3), 15-20.

Frankel, O. H. and E. Bennett (1970), *Genetic Resources in Plants — their Exploration and Conservation,* IBP Handbook No. 11, Blackwell, Oxford.

Landsberg, H. E. (1980), Bicentenary of International Meteorological Observations, *WMO Bulletin,* 29 (3), 263-7.

SCOPE (1971), *Global Environmental Monitoring,* SCOPE, Stockholm.

Stratton, F. J. M. (Editor) (1946), *Reports of Proceedings of Fourth ICSU General Assembly*, Cambridge University Press, Cambridge.

WMO-ICSU (1980), *The Global Weather Experiment 1978-79*, WMO, Geneva.

# Index[1]

## GENERAL INDEX



1. Page references to Part I are in roman type, to Part II in *italics*. In using this index, readers should refer to particular localities as well as countries and regions.

# GEOGRAPHICAL INDEX